Nankai Symposium on Mathematical Dialogues

Yang-Hui He · Mo-Lin Ge · Cheng-Ming Bai ·
Jiakang Bao · Edward Hirst
Editors

Nankai Symposium on Mathematical Dialogues

Celebrating the 110th Anniversary of the Birth of Prof. S.-S. Chern

Springer

Editors
Yang-Hui He
London Institute for Mathematical Science
London, UK

Merton College, University of Oxford
London, UK

Cheng-Ming Bai
Chern Institute of Mathematics
Nankai University
Tianjin, China

Edward Hirst
City St George's, University of London
London, UK

IMECC, University of Campinas
São Paulo, Brazil

Mo-Lin Ge
Chern Institute of Mathematics
Nankai University
Tianjin, China

Jiakang Bao
City St George's, University of London
London, UK

IPMU, University of Tokyo
Tokyo, Japan

ISBN 978-981-19-2327-2 ISBN 978-981-19-2328-9 (eBook)
https://doi.org/10.1007/978-981-19-2328-9

Mathematics Subject Classification: 00A79, 11-XX, 14-XX, 92BXX, 81TXX, 81T30

This work was supported by Mathematical Institute, University of Oxford.

This Springer imprint is published by the registered company Springer Nature Singapore Pte Ltd.
The registered company address is: 152 Beach Road, #21-01/04 Gateway East, Singapore 189721, Singapore

Preface

The Chern Institute of Mathematics is one of the premium research centres for mathematics. It is named after **Prof. S.-S. Chern**, one of the greatest and most influential mathematicians of the 20th century (there are many key concepts in geometry that bear his name, such as Chern Classes and Chern-Weil Theory, etc.). Professor Chern was an undergraduate alumnus of Nankai University, China, in the early 1900s and spent the last years of his life there as Distinguished Professor, after many years as Director of the Mathematical Sciences Research Institute (MSRI) at UC Berkeley.

A strong proponent for the dialogue between various branches of mathematics and theoretical physics, Prof. Chern had always wished for a regular international symposium in this subject. The symposium's cover image is a copy of the mural in the entrance to the Chern Institute: a Mathematical Dialogue between Prof. Chern and **Prof. C.-N. Yang** (Nobel Laureate 1957), another famous alumnus of Nankai. The conversations between Prof. Chern and Yang played an important role in the interactions between modern physics and modern mathematics.

When Prof. Chern passed away in 2004, and the Mathematical Institute at Nankai was renamed the Chern Institute, the then director of the Mathematical Physics group, Prof. M.-L. Ge had many conversations with the current director Prof. C.-M. Bai to launch such a series of conferences. Professor Y.-H. He, whose field of expertise is in the interface between geometry, mathematical physics and machine learning, joined Nankai University as Chang-Jiang Chair in 2010 whilst holding his main appointments in the UK. The position allowed him to spend 1–2 months each year in Nankai; this was seen as an opportunity by Prof. Ge and Bai to initiate this series.

In 2017, the inaugural "Nankai Symposium in Physics, Geometry and Number Theory" was held in August at the Chern Institute, organised by Prof. Ge, Bai and He. Over 50 participants from more than 10 countries stayed in the Chern Institute for a week, interacted extensively and gave many interesting talks ranging from gauge theories to string compactifications, from modular forms to complex geometry. Several participants also contributed to the book "*Topology and Physics*", C. N. Yang, M.-L. Ge, Y.-H. He, Editors (World Scientific, 2019). The symposium was a definite success, so much so that there were ideas of extending it beyond dialogues between mathematical physicists and geometers/number theorists.

The funding for the 2nd Symposium had long been secured. Due to the COVID-19 pandemic, it was inevitably delayed. However, by kind permission of the Chern Institute, it was held online between August 2–13, 2021, with Prof. Bai and Ge organising from Nankai. Meanwhile, Mr. J. Bao, Prof. Y.-H. He and Mr. E. Hirst met in-person to coordinate each talk on Zoom from the London Institute for Mathematical Sciences, which occupies the inspiring space that once belonged to Michael Faraday in the Royal Institution of Great Britain. The timing fortuitously coincided with the 110th anniversary of the birth of Prof. S.-S. Chern.

In the course of the intense two weeks of only plenary talks so as to truly encourage every possible dialogue, each day consisted of eight 45-minute talks, which a special session dedicated to short-presentations for young researchers such as Ph.D. students and early postdocs. A total of some 80 talks with wide range of mathematical topics were given, making the conference a truly unique experience. The subject matter ranged from AI technology to number theory, from string theory to representation theory, from engineering in particle colliders to topology, from cosmology to algebraic geometry, etc., all linked by the spirit of "mathematical dialogues". One highlight of the conference is the Special Public Lecture by Prof. Sir Roger Penrose, who just received his Nobel Prize in physics earlier in 2021. He gave a wonderful lecture, paying tribute to Prof. Chern, on his ground-breaking research on singularity theorems which lead to predicting the existence of black holes.

The details of the conference can be found online at

https://sites.google.com/view/cim-dialogues/

and all talks have been recorded and posted at

https://www.youtube.com/channel/UCMDH7Wl391yRGufJjCC981Q

It is our sincere hope this Dialogues series will continue, and it will be in person at the Chern Institute next time.

London, UK	Yang-Hui He
Tianjin, China	Mo-Lin Ge
Tianjin, China	Cheng-Ming Bai
London, UK	Jiakang Bao
London, UK	Edward Hirst

Acknowledgments We are grateful to the Chern Institute for hosting the conference remotely, especially for arranging a regular face-to-face colloquium for the Penrose Lecture. We thank the London Institute for Mathematical Sciences for providing the beautiful physical space where much of the organising team could meet in-person during COVID-19 to discuss and to coordinate. Special thanks are due to Dr. Venkatesh Veeraraghavan and Ms. Elli Heyes for providing the careful review and editorial work for the Proceedings, as well as Mr. Suvajit Majumder and Ms. Victoria Harrison for invaluable help during the conference.

Contents

Space-Time Singularities and Cyclic Geometry: Public Lecture Commemorating Professor S.-S. Chern's 110th

Professor Sir Roger Penrose, FRS, OM, Kt.

Yang-Hui He

Abstract When too much material collapses into too small a region, the result is a black hole, and internal space-time singularities are expected, where space-time curvatures appear to diverge to infinity. The mathematics works both ways in time yet the Big-Bang singularity was of a completely different character, and better understood in terms of conformal geometry, leading to a cyclic picture of the universe, now with considerable observational support. The mathematics of this picture leads to intriguing speculations involving different cyclic geometrical ideas.

On August 7th, 2021, Professor Sir Roger Penrose gave a Special Public Lecture in Honour of Professor Shiing-Shen Chern. The timing is very opportune in so many ways. The lecture commemorates the 110th year of Professor Chern's birthday. Professor Penrose just received his Nobel Prize in Physics earlier in the year for his theoretical prediction of blackholes, a topic which he covered in his lecture. Importantly, the following day, Sir Roger celebrated his 90th birthday.

The Public Lecture was received with great enthusiasm: many hundreds attended online and a special in-person lecture hall was arranged in the Chern Institute where a packed room watched it together. The lecture lasted 1 h, with discussions that went on for another hour, including a singing of "Happy Birthday" to Professor Penrose. The entire event was recorded and can be watched here:

https://www.youtube.com/watch?v=25HjUmn0KCE&t=111s

Y.-H. He (✉)
London Institute, Royal Institution of Great Britain, London, UK
e-mail: yang-hui.he@merton.ox.ac.uk

Merton College, University of Oxford, Oxford, UK

City St George's, University of London, London, UK

Nan Kai University, Tian Jin, China

Y.-H. He et al. (eds.), *Nankai Symposium on Mathematical Dialogues*,
https://doi.org/10.1007/978-981-19-2328-9_1

Vanishing Yukawa Couplings and the Geometry of String Theory Models

Lara B. Anderson, James Gray, Magdalena Larfors, and Matthew Magill

Abstract We provide an overview of recent work which aims to understand patterns of vanishing Yukawa couplings that arise in models of particle physics derived from string theory. These patterns are seemingly linked to a plethora of different geometrical structures and our understanding of the subject has yet to be consolidated in a unified framework. This short note is based upon a talk that was given by one of the authors at the Nankai Symposium on Mathematical Dialogues. Therefore it is aimed at a mathematical audience of mixed academic background.

1 Introduction

Yukawa couplings are parameters that appear in theories of physics which describe how strongly certain particles interact. In many string theory constructions these Yukawa couplings are found to vanish in a somewhat mysterious fashion [1–17]. These vanishings are linked to interesting geometrical structures associated to those theories. In this brief note we review this topic, in a manner aimed at an audience of geometers of mixed academic background, and highlight some open questions in this setting.

L. B. Anderson · J. Gray (✉)
Department of Physics, Robeson Hall, Virginia Tech, Blacksburg, VA, USA
e-mail: jamesgray@vt.edu

L. B. Anderson
e-mail: lara.anderson@vt.edu

M. Larfors · M. Magill
Department of Physics and Astronomy, Uppsala University, Uppsala, Sweden
e-mail: magdalena.larfors@physics.uu.se

M. Magill
e-mail: matthew.magill@physics.uu.se

M. Larfors
Department of Mathematical Sciences, Centre for Particle Theory, Durham University, Durham, UK

Y.-H. He et al. (eds.), *Nankai Symposium on Mathematical Dialogues*,
https://doi.org/10.1007/978-981-19-2328-9_2

We will consider particle physics models derived from heterotic string theory. In detailing such a model two pieces of geometrical data must be specified. These are a Calabi-Yau threefold X and a holomorphic poly-stable vector bundle of zero slope V_X over that threefold base whose structure group is a subgroup of E_8 [18]. Given this data, the particle content of the resulting physical theory, and the Yukawa couplings between those particles, can be computed in a quasi-topological fashion.

The particle content of a heterotic model is given by various cohomology groups of associated bundles to V_X [18]. As a simple example, in the case of an $SU(3)$ vector bundle, families of matter fields in the particle physics theory are in one to one correspondence with elements of $H^1(V_X)$ and anti-families of matter fields with elements of $H^1(V_X^\vee)$. Superpotential Yukawa couplings are numbers describing how three of the families (or anti-families) in such a theory interact. As such they are given by a map which takes the following form (in this simple setting) [8].

$$H^1(V_X) \otimes H^1(V_X) \otimes H^1(V_X) \to H^3(\wedge^3 V_X) = H^3(\mathcal{O}) = \mathbb{C} \tag{1}$$

In the last two equalities in the above expression we have used that the bundle has $SU(3)$ structure group and that $h^3(\mathcal{O}) = 1$ on a Calabi-Yau threefold. If these cohomology groups are represented in terms of bundle valued forms $\omega_i \in H^1(V_X)$ for $i = 1, \ldots, h^1(V_X)$ then this map can be represented by the following integral [18].

$$\int_X \mathrm{tr}(\omega_i \wedge \omega_j \wedge \omega_k) \wedge \Omega \tag{2}$$

Here the tr indicates taking the $SU(3)$ invariant combination and Ω is the nowhere vanishing holomorphic three-form on X. A similar, but somewhat more complex, discussion to that given here exists for the case of some other structure groups besides $SU(3)$ [8, 14, 18].

In building models of phenomenological particle physics, a generic guideline is that, unless there is a good reason for a Yukawa coupling to vanish, then one should not set it to zero. After all, if a number could take any value in the complex plane it seems improbable that it would turn out to be precisely and exactly zero by chance. The canonical example of a 'good reason' would be a symmetry that forbids the coupling from appearing in the theory. The surprise seen in string theory models (see [1, 2, 12, 13, 17, 19] for some examples) is that many of these couplings do vanish, even though there is no known associated symmetry that would account for such a feature. There are several possible reasons that this phenomenon might occur. One is that there may be a, previously unknown, symmetry for such systems that has thus far been missed. If such a symmetry did exist it would be of great interest to the subject of string theory compactifications to understand it. Another possibility is that for these vanishings the 'good reason' may be associated to quasi-topological properties of the map (1). This too would be exciting, in that it would provide a signature of the higher dimensional nature of the theory. From the point of view of the four-dimensional particle physics model these vanishings would be seemingly unexplained, and only the higher dimensional avatar would provide a reason as to why this effect had occurred.[1]

2 Vanishing Couplings and Geometrical Structures

In this short note we will focus on just two types of effect that can lead to vanishing Yukawa couplings.

- Constraints descending from fibration structures of X
- Constraints descending from the possible embeddings of X into some larger ambient space.

Let us start by considering the case where vanishing Yukawa couplings are caused by the existence of a torus fibration structure of X. A discussion of this situation can be found in [12] and we summarize some of those results here. In such a case, we have a projection map from X to some two dimensional base of the fibration B_2.

$$X \xrightarrow{\pi} B_2 \tag{1}$$

For simplicity we will consider the case where V_X pushes down to a bundle on B_2. In such a case we have that

$$H^1(X, V_X) = \bigoplus_{p,q}^{p+q=1} H^p(B_2, R^q\pi_* V_X) \tag{2}$$

This implies that we have a (p, q) grading on $H^1(X, V_X)$, which we will denote by

$$H^p(B_2, R^q\pi_* V_X) := (p, q|V_X)\ . \tag{3}$$

This grading is preserved under mappings such as (1).

$$(p_1, q_1|V_X) \otimes (p_2, q_2|V_X) \otimes (p_3, q_3|V_X) \to (p_1 + p_2 + p_3, q_1 + q_2 + q_3| \wedge^3 V_X) \tag{4}$$

The target in (1) is (2, 1) graded: $H^3(X, \mathcal{O}) = (2, 1|\mathcal{O})$. Thus for a non-zero Yukawa coupling we require three matter fields with grading (p_i, q_i) such that $p_1 + p_2 + p_3 = 2$ and $q_1 + q_2 + q_3 = 1$. In any other case the coupling will be zero. In the language of forms, via the condition (2), this is simply the statement that we require the form $\mathrm{tr}(\omega_i \wedge \omega_j \wedge \omega_k)$ to have two legs on the base and one on the fiber. This

[1] For particle physics experts, it should be noted that the couplings being discussed here are high energy superpotential parameters. If there is truly no symmetry involved in this phenomenon, then this effect would provide a seemingly unexplained set of vanishings which would appear in the initial conditions for a computation of running couplings. This could lead to unexplained textures at low energies.

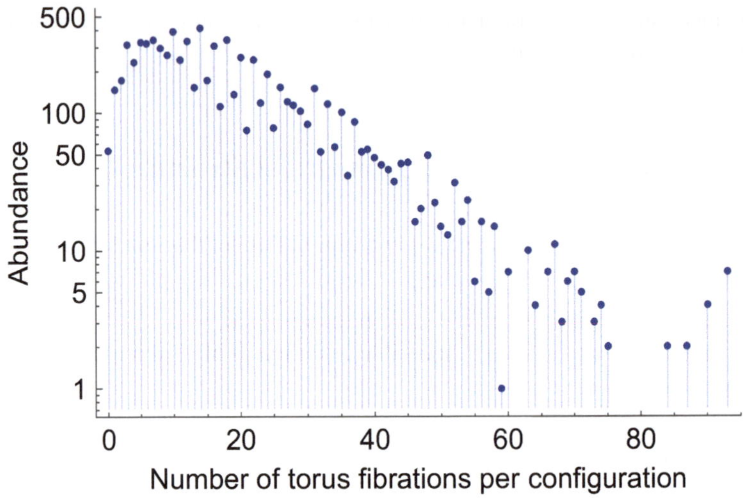

Fig. 1 The number of Calabi-Yau threefolds (abundance), described as a complete intersection in products of projective spaces, that admit a given number of inequivalent torus fibrations. Figure taken from [23]

can then be paired with Ω, which has a similar structure, except that the legs are holomorphic rather than anti-holomorphic, to form a top form which can be integrated to give a non-zero value of the coupling.

The above discussion is for the simple case where V_X pushes down to a bundle on B_2. We have focussed on this situation in the interests of brevity in this short note. However, a somewhat modified expression analogous to (4) does exist in more general cases, as is discussed in [12].

The reason that constraints on Yukawa couplings due to fibrations are so important is that such substructures are ubiquitous in the known set of Calabi-Yau threefolds (see [20–29] for studies of this in the physics literature and [30] for a recent discussion of this point in the mathematics literature). Indeed, in cases that have been analyzed in complete generality most Calabi-Yau manifolds are torus fibered in multiple different manners (see Fig. 1) [20, 21, 23, 31]. Each of these different fibrations will lead to a different pattern of vanishing Yukawa couplings, all of which must then be manifested in the resulting physical theory.

Note that one would expect other fibrations, such as $K3$ fibrations, to offer similar constraints, although to our knowledge this has not been systematically investigated in the literature to date. Such fibrations are also ubiquitous in the known set of Calabi-Yau threefolds, with most manifolds being multiply fibered by every smaller dimensional Calabi-Yau n-folds in cases that have been studied (see [23] and Fig. 2).

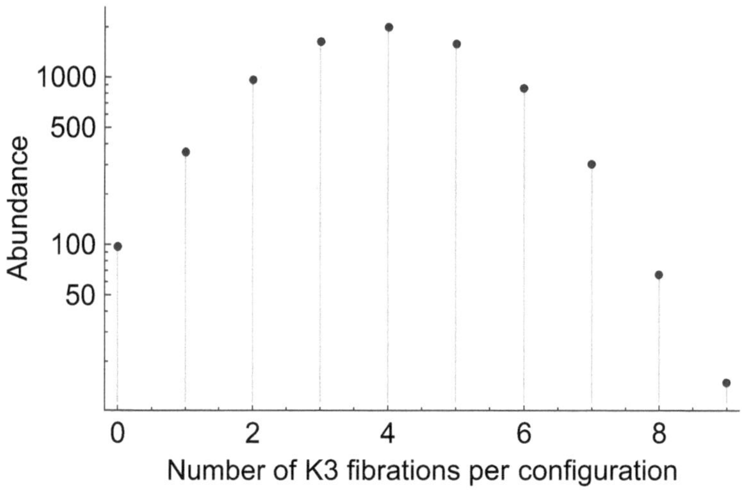

Fig. 2 The number of Calabi-Yau threefolds (abundance), described as a complete intersection in products of projective spaces, that admit a given number of inequivalent K3 fibrations. Figure taken from [23]

Another source of vanishing Yukawa couplings stems from the description of Calabi-Yau manifolds as sub-loci inside a larger ambient space. The simplest form of this effect was described, and analyzed in a differential form associated to (2), in [1, 2]. It was then analyzed in an algebraic language associated to (1) and generalized in [17].

To see the basic effect, consider a Calabi-Yau threefold X embedded in an ambient space A. For simplicity we will consider X to be defined as a hypersurface, with normal bundle $\mathcal{N}$, although the generalization to complete intersections is straight forward and included in the above references. Initially, we will also simplify the discussion to the case where the bundle V_X on X is the restriction of a bundle V on the ambient space. The more general case where V_X extends to a sheaf was considered in [17] and we will return to that situation shortly. In the case where V is a bundle we have a short exact Koszul sequence as follows.

$$0 \to \mathcal{N}^\vee \otimes V \to V \to V|_X \to 0 \tag{5}$$

Studying the associated long exact sequence in cohomology, we find that the first cohomology of V_X can be expressed in the following manner.

$$H^1(V_X) = H^1(V|_X) = \text{coker}\left\{H^1(\mathcal{N}^\vee \otimes V) \to H^1(V)\right\} + \text{ker}\left\{H^2(\mathcal{N}^\vee \otimes V) \to H^2(V)\right\} \tag{6}$$

Now, as in (2), we once again have a grading. Following [1, 2], a field is referred to as 'type i' if it descends from the ith ambient space cohomology. So for example, the fields associated to the coker in (6) are of type 1 while those associated to the ker are of type 2. The authors of [1, 2] then go on to prove that the Yukawa couplings between three fields will vanish if the sum of their types is less than the dimension of the ambient space A.

This condition can already be restrictive when considering a description of X in terms of just a single ambient space. For example, in [19], a study was performed of how many Yukawa couplings, allowed by every known symmetry, were forbidden by this effect in the largest data set of particle physics standard models that has been created using the construction under discussion [32–34]. There it was shown that this effect lead to unexpected vanishings of Yukawa couplings in 30% of the models in the data set.

The vanishings due to a single ambient space embedding of X, however, are not the full story in considering the implications of this phenomenon for Yukawa couplings in string theory models. Every Calabi-Yau manifold can be embedded in a huge variety of ambient spaces, and each of these descriptions of the manifold can potentially lead to different constraints on the Yukawa couplings. For example, the following four configuration matrices all describe the same Calabi-Yau threefold.

$$\left[\begin{array}{c|cccccccc} \mathbb{P}^1 & 1&1&0&0&0&0&0&0 \\ \mathbb{P}^1 & 0&0&1&0&0&0&1&0 \\ \mathbb{P}^1 & 1&0&0&1&0&0&0&0 \\ \mathbb{P}^1 & 0&0&0&0&1&1&0&0 \\ \mathbb{P}^1 & 1&0&0&0&0&0&0&1 \\ \mathbb{P}^2 & 0&1&0&0&0&0&1&1 \\ \mathbb{P}^2 & 0&0&1&0&1&0&0&1 \\ \mathbb{P}^2 & 0&0&0&1&0&2&0&0 \end{array}\right] = \left[\begin{array}{c|ccccc} \mathbb{P}^1 & 1&1&0&0&0 \\ \mathbb{P}^1 & 0&0&2&0&0 \\ \mathbb{P}^1 & 0&0&0&0&2 \\ \mathbb{P}^1 & 1&0&0&1&0 \\ \mathbb{P}^1 & 1&0&0&1&0 \\ \mathbb{P}^3 & 0&1&1&1&1 \end{array}\right]$$
$$= \left[\begin{array}{c|cccccc} \mathbb{P}^1 & 1&1&0&0&0&0 \\ \mathbb{P}^1 & 0&0&1&1&0&0 \\ \mathbb{P}^1 & 0&0&0&0&2&0 \\ \mathbb{P}^1 & 0&0&1&0&0&1 \\ \mathbb{P}^1 & 0&0&1&0&0&1 \\ \mathbb{P}^4 & 1&1&0&1&1&1 \end{array}\right] = \left[\begin{array}{c|ccccccccc} \mathbb{P}^1 & 0&0&0&1&1&0&0&0&0 \\ \mathbb{P}^1 & 0&0&0&0&0&1&1&0&0 \\ \mathbb{P}^1 & 1&0&0&0&0&0&0&1&0 \\ \mathbb{P}^1 & 0&1&0&0&0&0&0&0&1 \\ \mathbb{P}^1 & 0&0&1&0&0&0&0&0&1 \\ \mathbb{P}^5 & 0&0&0&1&1&1&1&1&1 \\ \mathbb{P}^2 & 1&1&1&0&0&0&0&0&0 \end{array}\right] \tag{7}$$

In each of these matrices, the first column gives a series of projective spaces, the direct product of which is A. As we can see, in this simple example all of the ambient spaces are different. The remaining columns in each matrix each denotes the multi-degree of a defining polynomial of the variety with respect to the homogeneous coordinates of the projective ambient factors. All of the different constraints on couplings descending from such different descriptions of X must be realized in the physics of the compactification on that manifold, which should not depend on the description being used of the threefold.

To finish our discussion let us return briefly to the case where the bundle V_X descends from a sheaf, not a bundle, on the ambient space A in some description. In such a situation, the exact Koszul sequence (such as (5)) is replaced with a sequence which is does not terminate on the left. The above discussion can then be generalized by using the following result [17].

Given any exact sequence,

$$\ldots \to \mathcal{F}_2 \to \mathcal{F}_1 \to \mathcal{F}_0 \to V_X \to 0 \tag{8}$$

we can define a notion of 'type' for elements of $H^1(V_X)$. Namely, we say that such a cohomology element has a type $\tau = i$ if it descends from an element of $H^i(\mathcal{F}_{i-1})$. Then, if $H^3(\wedge^3 \mathcal{F}_0) = 0$, all Yukawa couplings between three type $\tau = 1$ fields will vanish. If, in addition, $H^4(\mathcal{F}_1 \otimes \wedge^2 \mathcal{F}_0) = 0$ then all Yukawa couplings between one type $\tau = 2$ and two type $\tau = 1$ fields will also vanish.

This result was used in [17] to show that, even for a given ambient space, different sheaf lifts of V_X can lead to different constraints on the Yukawa couplings. As a simple example, consider the Calabi-Yau threefold given by the following configuration matrix.

$$X = \left[\begin{array}{c|cc} \mathbb{P}^1 & 1 & 1 \\ \mathbb{P}^1 & 1 & 1 \\ \mathbb{P}^1 & 1 & 1 \\ \mathbb{P}^1 & 1 & 1 \\ \mathbb{P}^1 & 1 & 1 \end{array}\right] \tag{9}$$

A simple sum of line bundles $V_X = \mathcal{O}_X(2, -2, 0, 0, 0) \oplus \mathcal{O}_X(-1, 1, 0, 0, 0)^{\oplus 2}$ can descend from the ambient space of this example either from a bundle,

$$V_{\text{bundle}} = \mathcal{O}(2, -2, 0, 0, 0) \oplus \mathcal{O}(-1, 1, 0, 0, 0)^{\oplus 2}, \tag{10}$$

or from a sheaf

$$V_{\text{sheaf}} = \mathcal{O}(2, -2, 0, 0, 0) \otimes \mathcal{I}_L \oplus \mathcal{O}(-1, 1, 0, 0, 0)^{\oplus 2}. \tag{11}$$

In these expressions the arguments in the line bundles indicate the coefficients in an expansion of the first Chern class of that object in a basis of Kähler forms associated to the ambient space factors of (9) (restricted to X if necessary). In defining the ideal sheaf $\mathcal{I}_L$ we have defined a curve given by the following configuration matrix.

$$L = \left[\begin{array}{c|cccc} \mathbb{P}^1 & 1 & 1 & 1 & 1 \\ \mathbb{P}^1 & 1 & 1 & 1 & 1 \\ \mathbb{P}^1 & 1 & 1 & 1 & 1 \\ \mathbb{P}^1 & 1 & 1 & 1 & 1 \\ \mathbb{P}^1 & 1 & 1 & 1 & 1 \end{array}\right] \tag{12}$$

The ideal sheaf itself is then defined via the usual short exact sequence.

$$0 \to \mathcal{I}_L \to \mathcal{O} \to \mathcal{O}_L \to 0 \tag{13}$$

The constraints on Yukawa couplings descending from Koszul sequences for (10) and (11) are different. In particular, in this example, the sheaf lift of V_X leads to no constraints on the Yukawa couplings at all, whereas the bundle lift leads to every coupling being forbidden. Thus, different lifts of the bundle can lead to different constraints on the Yukawa couplings, even in a single ambient space description of X.

3 Conclusions

In this short note we have seen a plethora of geometrical structures that can lead to patterns of vanishing Yukawa couplings in string theory models. Most known Calabi-Yau threefolds admit multiple fibration structures, each of which can lead to such features in the physical theory. In addition Calabi-Yau manifolds can be embedded in a wide variety of ambient spaces, and most of these embeddings will lead to a pattern of vanishing Yukawa couplings. Further, even for a given ambient embedding, different extensions of the gauge bundle to the ambient space can lead to different patterns of vanishing Yukawa couplings. This entire plethora of possible vanishing patterns of Yukawa couplings must *all* be realized in the associated particle physics theory. As such, it is common for Yukawa couplings to vanish in these constructions, even if there is no apparent symmetry that would cause these numbers to be zero. In fact, there are other effects that can also lead to vanishing Yukawa couplings that we have not had space to cover in this short note. For example, structures in the Kähler cone associated to the poly-stability of the vector bundle V_X [15, 35–37] can also lead to zero couplings [15, 16, 38].

In studying these phenomena one is left with the inescapable feeling that something is being missed. While it is true that all of these disparate causes do lead to vanishing couplings, one might expect that this phenomenon could be understood in a more unified manner. This could either be in terms of a symmetry that has thus far been missed in these constructions, or something more exotic. For example, perhaps there is some unified quasi-topological framework which encapsulates all of the possible information about vanishing couplings in a manner that doesn't refer to ancillary structures such as particular ambient spaces.

Whether there is a previously unknown symmetry of these string theory constructions, or a more exotic mechanism that predicts otherwise unexpected vanishings of Yukawa couplings in particle physics models coming from string theory, it seems that there is an exciting structure underlying these models which is waiting to be discovered.

Acknowledgements The work of LA and JG is supported by the NSF grant PHY-2014086. M.L., M.M. and R.S. are funded in part by Vetenskapsrådet, under grant number 2020-03230. The authors would like to thank the organizers of the Nankai Symposium on Mathematical Dialogues where the talk upon which this short note is based was presented.

References

1. S. Blesneag, E. I. Buchbinder, P. Candelas and A. Lukas, "Holomorphic Yukawa Couplings in Heterotic String Theory," JHEP **01**, 152 (2016) [arXiv:1512.05322 [hep-th]].
2. S. Blesneag, E. I. Buchbinder and A. Lukas, "Holomorphic Yukawa Couplings for Complete Intersection Calabi-Yau Manifolds," JHEP **01**, 119 (2017) [arXiv:1607.03461 [hep-th]].
3. A. Strominger, "Yukawa Couplings in Superstring Compactification," Phys. Rev. Lett. **55**, 2547 (1985).
4. B. R. Greene, K. H. Kirklin, P. J. Miron and G. G. Ross, "A Three Generation Superstring Model. 1. Compactification and Discrete Symmetries," Nucl. Phys. B **278** (1986) 667.
5. B. R. Greene, K. H. Kirklin, P. J. Miron and G. G. Ross, "A Three Generation Superstring Model. 2. Symmetry Breaking and the Low-Energy Theory," Nucl. Phys. B **292** (1987) 606.
6. P. Candelas, "Yukawa Couplings Between (2,1) Forms," Nucl. Phys. B **298**, 458 (1988).
7. J. Distler, B. R. Greene, K. Kirklin and P. Miron, "Evaluation of 27-bar**3 Yukawa Couplings in a Three Generation Superstring Model," Phys. Lett. B **195**, 41 (1987).
8. J. Distler and B. R. Greene, "Aspects of (2,0) String Compactifications," Nucl. Phys. B **304** (1988) 1.
9. B. R. Greene, K. H. Kirklin, P. J. Miron and G. G. Ross, "27**3 Yukawa Couplings for a Three Generation Superstring Model," Phys. Lett. B **192**, 111 (1987).
10. P. Candelas and X. de la Ossa, "Moduli Space of Calabi-Yau Manifolds," Nucl. Phys. B **355**, 455 (1991).
11. J. Distler and S. Kachru, "Duality of (0,2) string vacua," Nucl. Phys. B **442**, 64 (1995) [hep-th/9501111].
12. V. Braun, Y. H. He and B. A. Ovrut, "Yukawa couplings in heterotic standard models," JHEP **0604**, 019 (2006) [hep-th/0601204].
13. V. Bouchard, M. Cvetic and R. Donagi, "Tri-linear couplings in an heterotic minimal supersymmetric standard model," Nucl. Phys. B **745**, 62 (2006) [arXiv:hep-th/0602096].
14. L. B. Anderson, J. Gray, D. Grayson, Y. H. He and A. Lukas, "Yukawa Couplings in Heterotic Compactification," Commun. Math. Phys. **297**, 95-127 (2010) [arXiv:0904.2186 [hep-th]].
15. L. B. Anderson, J. Gray and B. Ovrut, "Yukawa Textures From Heterotic Stability Walls," JHEP **1005**, 086 (2010) [arXiv:1001.2317 [hep-th]].
16. E. I. Buchbinder, A. Constantin and A. Lukas, "Non-generic Couplings in Supersymmetric Standard Models," Phys. Lett. B **748**, 251 (2015) [arXiv:1409.2412 [hep-th]].
17. L. B. Anderson, J. Gray, M. Larfors, M. Magill and R. Schneider, "Generalized Vanishing Theorems for Yukawa Couplings in Heterotic Compactifications," JHEP **05**, 085 (2021) [arXiv:2103.10454 [hep-th]].
18. M. B. Green, J. H. Schwarz and E. Witten, "Superstring Theory. Vol. 2: Loop Amplitudes, Anomalies and Phenomenology,"
19. J. Gray and J. Wang, "Jumping Spectra and Vanishing Couplings in Heterotic Line Bundle Standard Models," JHEP **11**, 073 (2019) [arXiv:1906.09373 [hep-th]].
20. J. Gray, A. S. Haupt and A. Lukas, "Topological Invariants and Fibration Structure of Complete Intersection Calabi-Yau Four-Folds," JHEP **1409**, 093 (2014) [arXiv:1405.2073 [hep-th]].
21. L. B. Anderson, X. Gao, J. Gray and S. J. Lee, "Multiple Fibrations in Calabi-Yau Geometry and String Dualities," JHEP **10**, 105 (2016) [arXiv:1608.07555 [hep-th]].
22. L. B. Anderson, X. Gao, J. Gray and S. J. Lee, "Tools for CICYs in F-theory," JHEP **11**, 004 (2016) [arXiv:1608.07554 [hep-th]].
23. L. B. Anderson, X. Gao, J. Gray and S. J. Lee, "Fibrations in CICY Threefolds," JHEP **10**, 077 (2017) [arXiv:1708.07907 [hep-th]].
24. L. B. Anderson, J. Gray and B. Hammack, "Fibrations in Non-simply Connected Calabi-Yau Quotients," JHEP **08**, 128 (2018) [arXiv:1805.05497 [hep-th]].
25. W. Taylor, "On the Hodge structure of elliptically fibered Calabi-Yau threefolds," JHEP **08**, 032 (2012) [arXiv:1205.0952 [hep-th]].
26. S. B. Johnson and W. Taylor, "Calabi-Yau threefolds with large $h^{2,1}$," JHEP **10**, 023 (2014) [arXiv:1406.0514 [hep-th]].

27. Y. C. Huang and W. Taylor, "Comparing elliptic and toric hypersurface Calabi-Yau threefolds at large Hodge numbers," JHEP **02**, 087 (2019) [arXiv:1805.05907 [hep-th]].
28. Y. C. Huang and W. Taylor, "On the prevalence of elliptic and genus one fibrations among toric hypersurface Calabi-Yau threefolds," JHEP **03**, 014 (2019) [arXiv:1809.05160 [hep-th]].
29. Y. C. Huang and W. Taylor, "Fibration structure in toric hypersurface Calabi-Yau threefolds," JHEP **03**, 172 (2020) [arXiv:1907.09482 [hep-th]].
30. S. Filipazzi, C. D. Hacon and R. Svaldi, "Boundedness of elliptic Calabi-Yau threefolds," [arXiv:2112.01352 [math.AG]].
31. F. Rohsiepe, "Fibration structures in toric Calabi-Yau fourfolds," [arXiv:hep-th/0502138 [hep-th]].
32. L. B. Anderson, J. Gray, A. Lukas and E. Palti, "Two Hundred Heterotic Standard Models on Smooth Calabi-Yau Threefolds," Phys. Rev. D **84**, 106005 (2011) [arXiv:1106.4804 [hep-th]].
33. L. B. Anderson, J. Gray, A. Lukas and E. Palti, "Heterotic Line Bundle Standard Models," JHEP **06**, 113 (2012) [arXiv:1202.1757 [hep-th]].
34. L. B. Anderson, A. Constantin, J. Gray, A. Lukas and E. Palti, "A Comprehensive Scan for Heterotic SU(5) GUT models," JHEP **01**, 047 (2014) [arXiv:1307.4787 [hep-th]].
35. E. R. Sharpe, "Kahler cone substructure," Adv. Theor. Math. Phys. **2**, 1441-1462 (1999) [arXiv:hep-th/9810064 [hep-th]].
36. L. B. Anderson, J. Gray, A. Lukas and B. Ovrut, "The Edge Of Supersymmetry: Stability Walls in Heterotic Theory," Phys. Lett. B **677**, 190-194 (2009) [arXiv:0903.5088 [hep-th]].
37. L. B. Anderson, J. Gray, A. Lukas and B. Ovrut, "Stability Walls in Heterotic Theories," JHEP **09**, 026 (2009) [arXiv:0905.1748 [hep-th]].
38. M. Kuriyama, H. Nakajima and T. Watari, "Theoretical Framework for R-parity Violation," Phys. Rev. D **79**, 075002 (2009) [arXiv:0802.2584 [hep-ph]].

Calabi-Yau Genus-One Fibrations and Twisted Dimensional Reductions of F-Theory

Lara B. Anderson, James Gray, and Paul-Konstantin Oehlmann

Abstract In this brief note we explore the space of genus one and elliptic fibrations within CY manifolds, their organizing principles, and how they relate to the set of all CY manifolds. We provide examples of genus one fibered manifolds that exhibit different Hodge numbers—and physically lead to different gauge groups—than their Jacobian fibrations. We suggest a physical mechanism for understanding this difference in twisted circle reductions of 6-dimensional compactifications of F-theory.

1 Introduction

Much recent work has been devoted to mapping out the space of string vacua, in the context of the string Swampland program and beyond. Playing a central role in this study, string compactifications sweep out a rich landscape of possible solutions. In this context the question of *which effective field theories* can arise is intrinsically linked to the question of *which geometries* can serve as backgrounds for string dimensional reductions? In this work we will explore briefly some of what is understood about one of the simplest classes of such of string geometries—Calabi-Yau (CY) manifolds and how attempts to bound/characterize Calabi-Yau geometry can shed light on the space of possible M-/F-theory vacua in 6-/5-dimensions.

Since the very first examples of string compactifications [1], Calabi-Yau geometries have played an important role. Even in this well-studied context though, there remain many open questions about CY geometries—including whether the class of CY n-folds is finite for $n > 2$? It is also unknown whether the possible values of topological invariants (i.e. for a CY 3-fold, independent Hodge numbers $h^{1,1}$, $h^{2,1}$, Chern classes $c_2(X)$, etc) are finite/bounded.

L. B. Anderson · J. Gray (✉)
Physics Department, Robeson Hall, Virginia Tech, Blacksburg, VA, USA
e-mail: jamesgray@vt.edu

P.-K. Oehlmann
Department of Physics and Astronomy, Uppsala University, Uppsala, Sweden

Y.-H. He et al. (eds.), *Nankai Symposium on Mathematical Dialogues*,
https://doi.org/10.1007/978-981-19-2328-9_3

One path towards resolving these questions involves the study of genus-one fibered CY manifolds. In particular, the set of genus one fibered CY 3-folds has been proven to be finite [3, 5, 6], and recent progress [2, 4] has given evidence of boundedness for genus one fibered CY 4- and 5-folds as well. Moreover, recent analysis of existing datasets of CY 3- and 4-folds has demonstrated that the vast majority of all CY manifolds exhibit genus-one fibrations [7–16]. This apparent genericity of fibration structures provides hope that a classifications of CY genus-one fibrations may lead to tools which could be used to establish the finiteness of the set of *all* CY n-folds.

Returning to the question of string effective field theories, elliptic and genus one fibered CY geometries are natural backgrounds for F-/M-theory in 6- and 5-dimensions [17–19]. As a result, more complete characterization of these CY geometries makes it possible to attempt to systematically map out a subspace of the space of M/F-theory vacua. In particular, the close relationship between F-/M-theory under a circle (S^1) reduction of the 6-dimensional theory has yielded much information about the possible structure of such theories and inspired many recent classifications of both EFTs and SCFT limits.

In this brief work, we will highlight one aspect of genus-one/elliptic CY geometry that has remained relatively unexplored and the consequences for M/F-theory on elliptically and genus one fibered Calabi-Yau threefolds. In particular, we are interested mathematically in the question of *how different can a genus one fibered Calabi-Yau manifold and its nearest elliptic cousin—its Jacobian, $J(X)$, be?* Genus one fibered geometries remain much less explicitly explored than elliptically fibered manifolds and the notion of a Jacobian fibration plays a key role in characterizing genus one fibrations. As we will review below, many different genus one fibered manifolds can be related to a single elliptic fibration and the set of manifolds related in this way is called a group of "CY torsors" (see e.g. [38]). Phrasing the question above more broadly, we are interested in how distinct are the manifolds/elements in the group of CY torsors? A physical application of this question arises in the context of S^1 reductions of 6-dimensional F-theory effective theories which include flux along the circle and the possibility of *different non-Abelian gauge groups* in the 6- and 5-dimensional theories. This will include an F-/M-theory realization of Dynkin diagrams associated to twisted affine Lie groups and automorphisms that "fold" them [42]. In this context we hope to map out a relatively unexplored corner of this string effective field space. This study is soon to appear [20] and we provide only a sketch of some of these questions and an illustrative example in the following sections.

2 Elliptically and Genus-One fibered Calabi-Yau Geometry and the Basic Ingredients of F-Theory

In this section we will briefly review some essential results about elliptic/genus one fibered CY manifolds and their consequences for F-/M-theory. In particular, we shall refer to a *genus one fibered* CY n-fold as one that can be described via a surjection

$\pi : X \rightarrow B_{n-1}$ such that the generic fiber is $\pi^{-1}(b) \simeq \mathbb{E}$ (i.e. a genus one complex curve or T^2 as a real manifold) for general points $b \in B_{n-1}$. In general such fibrations do not admit sections, only multi-sections (i.e. m-sheeted covers of the base manifold B_{n-1}). In the case that the fibration *does* admit a section, the generic fibers are elliptic curves and we refer to the fibration as an *elliptic fibration*. It is this latter class of manifolds that plays a key role in F-theory in 6- and 4-dimensions in which the elliptic fibers of a CY 3 or 4-fold respectively parameterize the behavior of the axio-dilaton of the (strongly coupled) type IIB theory.

As described above, recent work has attempted to map out how generic genus-one fibered CY 3-folds/4-folds are within the datasets of known CY manifolds. Overwhelmingly, the available datasets have shown that the presence of fibration structures (i.e. CY m-fold fibrations within CY n-folds with $m < n$) is ubiquitous for large[1] $h^{1,1}$. Moreover, a given CY manifold not only appears to be describable as a fibration, generic CY manifolds appear to admit *multiple* such descriptions. In [9, 15], *multiply elliptically fibered* (or genus one fibered in the case without section) CY n-folds were described as manifolds admitting multiple descriptions of the form $\pi_i : X_n \longrightarrow B^{(i)}_{n-1}$ with elliptic fiber $\mathbb{E}_{(i)b} = \pi^{-1}(b \in B^{(i)}_{n-1})$ (denoted succinctly by $\pi_i : X_n \xrightarrow{\mathbb{E}_{(i)}} B^{(i)}_{n-1}$). That is,

$$\begin{array}{ccccc} & & X_n & & \\ & \swarrow^{\mathbb{E}_{(1)}} & \downarrow^{\mathbb{E}_{(2)}} & \searrow^{\mathbb{E}_{(i)}} & \\ B^{(1)}_{n-1} & & B^{(2)}_{n-1} \cdots & & B^{(i)}_{n-1} \end{array} \tag{2.1}$$

It is important to note that for each fibration, π_i, the structure of the singular fibers, discriminant locus, and number of rational sections can all be different, as can the topology of the base manifolds $B^{(i)}_{n-1}$. This can lead to a rich array of consequences for string dualities. Initial steps to explore such prolific fibration structures and their physical consequences were taken in [7, 9, 15] and the results from [9] for one dataset are shown in Fig. 1.

Elliptically fibered manifolds exhibit a simple characterization mathematically in that each such manifold is birational to a specified minimal form, a so-called *Weierstrass model* [21]. Schematically, a Weierstrass model for an elliptic fibration is built as a hypersurface constraint (with cubic fiber) within a projectivization of three line bundles $\mathbb{P}(\mathcal{O} \oplus \mathcal{L}^2 \oplus \mathcal{L}^3)$ with $\mathcal{L} = K^{-1}_{B_{n-1}}$ in the CY case. This minimal form is achieved by blowing down all components of reducible fibers which do not intersect the zero-section to the elliptic fibration, leaving a generically singular, but simple CY variety.

Another useful property of elliptic fibrations is that the Picard lattice/Kähler forms in such a space can be sorted according to their vertical/horizontal form relative to the elliptic fibration. Namely, each divisor belongs to one of four types [22]: (1)

[1] In fact, fibrations seem to occur for all manifolds with $h^{1,1} > 19$ in known CY datasets where classifications have been undertaken (see e.g. [9]).

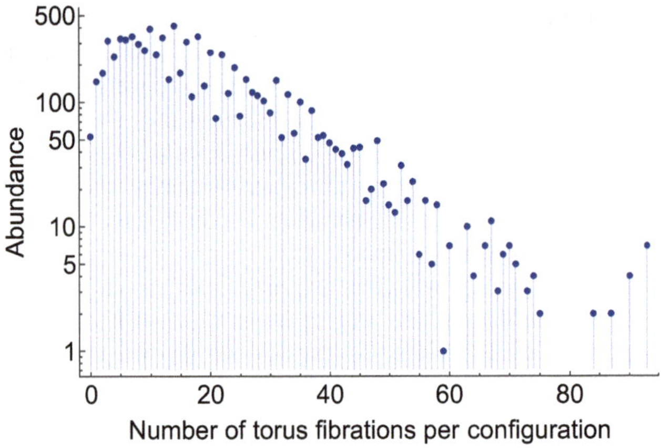

Fig. 1 Distribution of obvious torus fibrations in the set of CY 3-folds described as complete intersections in products of projective spaces from [9]. The values lie in the range 0–93. The search in [9] yielded 139, 597 fibrations in total

The divisors of the base, B_{n-1} pulled back to X, (2) "fibral" divisors associated to blow-ups of singular elliptic fibers, (3) the zero-section and (4) the free part of the so-called Mordell-Weil group of rational sections (see e.g. [23]). Within F-theory in 6-dimensional theories the number of "fibral" divisors determine the rank of a non-Abelian gauge group, while the rank of the Mordell-Weil group determines the number of Abelian factors in the effective theory. Elliptic CY fibrations have played a central role in F-theory for the past several decades and many systematic datasets have been constructed [24–27].

In contrast, the set of genus-one fibered CY manifolds has remained more obscure until recently. The absence of a zero-section to the fibration makes it difficult to interpret these geometries directly as F-theory backgrounds in any dimension. However, work in recent years has characterized explicitly how they can be viewed as solutions of M-theory in one dimension lower. The precise relationship between M-theory on a genus one fibered manifold in $(m-1)$-dimensions and m-dimensional solutions of F-theory on its elliptically fibered "cousin" has been very well studied in the context of Abelian gauge symmetries (see e.g. [28–34])

From a mathematical point of view, given a genus one fibered CY manifold, X, the elliptic CY geometry most closely resembling it is the so-called "Jacobian fibration" associated to X. The Jacobian fibration, $J(X)$ is the fiber-wise Jacobian of each genus one fiber of X. By construction, the Jacobian of a genus-one fibered CY manifold is an elliptically fibered manifold which shares the same discriminant locus to the fibration (and the same j-invariant associated to each fiber). In general, for CY threefolds $J(X)$ is birational to X (and shares the same value for $h^{2,1}$) and furthermore, multiple distinct genus one fibered manifolds can share *the same* Jacobian. The set of such geometries (together with information about the form the birational mapping takes) form a group of "CY Torsors" [38]. In this context the Tate-

Schaferavich group and the Weil-Chatelet group have played roles in the literature (see e.g. [43]) as examples of CY torsors.

While many systematic constructions of datasets of elliptically fibered CY manifolds exist (see e.g. [24]) no such systematic dataset has been built for genus-one fibered manifolds. One step in this direction would be to understand how to build *all* genus one fibered manifolds in the same CY torsor group—i.e. given a CY elliptic fibration, X_0, is it possible to construct[2] all genus one fibered manifolds X_i which have X_0 as their Jacobian: $J(X_i) = X_0$? To this end, in the next sections we will explore an explicit example in which a genus one fibered manifold and its Jacobian are quite different indeed and the new physics that this must involve in F-/M-theory.

3 S^1-Dimensional Reductions of F-Theory in 6-Dimensions and 5-Dimensional M-Theory and Twisted Boundary Conditions

Let us begin by briefly reviewing a now well-established story involving F-theory circle reductions with *Abelian gauge groups* and M-theory on genus one fibered Calabi-Yau manifolds (see [34] for a review). We will outline a few key ideas here and leave a more detailed discussion for [20]. As initially explored in [29], a 5-dimensional compactification of M-theory on a genus one fibered Calabi-Yau threefold, X is closely related to a 6-dimensional compactification of F-theory on the Jacobian of X, the elliptically fibered manifold $J(X)$ described above.

In terms of physics, a compactification of F-theory *requires* the existence of a section in addition to a T^2 fiber and thus, we can consider defining F-theory on a given elliptically fibered CY threefold. However, this single compactification of F-theory can give rise to a number of distinct 5-dimensional compactifications of M-theory via circle reduction with flux (and possibly matter field vevs) [44, 45]. In particular, if the 6-dimensional effective theory arising from F-theory is Abelian, then it is possible to consider a circle flux (i.e. discrete holonomy), $\xi = \int_{S^1} A$. Different choices of ξ lead to different compactifications of M-theory on (possibly) distinct genus-one manifolds which are different elements of the group of CY torsors[3] (note that the reduction with $\xi = 0$ leads to an M-theory compactification on the elliptically fibered manifold $J(X)$ itself) [29, 30, 33]. The set of these 5-dimensional theories are connected by geometric transitions in the underlying genus one fibered manifolds. This framework has been used to gain a deep understanding of discrete symmetries in F-/M-theory (see [31] for links to discrete torsion in genus-one/elliptically fibered CY geometries).

In the context of the questions raised here about Jacobian fibrations and groups of CY torsors, we are interested in manifolds that also contain fibral divisors. That

[2] It should be noted that the inverse problem—constructing Jacobians, given an explicit description of the genus one fiber—is better understood. See e.g. [35–37].

[3] In simple cases the Tate-Shaferavich group.

is, compactifications of F-theory in 6-dimensions which include non-Abelian gauge group factors, in addition to the possible existence of multiple rational sections, i.e. symmetry groups of the form $G \times U(1)$ (possibly with discrete group quotients or multiple Abelian factors as well). In this case, the circle reductions can involve flux from gauge fields in the Cartan subgroup of G as well as the $U(1)$ factors. And as we will see, a new feature that is possible in this context is the presence of boundary conditions which can mix these different fluxes together in non-trivial ways. As an action on gauge fields this is schematically of the form

$$A^B(s) = \sigma(A^B(s + 2\pi)) \tag{3.1}$$

where σ is a discrete, outer automorphism of $G \times U(1)$. These boundary conditions can reduce the 5-dimensional symmetry down to a subgroup.

Such non-trivial boundary conditions (and fluxes) have been studied in a variety of other contexts under the name of "twisted" circle reductions (see e.g. [49–51]) but has not been previously implemented in F-theory. We find that in the presence of these boundary conditions we generate a new set of 5-dimensional vacua associated to M-theory compactifications on genus-one fibered manifolds with different non-Abelian gauge groups of the form $H_i \times U(1)$ (or $H_i \times \mathbb{Z}_n$) where $H_i \subseteq G$ is a sub-group. Geometrically, these geometries are novel in that the genus one fibered manifold and its Jacobian can realize *different* non-Abelian gauge groups, despite the fact that they share a discriminant locus (see also [28, 39–41]).

In [20] we find that the possible groups H, G that can arise are highly constrained and the discrete boundary conditions allowed closely relate to "foldings" of the Dynkin diagrams (see Fig. 2). Moreover, in the context of the genus one fibered geometries considered here we find that the discrete automorphisms in play are in fact not simply those of the Dynkin diagram associated to the Lie group G itself, rather we find geometric realizations of the folding of the Dynkin diagrams associated to affine groups $\tilde{G}$). This is perhaps to be expected since it is the affine Dynkin diagrams which are realized as components of the resolved elliptic fibrations. To illustrate this conceretely, we turn now to a simple example.

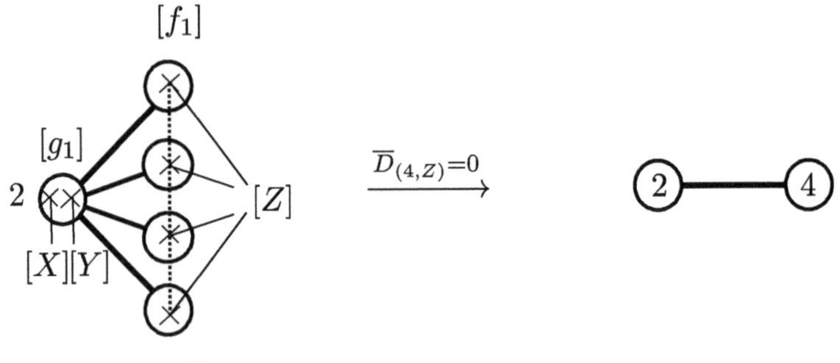

Fig. 2 Depiction of an $\mathfrak{a}_2^{(2)}$ fiber. The two two-sections X, Y intersect the middle node. All four outer roots are identified by the four-section Z. At the intersection with the monodromy divisor the fiber degenerates to a multiplicity two fiber

4 An Illustrative Example

In this section we illustrate a genus one geometry that exhibits one of the simplest twisted affine fibrations possible. This CY 3-fold is a genus one fibration $\pi : X \to \mathbb{P}^2$ and is described by the following reflexive polytope

$$\begin{array}{|c|c|} \hline X & (-1,1,0,0) \\ Y & (-1,-1,0,0) \\ Z & (1,0,0,0) \\ \hline f_1 & (1,0,1,0) \\ g_1 & (1,0,2,0) \\ z_1 & (-3,0,-1,-1) \\ z_2 & (0,0,0,1) \\ \hline \end{array} \tag{4.1}$$

which allows the computation of the Hodge numbers via the Batyrev formula as $(h^{1,1}, h^{2,1}) = (3, 107)$. The genus-one fiber is of quartic type and admits an affine $A_2^{(2)}$ fiber whose associated finite group is $SU(2)$ given as

$$p = d_1X^4 + d_2X^3Y + d_3X^2Y^2 + d_4XY^3 + d_5Y^4 + d_6f_1g_1X^2Z + d_7f_1g_1XYZ + d_8f_1g_1Y^2Z + f_1Z^2 \tag{4.2}$$

where f_1, g_1 are the fibral divisors that resolve the $A_2^{(2)}$ type of fiber and the polynomials d_i are valued in the $\mathbb{P}^2$ base. The Stanley-Reisner ideal is determined by

$$\mathcal{SRI} : \{Zg_1, YXZ, YXf_1, g_1z_1z_2, f_1z_1z_2\}\,. \tag{4.3}$$

Here we note, that the divisor $f_1 = 0$ contains four components that are intersected by the four-section $Z = 0$ given as the discriminant locus

$$p_{|Z=0} = d_1X^4 + d_2X^3Y + d_3X^2Y^2 + d_4XY^3 + d_5Y^4\,, \tag{4.4}$$

$$p_{|f_1=0} = d_1X^4 + d_2X^3Y + d_3X^2Y^2 + d_4XY^3 + d_5Y^4\,. \tag{4.5}$$

These four-components are interchanged along the discriminant locus $Disc(p_{|Z=0})$. The components in general intersect as a $A_2^{(2)}$ form fiber, and notably over $f_1 = Disc(p_{|Z=0}) = 0$, the four components degenerate to a single $\mathbb{P}^1$ and hence the fiber admits the multiplicities (4, 2) and therefore a multiple fiber [46, 47]. The monodromy which reduces the symmetry to $A_2^{(2)}$ is controlled by the following locus in the base:

$$\overline{D}_{4,Z} = -4(d_3^2 - 3d_2d_4 + 12d_1d_5)^3 + (2d_3^3 - 9d_3(d_2d_4 + 8d_1d_5) + 27(d_1d_4^2 + d_2^2d_5))^2\,, \tag{4.6}$$

See Fig. 2 for an illustration of the "folding" of this affine Dynkin diagram realized in the fiber.

Notably, such a four-section does not exist in the generic Weierstrass or Tate model. Indeed, upon shrinking the g_1 component to a singular genus-one fibration and taking the Jacobian mapping to produce an elliptic fibration in Weierstrass form [35] one finds the Weierstrass coefficients

$$f = \frac{1}{48} f_1^2 (-(4d_3 - d_7^2 f_1 + 4d_6 d_8 f_1)^2 - 24(-2d_2 d_4 + 8d_1 d_5 - 2d_5 d_6^2 f_1 + d_4 d_6 d_7 f_1 + d_8(-2d_3 d_6 + d_2 d_7 - 2d_1 d_8) f_1)) . \tag{4.7}$$

$$\begin{aligned} g = &\frac{1}{864} f_1^3 (36(-2d_2 d_4 + 8d_1 d_5 - 2d_5 d_6^2 f_1 + d_4 d_6 d_7 f_1 + d_8(-2d_3 d_6 + d_2 d_7 - 2d_1 d_8) f_1) \\ &\times (-4d_3 + (d_7^2 - 4d_6 d_8) f_1) + (-4d_3 + (d_7^2 - 4d_6 d_8) f_1)^3 \\ &+ 216((d_4^2 - 4d_3 d_5) d_6^2 f_1 + 2d_2 d_6 (2d_5 d_7 - d_4 d_8) f_1 + d_2^2(-4d_5 + d_8^2 f_1) \\ &- 4d_1 (d_4^2 - 4d_3 d_5 + d_5 d_7^2 f_1 - d_4 d_7 d_8 f_1 + d_3 d_8^2 f_1))). \end{aligned} \tag{4.8}$$

Here we find the fiber to be of D_4^* non-split form which corresponds to a G_2 gauge group in 6-dimensions [52, 53]. Such a fiber can be resolved via two more resolution divisors [54]. Thus this example illustrates the differences described above between CY torsors in that the G_2 gauge symmetry realized in the Jacobian is larger than the $SU(2)$ symmetry found in the genus one fibered manifold.

5 Summary

In this work we have given a brief illustration of the important role played by elliptic/genus-one fibrations both in mathematically characterizing Calabi-Yau manifolds as well as mapping out spaces of string solutions in various dimensions. In particular, we have provided explicit examples of genus one fibered manifolds exhibiting different singular fibers—and hence corresponding to different physical gauge groups—compared to their elliptic Jacobian counterparts. In this setting we expect these genus-one fibered geometries to correspond to the twisted dimensional reductions of F-theory described in Sect. 3. However, there remain a number of interesting questions to be addressed in this context, including (1) How is an outer automorphism of the form shown (3.1) linked to the apparent folding of the *affine* Dynkin diagrams which are realized in the elliptic fibers? (2) How are the necessary discrete symmetries realized explicitly in the CY geometry? (3) How are the CY torsors appearing in this setting linked by geometric transitions (i.e. Higgsing transitions in the 5D theory)? We will address these and other questions in soon to appear work [20]. For now we hope that the brief discussion above illustrates interesting open questions—both physical and mathematical—remaining within the study of CY fibrations.

Acknowledgements We thank Luca Cassia, Markus Dierigl, Jonathan Mboyo Esole, Antonella Grassi, Jie Gu, Sheldon Katz, David Morrison, Nikhil Raghuram and Thorsten Schimannek, for illuminating discussions. The work of L.A. and J.G. is supported by the NSF grant PHY-2014086. The work of P.K.O. is supported by a Swedish grant of the Carl Trygger Foundation for Scientific

Research and the European Research Council (ERC) under the European Union's Horizon 2020 research and innovation program (grant agreement No. 851931). The authors would like to thank the organizers of the Nankai Symposium on Mathematical Dialogues where the talk inspiring this brief survey was presented.

References

1. P. Candelas, G. T. Horowitz, A. Strominger and E. Witten, "Vacuum Configurations for Superstrings," Nucl. Phys. B **258** (1985), 46–74.
2. C. Birkar, G. Di Cerbo and R. Svaldi, "Boundedness of elliptic Calabi-Yau varieties with a rational section," arXiv:2010.09769 [math.AG].
3. A. Grassi, "On minimal models of elliptic threefolds," Math. Ann. **290** (1991), 287.
4. G. Di Cerbo and R. Svaldi, "Birational boundedness of low-dimensional elliptic Calabi?Yau varieties with a section," Comp. Math. **8** (2021), 1766–1806.
5. M. Gross, "A Finiteness theorem for elliptic Calabi-Yau threefolds," arXiv:9305002 [math.AG].
6. , S. Filipazzi, C. Hacon, and R. Svaldi, "Boundedness of elliptic Calabi-Yau threefolds," arXiv:2112.01352 [math.AG].
7. L. B. Anderson, X. Gao, J. Gray and S. J. Lee, "Multiple Fibrations in Calabi-Yau Geometry and String Dualities," JHEP **10** (2016), 105 [arXiv:1608.07555 [hep-th]].
8. L. B. Anderson, X. Gao, J. Gray and S. J. Lee, "Tools for CICYs in F-theory," JHEP **11** (2016), 004 [arXiv:1608.07554 [hep-th]].
9. L. B. Anderson, X. Gao, J. Gray and S. J. Lee, "Fibrations in CICY Threefolds," JHEP **10** (2017), 077 [arXiv:1708.07907 [hep-th]].
10. L. B. Anderson, J. Gray and B. Hammack, "Fibrations in Non-simply Connected Calabi-Yau Quotients," JHEP **08** (2018), 128 [arXiv:1805.05497 [hep-th]].
11. L. B. Anderson, J. Gray, M. Karkheiran, P. K. Oehlmann and N. Raghuram, "$\mathbb{P}^1$-fibrations in F-theory and String Dualities," [arXiv:2109.11534 [hep-th]].
12. V. Braun, "Toric Elliptic Fibrations and F-Theory Compactifications," JHEP **01** (2013), 016 [arXiv:1110.4883 [hep-th]].
13. Y. C. Huang and W. Taylor, "On the prevalence of elliptic and genus one fibrations among toric hypersurface Calabi-Yau threefolds," JHEP **03** (2019), 014 [arXiv:1809.05160 [hep-th]].
14. Y. C. Huang and W. Taylor, "Fibration structure in toric hypersurface Calabi-Yau threefolds," JHEP **03** (2020), 172 [arXiv:1907.09482 [hep-th]].
15. J. Gray, A. S. Haupt and A. Lukas, "Topological Invariants and Fibration Structure of Complete Intersection Calabi-Yau Four-Folds," JHEP **09** (2014), 093 [arXiv:1405.2073 [hep-th]].
16. F. Rohsiepe, "Fibration structures in toric Calabi-Yau fourfolds," hep-th/0502138.
17. C. Vafa, "Evidence for F theory," Nucl. Phys. B **469** (1996), 403–418 [arXiv:hep-th/9602022 [hep-th]].
18. D. R. Morrison and C. Vafa, "Compactifications of F theory on Calabi-Yau threefolds. 1," Nucl. Phys. B **473** (1996), 74–92 [arXiv:hep-th/9602114 [hep-th]].
19. D. R. Morrison and C. Vafa, "Compactifications of F theory on Calabi-Yau threefolds. 2.," Nucl. Phys. B **476** (1996), 437–469 [arXiv:hep-th/9603161 [hep-th]].
20. L. B. Anderson, J. Gray and P .K. Oehlmann, To appear.
21. N. Nakayama, "On Weierstrass Models," Alg. Geom. and Comm. Alg., (1987) 405–431.
22. R. Wazir, "Arithmetic on elliptic threefolds," Compos. Math. **140** (2004), 567–580. arXiv:math/0112259 [math.NT].
23. J. H. Silverman, "The Arithmetic of Elliptic Curves," Graduate Texts in Mathematics **106** (Springer-Verlag New York, 2009).
24. D. R. Morrison and W. Taylor, "Toric bases for 6D F-theory models," Fortsch. Phys. **60** (2012), 1187–1216 [arXiv:1204.0283 [hep-th]].

25. G. Martini and W. Taylor, "6D F-theory models and elliptically fibered Calabi-Yau threefolds over semi-toric base surfaces," JHEP **06** (2015), 061 [arXiv:1404.6300 [hep-th]].
26. L. B. Anderson and W. Taylor, "Geometric constraints in dual F-theory and heterotic string compactifications," JHEP **08** (2014), 025 [arXiv:1405.2074 [hep-th]].
27. J. Halverson and W. Taylor, "$\mathbb{P}^1$-bundle bases and the prevalence of non-Higgsable structure in 4D F-theory models," JHEP **09** (2015), 086 [arXiv:1506.03204 [hep-th]].
28. V. Braun and D. R. Morrison, "F-theory on Genus-One Fibrations," JHEP **08** (2014), 132 [arXiv:1401.7844 [hep-th]].
29. D. R. Morrison and W. Taylor, "Sections, multisections, and U(1) fields in F-theory," [arXiv:1404.1527 [hep-th]].
30. L. B. Anderson, I. García-Etxebarria, T. W. Grimm and J. Keitel, "Physics of F-theory compactifications without section," JHEP **12** (2014), 156 [arXiv:1406.5180 [hep-th]].
31. C. Mayrhofer, E. Palti, O. Till and T. Weigand, "On Discrete Symmetries and Torsion Homology in F-Theory," JHEP **1506** (2015) 029 [arXiv:1410.7814 [hep-th]].
32. M. Cvetic and L. Lin, "The Global Gauge Group Structure of F-theory Compactification with U(1)s," JHEP **01** (2018), 157 [arXiv:1706.08521 [hep-th]].
33. M. Cvetic, R. Donagi, D. Klevers, H. Piragua and M. Poretschkin, "F-theory vacua with $\mathbb{Z}_3$ gauge symmetry," Nucl. Phys. B **898** (2015) 736 [arXiv:1502.06953 [hep-th]].
34. M. Cvetič and L. Lin, "TASI Lectures on Abelian and Discrete Symmetries in F-theory," PoS **TASI2017** (2018), 020 [arXiv:1809.00012 [hep-th]].
35. M. Artin, F. Rodriguez-Villegas, J. Tate, "On the Jacobians of Plane Cubics," Adv. Math. 198 (2005) 366–382.
36. V. Braun, T. W. Grimm and J. Keitel, "Complete Intersection Fibers in F-Theory," JHEP **03**, 125 (2015) [arXiv:1411.2615 [hep-th]].
37. J. Knapp, E. Scheidegger and T. Schimannek, "On genus one fibered Calabi-Yau threefolds with 5-sections," [arXiv:2107.05647 [hep-th]].
38. I. Dolgachev and M Gross, "Elliptic threefolds I: Ogg-Shafarevich theory". J. Algebraic Geom. 3.1 (1994), pp. 39–80. ISSN:1056-3911.
39. F. B. Kohl, M. Larfors and P. K. Oehlmann, "F-theory on 6D Symmetric Toroidal Orbifolds," [arXiv:2111.07998 [hep-th]].
40. F. Baume, M. Cvetic, C. Lawrie and L. Lin, "When rational sections become cyclic — Gauge enhancement in F-theory via Mordell-Weil torsion," JHEP **03**, 069 (2018) [arXiv:1709.07453 [hep-th]].
41. L. B. Anderson, J. Gray and P. K. Oehlmann, "F-Theory on Quotients of Elliptic Calabi-Yau Threefolds," arXiv:1906.11955 [hep-th].
42. V. G. Kac, "Infinite Dimensional Lie Algebras", Cambridge University Press (1994).
43. L. Bhardwaj, M. Del Zotto, J. J. Heckman, D. R. Morrison, T. Rudelius and C. Vafa, "F-theory and the Classification of Little Strings," Phys. Rev. D **93** (2016) no. 8, 086002 [arXiv:1511.05565 [hep-th]].
44. T. W. Grimm and A. Kapfer, "Anomaly Cancelation in Field Theory and F-theory on a Circle," JHEP **05** (2016), 102 [arXiv:1502.05398 [hep-th]].
45. E. Witten, "Phase transitions in M theory and F theory," Nucl. Phys. B **471** (1996) 195 [hep-th/9603150].
46. M. Gross," Elliptic Three-folds II: Multiple Fibers", 1997. Trans. Am. Math. Soc. 349, 3409–3468.
47. L. B. Anderson, A. Grassi, J. Gray and P. K. Oehlmann, "F-theory on Quotient Threefolds with (2,0) Discrete Superconformal Matter," JHEP **1806** (2018) 098 [arXiv:1801.08658 [hep-th]].
48. P. K. Oehlmann and T. Schimannek, "GV-Spectroscopy for F-theory on genus-one fibrations," JHEP **09**, 066 (2020) [arXiv:1912.09493 [hep-th]].
49. Y. Tachikawa, "On S-duality of 5d super Yang-Mills on S^1," JHEP **11** (2011), 123 [arXiv:1110.0531 [hep-th]].
50. H. C. Kim, S. S. Kim and K. Lee, "Higgsing and twisting of 6d D_N gauge theories," JHEP **10** (2020), 014 [arXiv:1908.04704 [hep-th]].

51. S. Kim, K. M. Lee, H. U. Yee and P. Yi, "The N = 1* theories on R**(1+2) x S1 with twisted boundary conditions," JHEP **08** (2004), 040 [arXiv:hep-th/0403076 [hep-th]].
52. S. Katz, D. R. Morrison, S. Schafer-Nameki and J. Sully, "Tate's algorithm and F-theory," JHEP **1108** (2011) 094 [arXiv:1106.3854 [hep-th]].
53. M. Bershadsky, K. A. Intriligator, S. Kachru, D. R. Morrison, V. Sadov and C. Vafa, "Geometric singularities and enhanced gauge symmetries," Nucl. Phys. B **481** (1996) 215 [hep-th/9605200].
54. M. Esole, P. Jefferson and M. J. Kang, "The Geometry of F_4-Models," [arXiv:1704.08251 [hep-th]].

Calabi-Yau Metrics, CFTs and Random Matrices

Anthony Ashmore

Abstract Calabi-Yau manifolds have played a key role in both mathematics and physics, and are particularly important for deriving realistic models of particle physics from string theory. Unfortunately, very little is known about the explicit metrics on these spaces, leaving us unable, for example, to compute particle masses or couplings in these models. We review recent progress in this direction on using numerical approximations to compute the spectrum of the Laplacian on these spaces. We give an example of what one can do with this new "data", giving a surprising link between Calabi-Yau metrics and random matrix theory.

1 Introduction and Summary

Calabi-Yau manifolds have a rich mathematical history and give a starting point for recovering realistic four-dimensional physics from string theory. Despite much study, there are still no explicit expressions for non-trivial Ricci-flat metrics on these spaces.[1] Viewed from the string worldsheet, these manifolds define interacting two-dimensional superconformal field theories (SCFTs). These theories come in families labelled by the complex structure and Kähler moduli of the underlying Calabi-Yau. In the large-volume limit, these theories are described by non-linear σ-models whose target space is simply the Calabi-Yau equipped with its Ricci-flat metric.

Recent advances in numerical methods give us access to both the Ricci-flat metric and the spectrum of the Laplacian on these spaces. The spectrum is a source of new non-BPS "data" which characterises CFT operators with low scaling dimension. In this talk, we show that the spectrum of the Laplacian, and thus the spectrum of operators in the corresponding CFT, averaged over complex structure moduli displays the hallmarks of chaos in the form of random matrix statistics. In our companion

[1] At least in six dimensions and higher—see [1–3] for progress on computing K3 metrics.

A. Ashmore (✉)
Department of Physics, Skidmore College, Saratoga Springs, NY 12866, USA

Sorbonne Université, CNRS, LPTHE, Paris, France
e-mail: aashmore@skidmore.edu

Y.-H. He et al. (eds.), *Nankai Symposium on Mathematical Dialogues*,
https://doi.org/10.1007/978-981-19-2328-9_4

paper [4], we also examine K3 CFTs and the spectra of field theories on genus-three Riemann surfaces, again finding chaos in their spectra.

Further work includes extending our analysis to the non-scalar spectrum, understanding the interplay of the explicit spectrum with the conformal [5–11] and geometric [12–15] bootstrap programmes, exploring whether RMT sheds light on interacting CFTs in an averaged sense or if it can be used to understand the "typical" properties of Calabi-Yau compactifications [16–22].

A point to emphasise is that one could imagine taking random metrics on these manifolds and finding similar random matrix statistics. Our surprising result is that one sees this chaotic behaviour even within the restricted class of Ricci-flat metrics on a Calabi-Yau hypersurface. A related observation is that one sees similar behaviour for constant negative curvature metrics on genus-three Riemann surfaces. In this context, there are mathematical theorems relating the behaviour of geodesics to chaotic particle motion on these manifolds [23–26]. This suggests that something similar should be true for Ricci-flat metrics, though this still appears to be an open question.

2 Chaos in 2D CFTs from Calabi-Yau σ-Models?

Even in two dimensions, very little is known about the spectrum of interacting CFTs. For CFTs at generic points in moduli space, one can study quantities protected by supersymmetry [27–34] or use modular invariance and the conformal bootstrap to constrain the spectrum in some way [5–10]. In general, however, there is no way to compute the spectrum of an interacting CFT.

There are large families of SCFTs defined by Calabi-Yau manifolds. For example, a $(2, 2)$ σ-model with a Calabi-Yau threefold target space X is known to flow to a $(2, 2)$ SCFT [35–37], with the complex structure and Kähler moduli of the Calabi-Yau metric determining certain exactly marginal couplings in the field theory [38–43].

The relation of the geometry of the Calabi-Yau to the spectrum of operators in the CFT was given first by Witten [27]. At low energies, the σ-model reduces to a quantum mechanics for a point particle moving on the target space, with the Hamiltonian given by the de Rham Laplacian, Δ, for the Ricci-flat metric. Thanks to this, one can study the low-lying operators of the CFT via the geometry of the Calabi-Yau. Primary operators $\{\mathcal{O}_i\}$ in the CFT are eigenstates of the Hamiltonian with fixed scaling dimension D_i:

$$H|\mathcal{O}_i\rangle = D_i|\mathcal{O}_i\rangle, \qquad D_i \geq 0,$$

In the large-volume limit, these operators correspond to the (p, q)-eigenforms of Δ, with scaling dimensions

$$D = \lambda + \frac{p+q}{2},$$

where λ is the eigenvalue of the corresponding eigenform. Since $\lambda \sim \mathrm{Vol}(X)^{-1/\dim_{\mathbb{C}} X}$, at large volume the light operators come from *scalar* eigenmodes of Δ.

In the spirit of "experimental" theoretical physics, we ask the following question:

> Given an ensemble of CFTs, do the scaling dimensions $\{D_i\}$ of primary operators display any interesting statistics? In particular, do they display signs of chaos?

This is obviously a difficult question to answer: we need families of generic, interacting CFTs, and we have to be able to compute the spectrum of these theories. The requirement of being generic and interacting immediately rules out looking at solvable or rational CFTs, and needing families of such theories means we cannot look at special points in moduli space or isolated CFTs, such as the Ising model.

Until recently, it was not possible to answer such a question, other than in free field theories [44–46]. The idea we pursue in this talk is to construct an ensemble of CFTs and compute their spectra numerically via the connection to Calabi-Yau geometries. We then analyse the spectra statistically and find that they display random matrix statistics, indicative of chaos in the underlying field theory.

3 Numerical Calabi-Yau Metrics and Their Spectra

Calabi-Yau (CY) manifolds are Kähler manifolds which admit Ricci-flat metrics. Thanks to Yau's proof [47] of the Calabi conjecture [48], we know that such metrics exist when $c_1(X) = 0$. This proof is in no way constructive, however, and so the metrics must be determined by other means.

There are now many methods to compute numerical approximations to Ricci-flat metrics on CY manifolds [49–61], which have been discussed many times in the literature [51, 62–72]. In this work, we use an algorithm due to Donaldson [51], based on the *algebraic metrics* ansatz of Tian [50]. One iteratively solves for a *balanced* metric on X, which is known to approach the Ricci-flat metric in a certain limit. In practice, this involves converting integrals over X to discrete sums—one uses Monte Carlo methods to implement this using a sampling method discussed in [63, 64].

3.1 *The Laplacian on a Calabi-Yau*

In the large-volume limit, low-lying operators in the CFT are determined by scalar eigenmodes on the CY. These eigenmodes satisfy

$$\Delta\phi_n = \lambda_n\phi_n, \qquad \Delta = \delta \mathrm{d}.$$

Given the data of an approximate CY metric, we can then calculate the spectrum $\{\lambda_n\}$ following [68, 73–75]. We do this by expanding the eigenmodes in a truncated, finite basis of functions $\{\alpha_A\}$, so that $\phi = \phi_A \alpha_A$, thus giving a finite-dimensional generalised eigenvalue problem,

$$\Delta_{AB}\phi_B = \lambda\, O_{AB}\phi_B,$$

where $\Delta_{AB} = \langle \alpha_A, \Delta\alpha_B \rangle$ and $O_{AB} = \langle \alpha_A, \alpha_B \rangle$ both depend on the approximate CY metric.

4 Calabi-Yau CFTs and Random Matrix Theory

A suitable ensemble of CFTs is provided by constructing a family of CY manifolds with varying complex structure moduli. For example, a generic quintic threefold X is defined by a quintic equation in $\mathbb{P}^4$ with coordinates $[z_0 : \ldots : z_4]$:

$$X \equiv \sum_{m,n,p,q,r} c_{mnpqr} z_m z_n z_p z_q z_r = 0.$$

This describes a 101-dimensional family of CYs. We sample the c_{mnpqr} randomly from the unit disk in the complex plane with a flat measure:

$$c_{mnpqr} \in \mathbb{C}, \qquad |c_{mnpqr}| < 1.$$

For each choice of complex structure moduli, we compute an approximate Ricci-flat metric and the spectrum of the Laplacian numerically. This gives us an ensemble of spectra which can be interpreted as an ensemble of scaling dimensions of operators in the large-volume CFT.

Since we are looking for signs of chaos in the spectrum of CFTs, it is useful to have some diagnostics of chaotic behaviour. In our case, we take random matrix statistics to be indicative of chaos in the spectrum [76–81].

4.1 *Random Matrix Theory and Spectral Statistics*

Random matrix theory (RMT) is the study of the distribution of eigenvalues of matrices whose entries are independent and identically distributed random variables. RMT governs many physical systems, such as nuclear physics [82–85], billiards [80, 86, 87], quantum many-body systems such as the SYK model [88–91], and toy models of black-hole physics and quantum gravity [92–96]. The latter suggests, via holography, that generic conformal field theories may also display signs of chaos.

We will compare the statistics of scaling dimensions of primary operators with the eigenvalue statistics of a random matrix theory, specifically the Gaussian orthogonal ensemble (GOE). This is the ensemble of eigenvalues of $N \times N$ real, symmetric matrices, where the entries of the matrices are independently drawn from a Gaussian distribution. We will focus on universal features of random matrix theory by *unfolding*, letting us focus on the statistics of fluctuations in the spectrum.

We measure RMT behaviour via three diagnostics: the nearest-neighbour level spacing $p_1(s)$, the number variance $\Sigma^2(L)$, and the spectral form factor $S(t, \beta)$. Together, these provide measures of both long- and short-range correlations in the spectrum [76, 79].

The nearest-neighbour level spacing is the probability of two consecutive eigenvalues being separated by a distance s. For the GOE in the large-N limit, it is

$$p_1(s) = \frac{\pi}{2} s\, \mathrm{e}^{-\pi s^2/4}.$$

The maximum of this distribution is away from $s = 0$, indicative of *eigenvalue repulsion*. The number variance is the fluctuation of the number of eigenvalues averaged over an interval of size L. For the GOE, it scales as

$$\Sigma^2(L) \sim \log L.$$

The logarithmic growth is an indication of *spectral rigidity*. The spectral form factor (SFF) (up to an overall normalisation) is the analytically continued thermal partition [97]

$$S(t, \beta) \sim \left| \sum_n \mathrm{e}^{-(\beta + 2\pi \mathrm{i} t)\lambda_n} \right|^2.$$

The spectral form factor displays three characteristic features: a dip with ringing for finite N, a ramp, and a plateau. We show these features for the GOE in Fig. 1.

4.2 Results

We generated 1,000 different samples of quintic threefolds of the form (4), with complex structure moduli given by random choices of c_{mnpqr}. For each sample, we keep the lowest-lying 100 eigenvalues of the Laplacian. We then repeat this for 1,000 different choices of the c_{mnpqr}, and analyse the resulting distribution of operator scaling dimensions.

We compare the distribution of scaling dimensions with the distribution of eigenvalues from a random matrix ensemble in Fig. 2. The SFF shows the characteristic dip with ringing, ramp and plateau. The ramp shows a smooth transition to the plateau region, suggesting GOE statistics. The nearest-neighbour level spacing and number variance also match those of the GOE.

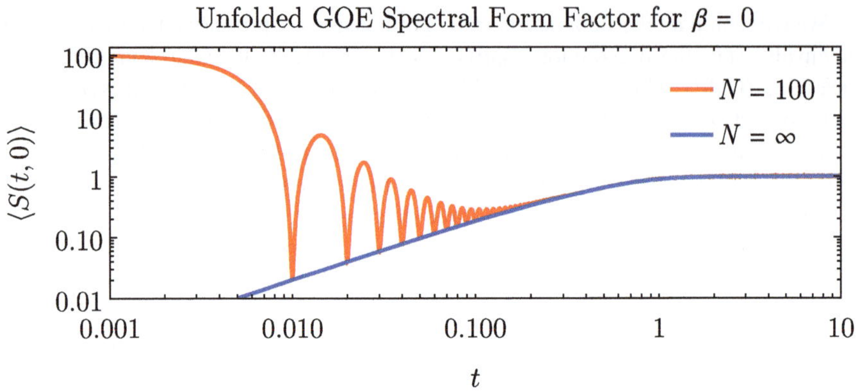

Fig. 1 Unfolded GOE spectral form factors for $\beta = 0$. The orange curve is computed from 100,000, 100×100 random symmetric matrices. The blue curve shows the SFF for $N \to \infty$ and $\beta \to 0$—only the ramp and plateau are present

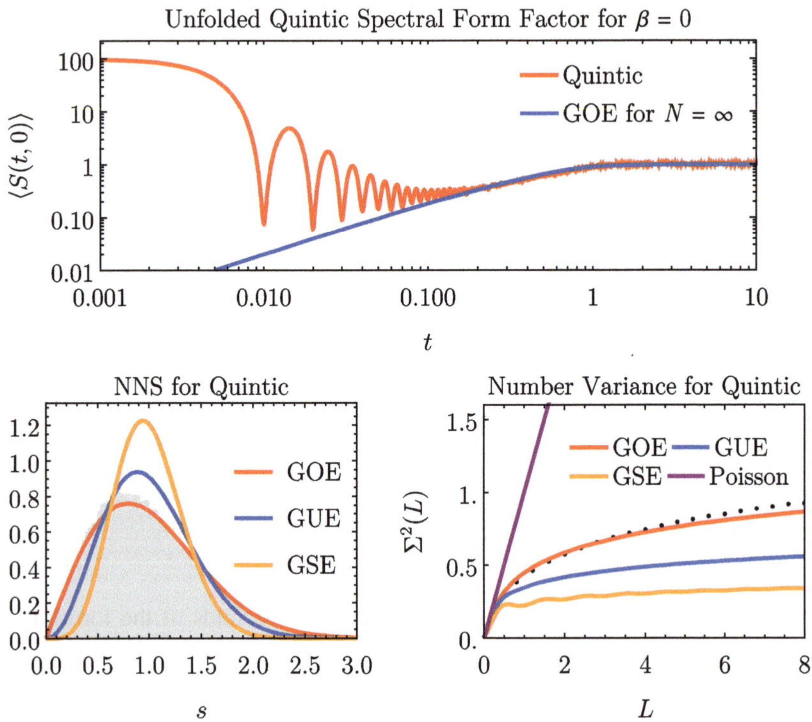

Fig. 2 The first row shows the unfolded spectral form factor for $\beta = 0$ for quintic Calabi-Yau threefolds. The orange curve shows an ensemble average consisting of 1,000 samples with varying complex structure, with each sample containing the 100 lowest-lying eigenvalues. The dip, ramp and plateau are all present. The second row shows the nearest-neighbour level spacings (NNS) and the number variance, together with fits to GOE. Other matrix ensembles are shown for comparison

Acknowledgements We thank Nima Afkhami-Jeddi and Clay Córdova for collaboration on the project which this article is based upon. AA is supported by the European Union's Horizon 2020 research and innovation program under the Marie Skłodowska-Curie grant agreement No. 838776. This work was completed in part with resources provided by the University of Chicago Research Computing Center.

References

1. S. Kachru, A. Tripathy, and M. Zimet, "K3 metrics from little string theory", arXiv:1810.10540 [hep-th].
2. S. Kachru, A. Tripathy, and M. Zimet, "K3 metrics", arXiv:2006.02435 [hep-th].
3. A. Tripathy and M. Zimet, "A plethora of K3 metrics", arXiv:2010.12581 [hep-th].
4. N. Afkhami-Jeddi, A. Ashmore, and C. Cordova, "Calabi-Yau CFTs and Random Matrices", arXiv:2107.11461 [hep-th].
5. S. Hellerman, "A Universal Inequality for CFT and Quantum Gravity", JHEP **08** (2011)130, http://dx.doi.org/10.1007/JHEP08(2011)130. arXiv:0902.2790 [hep-th].
6. C. A. Keller and H. Ooguri, "Modular Constraints on Calabi-Yau Compactifications", *Commun. Math. Phys***324** (2013)107–127, http://dx.doi.org/10.1007/s00220-013-1797-8. arXiv:1209.4649 [hep-th].
7. D. Friedan and C. A. Keller, "Constraints on 2d CFT partition functions", http://dx.doi.org/10.1007/JHEP10(2013)180 *JHEP***10** (2013)180, arXiv:1307.6562 [hep-th].
8. Y.-H. Lin, S.-H. Shao, D. Simmons-Duffin, Y. Wang, and X. Yin, "$\mathcal{N} = 4$ superconformal bootstrap of the K3 CFT", http://dx.doi.org/10.1007/JHEP05(2017)126 *JHEP***05** (2017)126, arXiv:1511.04065 [hep-th].
9. S. Collier, Y.-H. Lin, and X. Yin, "Modular Bootstrap Revisited", http://dx.doi.org/10.1007/JHEP09(2018)061 *JHEP***09** (2018)061, arXiv:1608.06241 [hep-th].
10. Y.-H. Lin, S.-H. Shao, Y. Wang, and X. Yin, "(2, 2) superconformal bootstrap in two dimensions", http://dx.doi.org/10.1007/JHEP05(2017)112 *JHEP***05** (2017)112, arXiv:1610.05371 [hep-th].
11. P. Kravchuk, D. Mazac, and S. Pal, "Automorphic Spectra and the Conformal Bootstrap", arXiv:2111.12716 [hep-th].
12. J. Bonifacio and K. Hinterbichler, "Unitarization from Geometry", http://dx.doi.org/10.1007/JHEP12(2019)165 *JHEP***12** (2019)165, arXiv:1910.04767 [hep-th].
13. J. Bonifacio and K. Hinterbichler, "Bootstrap Bounds on Closed Einstein Manifolds", http://dx.doi.org/10.1007/JHEP10(2020)069*JHEP***10** (2020)069, arXiv:2007.10337 [hep-th].
14. J. Bonifacio, "Bootstrap Bounds on Closed Hyperbolic Manifolds", arXiv:2107.09674 [hep-th].
15. J. Bonifacio, "Bootstrapping Closed Hyperbolic Surfaces", arXiv:2111.13215 [hep-th].
16. S. Ashok and M. R. Douglas, "Counting flux vacua", http://dx.doi.org/10.1088/1126-6708/2004/01/060*JHEP***01** (2004)060, arXiv:hep-th/0307049.
17. M. R. Douglas, B. Shiffman, and S. Zelditch, "Critical points and supersymmetric vacua", http://dx.doi.org/10.1007/s00220-004-1228-y*Commun. Math. Phys***252** (2004)325–358, arXiv:math/0402326.
18. M. R. Douglas, "The Statistics of string / M theory vacua", http://dx.doi.org/10.1088/1126-6708/2003/05/046*JHEP***05** (2003)046, arXiv:hep-th/0303194.
19. F. Denef and M. R. Douglas, "Distributions of flux vacua", http://dx.doi.org/10.1088/1126-6708/2004/05/072*JHEP***05** (2004)072, arXiv:hep-th/0404116.
20. F. Denef and M. R. Douglas, "Distributions of nonsupersymmetric flux vacua", http://dx.doi.org/10.1088/1126-6708/2005/03/061*JHEP***03** (2005)061, arXiv:hep-th/0411183.

21. J. Distler and U. Varadarajan, "Random polynomials and the friendly landscape", arXiv:hep-th/0507090.
22. D. I. Podolsky, J. Majumder, and N. Jokela, "Disorder on the landscape", http://dx.doi.org/10.1088/1475-7516/2008/05/024*JCAP***05** (2008)024, arXiv:0804.2263 [hep-th].
23. G. A. Hedlund, "On the metrical transitivity of the geodesics on closed surfaces of constant negative curvature", http://dx.doi.org/10.2307/1968495*Annals of Mathematics***35** 4, (1934)787–808.
24. E. Hopf, "Ergodic theory and the geodesic flow on surfaces of constant negative curvature", http://dx.doi.org/bams/1183533160*Bulletin of the American Mathematical Society***77** 6, (1971)863 – 877.
25. E. Hopf, "Statistik der geodätischen Linien in Mannigfaltigkeiten negativer Krümmung", *Ber. Verh. Sächs. Akad. Wiss.***91** (1939)261–304.
26. G. A. Hedlund, "Geodesic flows on closed riemann manifolds with negative curvature", *Proceedings of the Steklov Institute of Mathematics* 90, (1967).
27. E. Witten, "Constraints on Supersymmetry Breaking", http://dx.doi.org/10.1016/0550-3213(82)90071-2*Nucl. Phys. B***202** (1982)253.
28. T. Eguchi and A. Taormina, "Character Formulas for the $N = 4$ Superconformal Algebra", http://dx.doi.org/10.1016/0370-2693(88)90778-2*Phys. Lett. B***200** (1988)315.
29. T. Eguchi, H. Ooguri, A. Taormina, and S.-K. Yang, "Superconformal Algebras and String Compactification on Manifolds with SU(N) Holonomy", http://dx.doi.org/10.1016/0550-3213(89)90454-9*Nucl. Phys. B***315** (1989) 193–221.
30. S. Cecotti, P. Fendley, K. A. Intriligator, and C. Vafa, "A New supersymmetric index", http://dx.doi.org/10.1016/0550-3213(92)90572-S*Nucl. Phys. B***386** (1992)405–452, arXiv:hep-th/9204102.
31. T. Kawai, Y. Yamada, and S.-K. Yang, "Elliptic genera and N=2 superconformal field theory", http://dx.doi.org/10.1016/0550-3213(94)90428-6*Nucl. Phys. B***414** (1994)191–212, arXiv:hep-th/9306096.
32. R. Dijkgraaf, E. P. Verlinde, and H. L. Verlinde, "Counting dyons in N=4 string theory", http://dx.doi.org/10.1016/S0550-3213(96)00640-2*Nucl. Phys. B***484** (1997)543–561, arXiv:hep-th/9607026.
33. F. Benini, R. Eager, K. Hori, and Y. Tachikawa, "Elliptic genera of two-dimensional N = 2 gauge theories with rank-one gauge groups", http://dx.doi.org/10.1007/s11005-013-0673-y*Lett. Math. Phys***104** (2014)465–493, arXiv:1305.0533 [hep-th].
34. Y.-H. Lin, S.-H. Shao, Y. Wang, and X. Yin, "Supersymmetry Constraints and String Theory on K3", http://dx.doi.org/10.1007/JHEP12(2015)142*JHEP***12** (2015)142, arXiv:1508.07305 [hep-th].
35. C. M. Hull, "ULTRAVIOLET FINITENESS OF SUPERSYMMETRIC NONLINEAR SIGMA MODELS", http://dx.doi.org/10.1016/0550-3213(85)90317-7*Nucl. Phys. B***260** (1985)182–202.
36. L. Alvarez-Gaume, S. R. Coleman, and P. H. Ginsparg, "Finiteness of Ricci Flat $N = 2$ Supersymmetric σ Models", http://dx.doi.org/10.1007/BF01211757*Commun. Math. Phys***103** (1986)423.
37. D. Nemeschansky and A. Sen, "Conformal Invariance of Supersymmetric σ Models on Calabi-yau Manifolds", http://dx.doi.org/10.1016/0370-2693(86)91394-8*Phys. Lett. B***178** (1986)365–369.
38. D. Friedan, "Nonlinear Models in Two Epsilon Dimensions", http://dx.doi.org/10.1103/PhysRevLett.45.1057*Phys. Rev. Lett***45** (1980)1057.
39. L. Alvarez-Gaume and D. Z. Freedman, "Geometrical Structure and Ultraviolet Finiteness in the Supersymmetric Sigma Model", http://dx.doi.org/10.1007/BF01208280*Commun. Math. Phys***80** (1981)443.
40. L. Alvarez-Gaume and D. Z. Freedman, "Kahler Geometry and the Renormalization of Supersymmetric Sigma Models", http://dx.doi.org/10.1103/PhysRevD.22.846*Phys. Rev. D***22** (1980)846.

41. L. Alvarez-Gaume, D. Z. Freedman, and S. Mukhi, “The Background Field Method and the Ultraviolet Structure of the Supersymmetric Nonlinear Sigma Model”, http://dx.doi.org/10.1016/0003-4916(81)90006-3*Annals Phys.***134** (1981)85.
42. L. Alvarez-Gaume and P. H. Ginsparg, “Finiteness of Ricci Flat Supersymmetric Nonlinear Sigma Models”, http://dx.doi.org/10.1007/BF01229382*Commun. Math. Phys***102** (1985)311.
43. D. J. Gross and E. Witten, “Superstring Modifications of Einstein’s Equations”, http://dx.doi.org/10.1016/0550-3213(86)90429-3*Nucl. Phys. B***277** (1986)1.
44. N. Afkhami-Jeddi, H. Cohn, T. Hartman, and A. Tajdini, “Free partition functions and an averaged holographic duality”, http://dx.doi.org/10.1007/JHEP01(2021)130*JHEP***01** (2021)130, arXiv:2006.04839 [hep-th].
45. A. Maloney and E. Witten, “Averaging over Narain moduli space”, http://dx.doi.org/10.1007/JHEP10(2020)187*JHEP***10** (2020)187, arXiv:2006.04855 [hep-th].
46. N. Benjamin, C. A. Keller, H. Ooguri, and I. G. Zadeh, “Narain to Narnia”, arXiv:2103.15826 [hep-th].
47. S. T. Yau, “On the Ricci curvature of a compact Kähler manifold and the complex Monge-Ampère equation.I.”, *Commun. Pure Appl. Math.***31** 3, (1978)339–411.
48. E. Calabi and K. On, “On kähler manifolds with vanishing canonical class”, *Princeton Mathematical Series***12** (1957)78–89.
49. M. Headrick and T. Wiseman, “Numerical Ricci-flat metrics on K3”, http://dx.doi.org/10.1088/0264-9381/22/23/002*Class. Quant. Grav***22** (2005)4931–4960, arXiv:hep-th/0506129.
50. G. Tian, “On a set of polarized Kähler metrics on algebraic manifolds”, http://dx.doi.org/10.4310/jdg/1214445039*Journal of Differential Geometry***32** 1, (1990)99–130.
51. S. Donaldson, “Some numerical results in complex differential geometry”, http://dx.doi.org/10.4310/PAMQ.2009.v5.n2.a2*Pure and Applied Mathematics Quarterly***5** 2, (2009)571–618, arXiv:math/0512625 [math.DG].
52. M. R. Douglas, R. L. Karp, S. Lukic, and R. Reinbacher, “Numerical Calabi-Yau metrics”, http://dx.doi.org/10.1063/1.2888403*J. Math. Phys***49** (2008)032302, arXiv:hep-th/0612075.
53. V. Braun, T. Brelidze, M. R. Douglas, and B. A. Ovrut, “Calabi-Yau Metrics for Quotients and Complete Intersections”, http://dx.doi.org/10.1088/1126-6708/2008/05/080*JHEP***05** (2008)080, arXiv:0712.3563 [hep-th].
54. M. Headrick and A. Nassar, “Energy functionals for Calabi-Yau metrics”, http://dx.doi.org/10.4310/ATMP.2013.v17.n5.a1*Adv. Theor. Math. Phys***17** 5, (2013)867–902, arXiv:0908.2635 [hep-th].
55. A. Ashmore, Y.-H. He, and B. A. Ovrut, “Machine Learning Calabi-Yau Metrics”, http://dx.doi.org/10.1002/prop.202000068*Fortsch. Phys***68** 9, (2020)2000068, arXiv:1910.08605 [hep-th].
56. L. B. Anderson, M. Gerdes, J. Gray, S. Krippendorf, N. Raghuram, and F. Ruehle, “Moduli-dependent Calabi-Yau and SU(3)-structure metrics from Machine Learning”, http://dx.doi.org/10.1007/JHEP05(2021)013*JHEP***05** (2021)013, arXiv:2012.04656 [hep-th].
57. M. R. Douglas, S. Lakshminarasimhan, and Y. Qi, “Numerical Calabi-Yau metrics from holomorphic networks”, arXiv:2012.04797 [hep-th].
58. V. Jejjala, D. K. Mayorga Pena, and C. Mishra, “Neural Network Approximations for Calabi-Yau Metrics”, arXiv:2012.15821 [hep-th].
59. M. R. Douglas, “Holomorphic feedforward networks”, arXiv:2105.03991 [math.CV].
60. M. Larfors, A. Lukas, F. Ruehle, and R. Schneider, “Learning Size and Shape of Calabi-Yau Spaces”, arXiv:2111.01436 [hep-th].
61. A. Ashmore, L. Calmon, Y.-H. He, and B. A. Ovrut, “Calabi-Yau Metrics, Energy Functionals and Machine-Learning”, arXiv:2112.10872 [hep-th].
62. M. Headrick and T. Wiseman, “Numerical Ricci-flat metrics on K3”, http://dx.doi.org/10.1088/0264-9381/22/23/002*Class. Quant. Grav***22** (2005)4931–4960, arXiv:hep-th/0506129.
63. M. R. Douglas, R. L. Karp, S. Lukic, and R. Reinbacher, “Numerical Calabi-Yau metrics”, http://dx.doi.org/10.1063/1.2888403*J. Math. Phys***49** (2008)032302, arXiv:hep-th/0612075.
64. V. Braun, T. Brelidze, M. R. Douglas, and B. A. Ovrut, “Calabi-Yau Metrics for Quotients and Complete Intersections”, http://dx.doi.org/10.1088/1126-6708/2008/05/080*JHEP***05** (2008)080, arXiv:0712.3563 [hep-th].

65. M. Headrick and A. Nassar, "Energy functionals for Calabi-Yau metrics", http://dx.doi.org/10.4310/ATMP.2013.v17.n5.a1*Adv. Theor. Math. Phys***17** 5, (2013)867–902, arXiv:0908.2635 [hep-th].
66. A. Ashmore, Y.-H. He, and B. A. Ovrut, "Machine Learning Calabi–Yau Metrics", http://dx.doi.org/10.1002/prop.202000068*Fortsch. Phys***68** 9, (2020)2000068, arXiv:1910.08605 [hep-th].
67. W. Cui and J. Gray, "Numerical Metrics, Curvature Expansions and Calabi-Yau Manifolds", http://dx.doi.org/10.1007/JHEP05(2020)044*JHEP***05** (2020)044, arXiv:1912.11068 [hep-th].
68. A. Ashmore, "Eigenvalues and eigenforms on Calabi-Yau threefolds", arXiv:2011.13929 [hep-th].
69. L. B. Anderson, M. Gerdes, J. Gray, S. Krippendorf, N. Raghuram, and F. Ruehle, "Moduli-dependent Calabi-Yau and SU(3)-structure metrics from Machine Learning", http://dx.doi.org/10.1007/JHEP05(2021)013*JHEP***05** (2021)013, arXiv:2012.04656 [hep-th].
70. M. R. Douglas, S. Lakshminarasimhan, and Y. Qi, "Numerical Calabi-Yau metrics from holomorphic networks", arXiv:2012.04797 [hep-th].
71. V. Jejjala, D. K. Mayorga Pena, and C. Mishra, "Neural Network Approximations for Calabi-Yau Metrics", arXiv:2012.15821 [hep-th].
72. A. Ashmore, R. Deen, Y.-H. He, and B. A. Ovrut, "Machine Learning Line Bundle Connections", arXiv:2110.12483 [hep-th].
73. C. Iuliu-Lazaroiu, D. McNamee, and C. Saemann, "Generalized Berezin quantization, Bergman metrics and fuzzy Laplacians", http://dx.doi.org/10.1088/1126-6708/2008/09/059*JHEP***09** (2008)059, arXiv:0804.4555 [hep-th].
74. V. Braun, T. Brelidze, M. R. Douglas, and B. A. Ovrut, "Eigenvalues and Eigenfunctions of the Scalar Laplace Operator on Calabi-Yau Manifolds", http://dx.doi.org/10.1088/1126-6708/2008/07/120*JHEP***07** (2008)120, arXiv:0805.3689 [hep-th].
75. A. Ashmore and F. Ruehle, "Moduli-dependent KK towers and the swampland distance conjecture on the quintic Calabi-Yau manifold", http://dx.doi.org/10.1103/PhysRevD.103.106028*Phys. Rev. D***103** 10, (2021)106028, arXiv:2103.07472 [hep-th].
76. M. Mehta, http://dx.doi.org/10.1016/C2009-0-22297-5*Random Matrices*, vol. 142 of *Pure and Applied Mathematics*. Elsevier/Academic Press, 2004.
77. E. Brézin and S. Hikami, "Spectral form factor in a random matrix theory", http://dx.doi.org/10.1103/PhysRevE.55.4067*Phys. Rev. E***55** (1997)4067–4083.
78. T. A. Brody, J. Flores, J. B. French, P. A. Mello, A. Pandey, and S. S. M. Wong, "Random-matrix physics: spectrum and strength fluctuations", http://dx.doi.org/10.1103/RevModPhys.53.385*Rev. Mod. Phys***53** (1981)385–479.
79. T. Guhr, A. Muller-Groeling, and H. A. Weidenmuller, "Random matrix theories in quantum physics: Common concepts", http://dx.doi.org/10.1016/S0370-1573(97)00088-4*Phys. Rept***299** (1998)189–425, arXiv:cond-mat/9707301.
80. O. Bohigas, M. J. Giannoni, and C. Schmit, "Characterization of chaotic quantum spectra and universality of level fluctuation laws", http://dx.doi.org/10.1103/PhysRevLett.52.1*Phys. Rev. Lett***52** (Jan, 1984)1–4.
81. O. Bohigas, M. Giannoni, and C. Schmit, "Spectral properties of the laplacian and random matrix theories", http://dx.doi.org/10.1051/jphyslet:0198400450210101500*Journal de Physique Lettres***45** 21, (1984)1015–1022.
82. E. P. Wigner, "Characteristic vectors of bordered matrices with infinite dimensions", http://dx.doi.org/10.2307/1970079*Annals of Mathematics***62** 3, (1955)548–564.
83. N. Rosenzweig and C. E. Porter, ""Repulsion of energy levels" in complex atomic spectra", http://dx.doi.org/10.1103/PhysRev.120.1698*Phys. Rev***120** (Dec, 1960)1698–1714.
84. R. E. Trees, ""Repulsion of energy levels" in complex atomic spectra", http://dx.doi.org/10.1103/PhysRev.123.1293*Phys. Rev***123** (Aug, 1961)1293–1300.
85. R. Haq, A. Pandey, and O. Bohigas, "Fluctuation properties of nuclear energy levels: Do theory and experiment agree?", http://dx.doi.org/10.1103/PhysRevLett.48.1086*Physical Review Letters***48** 16, (1982)1086–1089.

86. M. C. Gutzwiller, http://dx.doi.org/10.1007/978-1-4612-0983-6*Chaos in Classical and Quantum Mechanics*. Springer New York, 1990.
87. N. Balazs and A. Voros, "Chaos on the pseudosphere", http://dx.doi.org/10.1016/0370-1573(86)90159-6*Physics Reports***143** 3, (1986)109–240.
88. S. Sachdev and J. Ye, "Gapless spin fluid ground state in a random, quantum Heisenberg magnet", http://dx.doi.org/10.1103/PhysRevLett.70.3339*Phys. Rev. Lett***70** (1993)3339, arXiv:cond-mat/9212030.
89. A. Kitaev, "Hidden correlations in the Hawking radiation and thermal noise", 1998. https://online.kitp.ucsb.edu/online/joint98/kitaev/.
90. A. Kitaev, "A simple model of quantum holography I", 2015. https://online.kitp.ucsb.edu/online/entangled15/kitaev/.
91. A. Kitaev, "A simple model of quantum holography II", 2015. https://online.kitp.ucsb.edu/online/entangled15/kitaev2/.
92. P. Saad, S. H. Shenker, and D. Stanford, "A semiclassical ramp in SYK and in gravity", arXiv:1806.06840 [hep-th].
93. A. M. García-García and J. J. M. Verbaarschot, "Spectral and thermodynamic properties of the Sachdev-Ye-Kitaev model", http://dx.doi.org/10.1103/PhysRevD.94.126010*Phys. Rev. D***94** 12, (2016)126010, arXiv:1610.03816 [hep-th].
94. Y.-Z. You, A. W. W. Ludwig, and C. Xu, "Sachdev-Ye-Kitaev model and thermalization on the boundary of many-body localized fermionic symmetry-protected topological states", http://dx.doi.org/10.1103/PhysRevB.95.115150*Phys. Rev. B***95** (Mar, 2017)115150, arXiv:1602.06964 [cond-mat.str-el].
95. J. S. Cotler, G. Gur-Ari, M. Hanada, J. Polchinski, P. Saad, S. H. Shenker, D. Stanford, A. Streicher, and M. Tezuka, "Black Holes and Random Matrices", http://dx.doi.org/10.1007/JHEP05(2017)118*JHEP***05** (2017)118, arXiv:1611.04650 [hep-th]. [Erratum: JHEP 09, 002 (2018)].
96. H. Gharibyan, M. Hanada, S. H. Shenker, and M. Tezuka, "Onset of Random Matrix Behavior in Scrambling Systems", http://dx.doi.org/10.1007/JHEP07(2018)124*JHEP***07** (2018)124, arXiv:1803.08050 [hep-th]. [Erratum: JHEP 02, 197 (2019)].
97. F. Haake, http://dx.doi.org/10.1007/978-3-642-05428-0*Quantum Signatures of Chaos*. Springer Series in Synergetics. Springer, Berlin, Heidelberg, 3 ed., 2010.

Why Do We Care about Mahler Measure in Physics?

Jiakang Bao

Abstract We give an overview on some recent developments of Mahler measure in supersymmetric gauge theories. It can be shown that maximization of Mahler measure for isoradial dimers is equivalent to a-maximization. This can be further generalized to non-isoradial cases by virtue of Seiberg duality. Mahler measure also enjoys nice properties under specular duality and hence encodes certain information of the master space. This is mainly based on the work [1].

1 Introduction

Mahler measure has been playing a crucial role in number theory and geometry since the paper [2] by Mahler. Given a Laurent polynomial $P(z) = P(z_1, \ldots, z_n) \in \mathbb{C}[z_1^{\pm 1}, \ldots, z_n^{\pm 1}]$, its Mahler measure is defined as[1]

$$m(P) := \int_0^1 \ldots \int_0^1 \log |P(\exp(2\pi i\theta_1), \ldots, \exp(2\pi i\theta_n))| \mathrm{d}\theta_1 \ldots \mathrm{d}\theta_n. \tag{1.1}$$

This *real* integral can be thought of as the geometric mean of the log-absolute value of P over the unit n-torus $\mathbb{T}^n$. For multivariate P, it is often hard to compute this integral. Nevertheless, we shall mainly contemplate Laurent polynomials of form $P(z) = k - p(z)$ where $p(z)$ has no constant terms and $|k| > \max_{|z| \in \mathbb{T}^n} |p(z)|$. In these cases, we may write the integral as

[1] In some literature, Mahler measure is referred to the exponential $\exp(m(P))$ while $m(P)$ is called the logarithmic Mahler measure. However, we will exclusively work with $m(P)$ and drop the word "logarithmic".

J. Bao (✉)
Department of Mathematics, City, University of London, London, UK

London Institute for Mathematical Sciences, Royal Institution, London, UK
e-mail: jiakang.bao.1@gmail.com

Y.-H. He et al. (eds.), *Nankai Symposium on Mathematical Dialogues*,
https://doi.org/10.1007/978-981-19-2328-9_5

$$m(P)=\mathrm{Re}\left(\frac{1}{(2\pi i)^n}\int_{|z_i|=1}\log(P(z_1,\dots,z_n))\frac{\mathrm{d}z_1}{z_1}\dots\frac{\mathrm{d}z_n}{z_n}\right), \tag{1.2}$$

and apply residue theorem after expanding $\log(k-p(z))$ formally in p. Then only the constant terms in the expansion will contribute. It would also be useful to introduce

$$u_0(k):=\frac{1}{(2\pi i)^n}\int_{|z_i|=1}\frac{1}{1-k^{-1}p(z_1,\dots,z_n)}\frac{\mathrm{d}z_1}{z_1}\dots\frac{\mathrm{d}z_n}{z_n} \tag{1.3}$$

so that

$$m(P)=\mathrm{Re}\left(\log k-\int_0^{k^{-1}}(u_0(t)-1)\frac{\mathrm{d}t}{t}\right). \tag{1.4}$$

In this note, we are interested in the cases when $n=2$, and we shall use $(z_1,z_2)=(z,w)$ to denote the two variables.

In fact, following its definition, $u_0(k)$ is a period of a holomorphic 1-form on the curve $1-k^{-1}P=0$ [3]. As a result, it satisfies certain Picard-Fuchs equation. This yields the so-called modular Mahler measure with nome $q=\mathrm{e}^{2\pi i\tau}:=\mathrm{e}^{u_1/u_0}$, where $u_1(k)=u_0(k)\log k+v(k)$ is another solution to the Picard-Fuchs equation and $v(k)$ is a holomorphic function vanishing at $k=0$.

It is not difficult to see that $\frac{\mathrm{d}m(P)}{\mathrm{d}\log k}=u_0(k)$, and this Mahler flow equation reveals the monotonic behaviour of $m(P)$ [1].

Proposition 1.1 *The Mahler measure $m(P)$ increases strictly when $|k|$ increases from $\max_{|z|=|w|=1}|p(z,w)|$ to ∞.*

It is also conjectured that this increasing can be extended to $|k|=0$.[2] In particular, $k\to\infty$ is known to be a tropical limit [1]. The precise definition of a tropical limit is a bit involved for us to expound here. Roughly speaking, at k_{trop}, the amoeba $\mathcal{A}_P=\{(\log|z|,\log|w|)|P(z,w)=0\}$ retracts to a (finite) set of lines (aka its spine).[3]

2 Dimer Models

Given a bipartite graph $\mathcal{G}$, a perfect matching is a set of edges such that each vertex (alternatively coloured black or white) is incident to precisely one edge. The dimer model is then the study of (random) perfect matchings. Moreover, the bipartite graph we focus on here is always $\mathbb{Z}^2$-periodic. In other words, $\mathcal{G}$ quotiented by $\mathbb{Z}^2$, known as the fundamental domain, is embedded on a torus. The n-times enlarged fundamental domain is given by $\mathcal{G}_n:=\mathcal{G}/(n\mathbb{Z}^2)$.

---

[2] When $|k|<\max_{|z|=|w|=1}|p|$, we shall take the Mahler flow as the definition of u_0 though it may not describe the period any more.

[3] Hence, $k\to\infty$ is *a* tropical limit, but may or may not be the unique one. See Appendix A in [1] for an example. However, the large k limit is the only tropical limit we will consider here.

For each edge e, we may associate a real function $\mathcal{E}(e)$ called the energy. The edge weight is defined to be $\mathrm{e}^{-\mathcal{E}(e)}$. For any set M of edges, its energy reads $\mathcal{E}(M) := \sum_{e\in M} \mathcal{E}(e)$. We can then define the partition function as $Z(\mathcal{G}_1) := \sum_{M\in\mathcal{M}(\mathcal{G}_1)} \mathrm{e}^{-\mathcal{E}(M)}$, where $\mathcal{M}(\mathcal{G}_1)$ is the set of perfect matchings of $\mathcal{G}_1$.

Now, let γ_x and γ_y be the paths winding horizontally and vertically around the torus. We can multiply an edge weight by z (or z^{-1}) if γ_x crosses the edge with the black node on its left (or right). Likewise, it is multiplied by $w^{\pm 1}$ if γ_y crosses the edge. These edge weights give rise to the Kasteleyn matrix $K(z, w)$ with rows and columns representing the black and white nodes respectively.[4] Then we can obtain a Laurent polynomial $P(z, w) = \det(K(z, w))$ (called the characteristic/Newton polynomial). This also yields a lattice polygon whose points (m, n) are given by the monomials $z^m w^n$ in the Newton polynomial. This Newton polygon encodes combinatorial data of a toric Calabi-Yau (CY) threefold (or Gorenstein singularity to be precise), and hence also the name toric diagram.

The seminal work [5] showed that the partition function is given by $Z(\mathcal{G}_1) = \frac{1}{2}(-P(1,1) + P(1,-1) + P(-1,1) + P(-1,-1))$. More generally, one can write $Z(\mathcal{G}_n)$ in terms of the characteristic polynomial P_n of $\mathcal{G}_n$ according to Theorem 3.3 and Corollary 3.4 in [5]. In particular.[5]

Theorem 2.1 *The partition function Z per fundamental domain is given by*

$$\log Z := \lim_{n\to\infty} \frac{1}{n^2} Z(\mathcal{G}_n) = m(P). \tag{2.1}$$

As a result, the free energy of the statistical model is then computed by the Mahler measure. In terms of quiver gauge theory, this describes the associated gauged linear sigma model (GLSM) as the perfect matchings are in one-to-one correspondence with the GLSM fields. However, we should bear in mind that this is *not* the free energy for the 4d $\mathcal{N} = 1$ quiver gauge theory (which should be obtained directly from its Lagrangian).

Crystal melting A dimer is associated to a crystal melting model following [6–8]. Each perfect matching of $\mathcal{G}$ corresponds to a crystal configuration. This enumerates the non-commutative Donaldson-Thomas (NCDT) invariants of the corresponding CY_3. Physically, we have supersymmetric quantum mechanics with a single D6-brane wrapping the CY 3-fold, and the molten crystals count the D4-D2-D0 bound states therein. As argued in [5], the crystal configuration is exactly the shape of the Ronkin function in the thermodynamic limit. The Ronkin function is a generalization of Mahler measure defined as

$$R(x, y) := \frac{1}{(2\pi i)^2} \int_{|z|=\mathrm{e}^x, |w|=\mathrm{e}^y} \log P(z, w) \frac{\mathrm{d}z}{z} \frac{\mathrm{d}w}{w}. \tag{2.2}$$

---

[4] More precisely, the edges should also be multiplied by ± 1. See [4] for more details.

[5] To be precise, $\mathcal{E}(e)$ should all be non-negative so that the Riemann sum in computing $Z(\mathcal{G}_n)$ leads to the integral in the limit. In such case, the spectral curve $P = 0$ is known to be Harnack, and the integral in (1.2) is automatically real.

Incidentally, its projection on the plane is the amoeba $\mathcal{A}_P := \{(\log|z|, \log|w|)|P(z, w) = 0\}$. Each bounded (unbounded) complement of the amoeba known as the gas (solid) phase corresponds to a region where $R(x, y)$ has constant gradient (which is precisely the coordinates of the corresponding vertex in the toric diagram). The remaining part with non-trivial gradients projects to the region of the amoeba called the liquid phase. As discussed in [1], Mahler measure indicates certain phase transition along the Mahler flow. Moreover, in the large k limit, the area A of the bounded complementary regions of the amoeba scales as $A \sim \log^2 k$.

From [9], we also learn that the thermodynamic partition function of molten crystals is $Z_{\text{crystal}} \sim \exp\left(\int dxdy R(x, y)\right)$. The Ronkin function can then be used to show that Z_{crystal} is equal to the partition function of topological string theory in the classical limit. This reflects a unified description of Z_{BPS}, $Z_{\text{(NC)DT}}$ and Z_{top}, and is intimately related to the counting of black hole microstates.

3 Quiver Gauge Theories

The dimer model has a natural interpretation in the brane set-up, and is therefore also known as the brane tiling [10–12]. The dual graph is the quiver of the 4d $\mathcal{N} = 1$ gauge theory (which can be dimensionally reduced to the aforementioned quiver quantum mechanics). Such theory arises from D-branes probing the CY geometry encoded by the toric diagram.

The a-maximization Of particular interest here would be the dimers whose faces are all inscribed in circles of the same radii (which can be taken as 1). Such dimers are called isoradial and a natural choice of the edge weights would be $\sqrt{4 - l^2}$ where l is the length of an edge [13, 14]. For brevity, we shall call this the canonical weight choice. It turns out that with such canonical choice, k is equal to the lower bound in Proposition 1.1, and is therefore known as the isoradial limit [1].

As shown in Fig. 1, the R-charges R_I in the gauge theory are encoded by the rhombus angles θ_I in the brane tiling, where $2\theta_I = \pi R_I$.

Then the canonical edge weight can be written as $\sqrt{4 - l^2} = 2\sin(\theta_I)$.

In such cases, a remarkable formula for the (isoradial) Mahler measure was given in [15]:

$$m_{\text{iso}}(P) = \sum_I \left(\frac{\theta_I}{\pi} \log(2\sin(\theta_I)) + \frac{1}{\pi} \Lambda(\theta_I) \right), \tag{3.1}$$

where $\Lambda(x) = -\int_0^x \log(2\sin(t))dt$ is the Lobachevsky function. Using this formula, we are able to show that [1].

Theorem 3.1 *For toric quiver gauge theory, the maximization of m_{iso} is equivalent to a-maximization.*

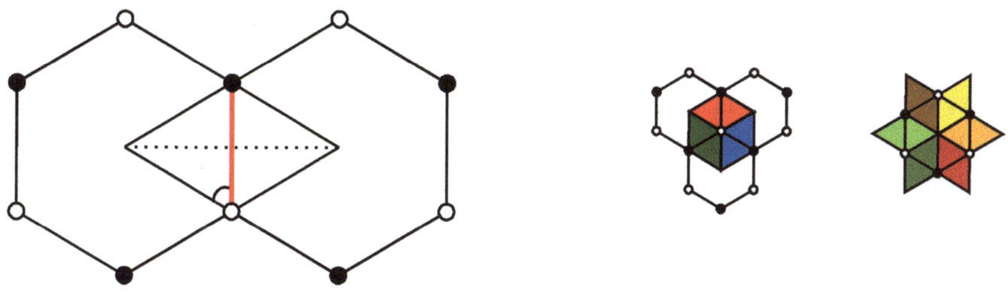

(a) (b)

Fig. 1 **a** The edge in red has length l. Its dashed dual edge is of length $\sqrt{4-l^2}$ which equals the weight of the edge. The rhombus angle is labelled by θ_I. The corresponding internal angle $2\theta_I$ of the rombus gives the R-charge physically. **b** The left plot indicates $\sum 2\theta_I = 2\pi$ which corresponds to each superpotential term while the right plot indicates $\sum(\pi - 2\theta_I) = 2\pi$ which correpsonds to (the fields connected to) each node in the quiver

Consequently, the concept of Mahler measure is now connected to volume minimization and K-semistability (for product test configurations) [16]. Moreover, $\Lambda(x)$ and m_{iso} also appear in the partition function on the 3-sphere when determining the R-symmetry for a 3d $\mathcal{N} = 2$ theory [17]. It is not hard to show that F-maximization can be translated to m_{iso}-maximization for some simple models. For future work, this naturally raises the question about whether Mahler measure could give hint on bridging the study of R-symmetry between odd and even dimensions.

Seiberg duality Given a toric diagram, there could be more than one brane tilings/quivers connected by Seiberg duality, and only one of them is isoradial. However, as observed in [1], the non-isoradial dimers can also be studied using the above methods and hence admit the maximization of Mahler measure. This can be understood from two equivalent viewpoints:

- We can still assign canonical weights to the edges. As the rhombus angles are not really well-defined for a non-isoradial dimer, we can directly express the canonical weights in light of R-charges as $2\sin(\pi R_I/2)$. This indicates that the canonical weight choice is universal for both isoradial and non-isoradial dimers.
- We can also think of the non-isoradial dimer being deformed so that it allows zero or negative edge lengths. Then it can be viewed as "isoradial". This means that (3.1) can also be applied to such case, and we still have m_{iso}-maximization.

It was then conjectured that the Newton polynomial and the Mahler measure are invariant under Seiberg duality.

Specular duality Mahler measure and $u_0(k)$ are actually closely related to (the coherent component of) the master space ${}^{\text{Irr}}\mathcal{F}^\flat = \mathbb{C}[q_i]//Q_F$, which is a symplectic quotient of the coordinate ring given by GLSM fields q_i and Q_F is the F-term charge matrix. As proven in [1], $u_0(k)$ is the generating function of it in the sense that the (possibly redundant) F-term relations (in terms of GLSM fields) are encoded at different orders of k.[6]

As specular duality is a duality that preserves the master space, Mahler measure has the same functions $f_n(q_i)$ in its expansion for specular duals. In other words, given specular dual theories a and b whose GLSM fields are mapped under $q_i^a \leftrightarrow q_i^b$, we have

$$m(P_a) = \log(k) - \sum_{n=2}^{\infty} \frac{f_n(q_i^a)}{nk^n}, \quad m(P_b) = \log(k) - \sum_{n=2}^{\infty} \frac{f_n(q_i^b)}{nk^n}, \tag{3.2}$$

where we also use $q_i^{a,b}$ to denote the weight for the corresponding perfect matching. This statement is also true for $u_0(k)$.

4 Other Applications and Outlook

It was observed in [18] that certain expansions in Mahler measure recovers the Gromov-Witten invariants of vanishing 4-cycles in the embedding CY geometry. The dictionary between such expansions and the instanton expansions of Yukawa couplings can be found in [19].

With the so-called maximally tempered coefficients of the Newton polynomial for reflexive polygons, it was also discovered in [19] that the Mahler measures are in one-to-one correspondence with the dessins d'enfants. It is conjectured that the parameter k would reproduce the Hauptmodul of some modular subgroup, which contains the modular subgroup associated to the corresponding dessin. This benefits from the modular property in the study of Mahler measure in [3]. Besides, different complex structures (namely $\tau_{R,G,B}$) on the torus are extrapolated by the Mahler flow [19]. It would then be natural to connect Mahler measure to j-invariants and even monstrous moonshine.

Mahler measure also appears in the context of topological strings [20]. It would be interesting to further study its connection to the volume conjecture and knot invariants.

Acknowledgements I am indebted to Yang-Hui He and Ali Zahabi for enjoyable collaborations. I would also like to express my gratitude to Cyril Closset, Sebastian Franco, Amihay Hanany, Jan Stienstra and Masahito Yamazaki for fruitful discussions. I am supported by a CSC scholarship.

---

[6] Here it would be more convenient to carry the (weights of) perfect matchings as variables in the expansion rather than assign any explicit weight choices as shown in (3.2).

References

1. J. Bao, Y.-H. He, and A. Zahabi, "Mahler Measure for a Quiver Symphony," http://dx.doi.org/10.1007/s00220-022-04404-y*Commun. Math. Phys* (2022) , arXiv:2108.13903 [hep-th].
2. K. Mahler, "On some inequalities for polynomials in several variables," *Journal of the London Mathematical Society***1** no. 1, (1962) 341–344.
3. F. R. Villegas, "Modular mahler measures i," in *Topics in number theory*, pp. 17–48. Springer, 1999.
4. P. Kasteleyn, "Graph theory and crystal physics," *Graph Theory and Theoretical Physics* (1967) 43–110.
5. R. Kenyon, A. Okounkov, and S. Sheffield, "Dimers and amoebae," arXiv:math-ph/0311005.
6. A. Okounkov, N. Reshetikhin, and C. Vafa, "Quantum Calabi-Yau and classical crystals," http://dx.doi.org/10.1007/0-8176-4467-9_16*Prog. Math***244** (2006) 597, arXiv:hep-th/0309208.
7. A. Iqbal, N. Nekrasov, A. Okounkov, and C. Vafa, "Quantum foam and topological strings," http://dx.doi.org/10.1088/1126-6708/2008/04/011*JHEP***04** (2008) 011, arXiv:hep-th/0312022.
8. H. Ooguri and M. Yamazaki, "Crystal Melting and Toric Calabi-Yau Manifolds," http://dx.doi.org/10.1007/s00220-009-0836-y*Commun. Math. Phys***292** (2009) 179–199, arXiv:0811.2801 [hep-th].
9. H. Ooguri and M. Yamazaki, "Emergent Calabi-Yau Geometry," https://doi.org/10.1103/PhysRevLett.102.161601*Phys. Rev. Lett.***102** (2009) 161601, arXiv:0902.3996 [hep-th].
10. A. Hanany and K. D. Kennaway, "Dimer models and toric diagrams," arXiv:hep-th/0503149.
11. S. Franco, A. Hanany, K. D. Kennaway, D. Vegh, and B. Wecht, "Brane dimers and quiver gauge theories," http://dx.doi.org/10.1088/1126-6708/2006/01/096*JHEP***01** (2006) 096, arXiv:hep-th/0504110.
12. S. Franco, A. Hanany, D. Martelli, J. Sparks, D. Vegh, and B. Wecht, "Gauge theories from toric geometry and brane tilings," http://dx.doi.org/10.1088/1126-6708/2006/01/128*JHEP***01** (2006) 128, arXiv:hep-th/0505211.
13. R. Kenyon, "The laplacian and dirac operators on critical planar graphs," http://dx.doi.org/10.1007/s00222-002-0249-4*Inventiones mathematicae***150** no. 2, (Nov, 2002) 409–439, arXiv:math-ph/0202018. http://dx.doi.org/10.1007/s00222-002-0249-4.
14. R. Kenyon and A. Okounkov, "Planar dimers and Harnack curves," arXiv:math/0311062.
15. B. de Tilière, "Partition function of periodic isoradial dimer models," *Probability theory and related fields***138** no. 3–4, (2007) 451–462, arXiv:math/0605583.
16. T. C. Collins and G. Szekelyhidi, "K-Semistability for irregular Sasakian manifolds," http://dx.doi.org/10.4310/jdg/1525399217*J. Diff. Geom***109** no. 1, (2018) 81–109, arXiv:1204.2230 [math.DG].
17. D. L. Jafferis, "The Exact Superconformal R-Symmetry Extremizes Z," http://dx.doi.org/10.1007/JHEP05(2012)159*JHEP***05** (2012) 159, arXiv:1012.3210 [hep-th].
18. J. Stienstra, "Mahler measure, Eisenstein series and dimers," in *Workshop on Calabi-Yau Varieties and Mirror Symmetry*, pp. 151–158. 2, 2005. arXiv:math/0502197.
19. J. Bao, Y.-H. He, and A. Zahabi, "Reflexions on Mahler: Dessins, Modularity and Gauge Theories," arXiv:2111.03655 [hep-th].
20. R. Dijkgraaf and H. Fuji, "The Volume Conjecture and Topological Strings," https://doi.org/10.1002/prop.200900067*Fortsch. Phys.***57** (2009) 825–856, arXiv:0903.2084 [hep-th].

References

Mirror Symmetry, Born Geometry and String Theory

The Meta Mirror Map

Per Berglund, Tristan Hübsch, and Djordje Minic

Abstract All known string theory models may be obtained as partial fermionization, projection and background Ansätze from the original, purely bosonic string theory. The latter theory in turn has been recently shown to describe a chirally and non-commutatively doubled and manifestly T-dual target spacetime. We show herein that this, so-called *metastring* theory automatically includes mirror symmetry.

Introduction: A closer look at the underlying structures in string theory shows that the familiar and (almost) independent left- and right-moving degrees of freedom span a chirally doubled phase space-like non-commutative and modular target spacetime $\mathscr{M}$ [1, 2]. This so called *Born geometry* is specified by (i) the symplectic structure ω, (ii) the bi-orthogonal metric η, and (iii) the doubly Lorentzian metric H. More precisely, this *metastring* target spacetime, $\mathscr{M}$, is an *almost symplectic and para-Hermitian manifold*[1] that has a compatible *foliation* [3]: Locally at every point, p, $\mathscr{M}$ is of the form $\mathscr{M}_p = M_p \times \widetilde{M}_p$, with local coordinates $(x^a, \tilde{x}_b)_p \in M_p \times \widetilde{M}_p$, where $x^a := x^a_{\scriptscriptstyle L} + x^a_{\scriptscriptstyle R}$ and $\tilde{x}^a := x^a_{\scriptscriptstyle L} - x^a_{\scriptscriptstyle R}$ are combinations of the zero-modes of the corresponding worldsheet scalar fields. For local diffeomorphisms (implemented by the Dorfman generalization of the Lie derivative) to be integrable to finite translations, the so-called "section condition" is imposed, which halves the spacetime akin to the quantum-mechanical restriction of the classical phase space, (p, q), in the coordinate

[1] Such spaces admit a signature-(d, d) metric η and a symplectic structure ω such that $K := \eta \cdot \omega$ is an *almost product structure,* $K^2 = +\mathbb{1}$. This η need not be flat, and *almost* means that ω need not be a closed 2-form. Owing to the almost symplectic structure, the existence of the corresponding flat (Bott) connection guarantees that a foliated space is everywhere locally a product of two half-dimensional affine spaces.

P. Berglund (✉)
Department of Physics and Astronomy, University of New Hampshire, Durham, NH, USA
e-mail: per.berglund@unh.edu

T. Hübsch
Department of Physics and Astronomy, Howard University, Washington D.C., USA
e-mail: thubsch@howard.edu

D. Minic
Department of Physics, Virginia Tech, Blacksburg, VA, USA
e-mail: dminic@vt.edu

Y.-H. He et al. (eds.), *Nankai Symposium on Mathematical Dialogues*,
https://doi.org/10.1007/978-981-19-2328-9_6

or momentum representation—or indeed any other $\pi := (\alpha p + \beta q)$-*polarization*, as familiar from Geometric Quantization program [4, 5].

Mirror Doubling: A hallmark of the above structure is that $\pi[T_{\mathscr{M}}] = (T \oplus T^*)_{\pi[\mathscr{M}]}$: a polarization of (the total space of) the tangent bundle on $\mathscr{M}$ is the generalized/doubled tangent bundle on that polarization of $\mathscr{M}$. In local coordinates of the π_x-polarization, $\pi_x[\mathscr{M}] = M_x$, elements of $T_{\mathscr{M}}$ are given as $v^a(x,\tilde{x})\,\partial_a + w_a(x,\tilde{x})\,\mathrm{d}x^a$, as $\tilde{x}$ are locally constant $(\tilde{\partial}^a, \mathrm{d}\tilde{x}_a \mapsto 0)$ [3]. In turn, the same element is $v^a(x,\tilde{x})\,\mathrm{d}\tilde{x}_a + w_a(x,\tilde{x})\,\tilde{\partial}^a$ in the $\pi_{\tilde{x}}$-polarization, $\pi_{\tilde{x}}[\mathscr{M}] = \widetilde{M}_{\tilde{x}}$, where x are locally constant and $\partial_a, \mathrm{d}x^a \mapsto 0$. Therefore, swapping the polarization $M_x \leftrightarrow \widetilde{M}_{\tilde{x}}$ explicitly identifies $T_{M_x} = T^*_{\widetilde{M}_{\tilde{x}}}$ and $T^*_{M_x} = T_{\widetilde{M}_{\tilde{x}}}$—which is the underlying premise of mirror symmetry. Displacing M and $\widetilde{M}$ for clarity, these relationships are sketched in Figure 1, and corroborate the argument presented in [6, Sect. 5.1]. The Born geometry of $\mathscr{M}$ explicitly includes mirror symmetry by virtue of having to *polarize* on one half of $\mathscr{M}$, such that polarizing on the complementary half swaps the corresponding tangent and cotangent spaces. This provides a non-commutative version of the *generalized Calabi-Yau space* framework [7] for the *entire* spacetime, resonating with the recent proposal [8].

Born Heterosis: The metastring intrinsic mirror symmetry applies also to an adaptation of the heterotic Ansatz [9], which relies on a partial *fermionization* and supersymmetry: On the worldsheet, every chiral boson maps non-locally and nonlinearly but exactly 1–1 to a (chiral) fermion [10, 11]; also [12–15]. A suitable superconformal action for the resulting collection of worldsheet scalar and fermion fields then induces supersymmetry in the target space, leading to the familiar result that all the various superstring models may be obtained from the bosonic string [16], all of which eliminate the tachyonic instability of the purely bosonic string; see also [17].

Re-bosonizing and adapting the original heterotic Ansatz [9, 16], we partition the 26 + 26 chiral bosons of the bosonic (meta)string as follows, by identifying $x^0, x^1, \tilde{x}^0, \tilde{x}^1$ in the light-cone gauge with the worldsheet (chirally doubled) coordinates:

WS	light-cone $\perp$-coordinates on M	light-cone $T_M = T^*_{\widetilde{M}}$-fiber	Simple pos. roots of E_8	
$x^0\ x^1$	$x^2\ x^3\ x^4\ x^5\ x^6\ x^7\ x^8\ x^9$	$x^{10} x^{11} x^{12} x^{13} x^{14} x^{15} x^{16} x^{17}$	$x^{18} x^{19} x^{20} x^{21} x^{22} x^{23} x^{24} x^{25}$	(1)
$\tilde{x}^0\ \tilde{x}^1$	$\tilde{x}^2\ \tilde{x}^3\ \tilde{x}^4\ \tilde{x}^5\ \tilde{x}^6\ \tilde{x}^7\ \tilde{x}^8\ \tilde{x}^9$	$\tilde{x}^{10} \tilde{x}^{11} \tilde{x}^{12} \tilde{x}^{13} \tilde{x}^{14} \tilde{x}^{15} \tilde{x}^{16} \tilde{x}^{17}$	$\tilde{x}^{18} \tilde{x}^{19} \tilde{x}^{20} \tilde{x}^{21} \tilde{x}^{22} \tilde{x}^{23} \tilde{x}^{24} \tilde{x}^{25}$	
	light-cone $\perp$-coordinates on $\widetilde{M}$	light-cone $T_{\widetilde{M}} = T^*_M$-fiber	Simple pos. roots of $\widetilde{E}_8$	

For $a = 10, \ldots, 17$, the chiral bosons $x^a, \tilde{x}^a$ are the preimages of 8 + 8 fermions, $\psi^a, \tilde{\psi}^a$, that are induced by supersymmetry to span the indicated (co)tangent bundles. The worldsheet action for these $x^a, \tilde{x}^a$ must reflect the re-bosonization of the supersymmetric structure of the action involving $\psi^a, \tilde{\psi}^a$. In turn, for $a = 18, \ldots, 25$, $x^a, \tilde{x}^a$ are compactified on two copies of the $\mathbb{R}^8/\Lambda(E_8)$ torus. Their dynamics is the (nonlocal!) re-bosonization of the structure specified for their fermionized counterparts, $x^a \mapsto \lambda^a$ and $\tilde{x}^a \mapsto \tilde{\lambda}^a$, adapting the standard description [9, 16]. The worldsheet actions are indeed much simpler (and local!) in the *fermionized picture* of

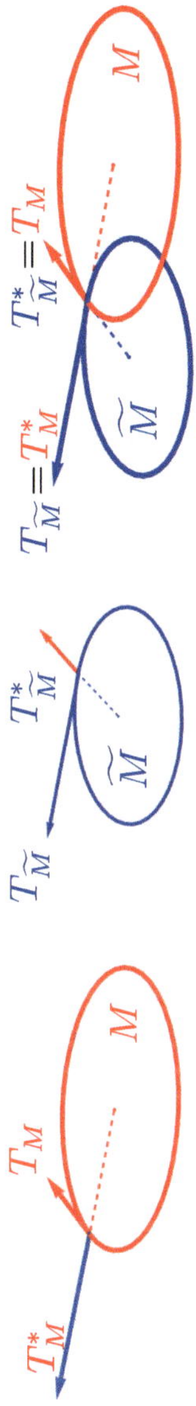

Fig. 1 The original (M, T_M, T^*_M), its mirror $(\widetilde{M}, T_{\widetilde{M}}, T^*_{\widetilde{M}})$, and both depicted together

$x^a, \tilde{x}^a$ for $a > 9$; the re-bosonized specification given here however highlights the implications of the metastring Born geometry.

The *non-commutative* metastring structure of $\mathscr{M}$ then has the following immediate implications:

1. The metastring metric, $\eta_{AB}|_{10...17}$, reproduces the canonical fiber-wise pairing of the two standard bundles: $\langle T_M, T^*_M \rangle = \langle T^*_{\tilde{M}}, T_{\tilde{M}} \rangle = 1$.
2. The metastring metric, $\eta_{AB}|_{18...25}$, implies a dual pairing between the E_8's simple positive roots and those of $\widetilde{E}_8$. This implies that the roots of $\widetilde{E}_8$ are η-canonically reciprocal to those of E_8—which identifies $\widetilde{E}_8$ as the *Langlands dual,* $E_8^\vee$ [18], i.e., the electro-magnetic dual [19, 20]! Being simply laced, $E_8^\vee \approx E_8$, and they are physically indistinguishable. The E_8-group elements are exponential functionals of the $\{x^{18}, \ldots, x^{25}\}$, which can be arranged to commute[1, 2] with the $\widetilde{E}_8$-group exponential functionals of the $\{\tilde{x}^{18}, \ldots, \tilde{x}^{25}\}$, realizing the $E_8 \times \widetilde{E}_8$ Yang-Mills gauge group as expected.
3. For $a = 2, \ldots, 9$, the diagram (1) identifies $x^a, \tilde{x}^a$ with the light-cone *base* coordinates, $x^{a+8} \mapsto \mathrm{d}x^a = \tilde{\partial}^a$ and $\tilde{x}^{a+8} \mapsto \partial_a = \mathrm{d}\tilde{x}_a$ with the light-cone *fibre* coordinates (as befits the preimages of fermions typically spanning the tangent and cotangent bundles), and $(x^{a+16}, \tilde{x}^{a+16}) \mapsto (\rho_a, \rho_{\tilde{a}})$ with the simple positive roots. Then:

   (a) All target spacetime fields are a priori bi-local: they *depend* on both x^μ and $\tilde{x}_\mu$.
   (b) They are *valued* as follows: $A_\mu{}^\alpha$ in $T^*_M \times E_8$ and $A^{\mu\tilde{\alpha}}$ in $T_M \times \widetilde{E}_8$, the latter of which define $A_\mu{}^{\tilde{\alpha}} := H_{\mu\nu} A^{\nu\tilde{\alpha}}$ in $T^\flat_M \times \widetilde{E}_8$ using the metastring H_{AB}-metric.

   These originally bi-local fields then give rise to both an E_8-gauge field *plus* the dual ("dark") field valued in the *same* E_8 algebra. With the analogous for the $\widetilde{E}_8$-valued gauge fields, we have both $A_\mu{}^\alpha(x)$, $A_\mu{}^{\tilde{\alpha}}(x)$ and their duals, $\widetilde{A}_\mu{}^\alpha(\tilde{x})$, $\widetilde{A}_\mu{}^{\tilde{\alpha}}(\tilde{x})$.
4. The metastring non-commutativity implies the correlation, $\frac{\lambda_4^2}{L^4} \int_x \mathfrak{F}^{\mu\nu} [\![A_\mu^\alpha(x), \widetilde{A}_\nu^{\tilde{\beta}}(\tilde{x})]\!] \eta_{\alpha\tilde{\beta}}$, between the *visible* E_8-fields and the *dark* $\widetilde{E}_8$-fields, as well as the analogous correlation term with $E_8 \leftrightarrow \widetilde{E}_8$ swapped [21].

The above statements in 3 insure that our Ansatz (1) reproduces the standard gauge field content of the heterotic string. The statements 1 and 2 however show that the Born geometry is not only perfectly aligned with the standard geometry and dynamics in the heterotic string theory, but additionally links the two copies of E_8 as each other's Langlands dual. Finally the correlation 4 is a straightforward but novel result [21].

Outlook: In conclusion, we comment on various ramifications of this new view of mirror symmetry. Our discussion of mirror symmetry and its relation to Born geometry of an intrinsically non-commutative and T-duality covariant formulation of string theory can be naturally related to the old observation that T-duality is deeply related to mirror symmetry [22]. In that case, as in our discussion, one should pay close attention to the issues of the local versus global formulation of mirror symmetry as T-duality. In particular, our phase space treatment of string theory, and mirror symmetry, as encapsulated in Born geometry, naturally relates to homological mirror symmetry involving the (derived) Fukaya categories (symplectic structure) on one side and the (derived) category of coherent sheaves (complex structure) [23]. Also, the world-sheet formulation of metastring theory should be related to the topological A and B model projections [24], so that these appear as gauge fixed versions of the metastring formulation that explicitly incorporates Born geometry. Finally, we would like to understand non-perturbative aspects of this approach in the context of a purely bosonic but non-commutative (and, in principle, non-associative) matrix-model like formulation; see [21].

Acknowledgements We would like to thank the organizers of the 2021 Nankai Chern symposium, and Yang-Hui He in particular for this extraordinary opportunity. DM thanks Laurent Freidel and Rob Leigh for numerous insightful discussions over many years on the topic of quantum gravity and string theory. PB would like to thank the CERN Theory Group for their hospitality over the past several years. TH is grateful to the Department of Physics, University of Maryland, College Park MD, and the Physics Department of the Faculty of Natural Sciences of the University of Novi Sad, Serbia, for the recurring hospitality and resources. DM is grateful to Perimeter Institute for hospitality and support. The work of PB is supported in part by the Department of Energy grant DE-SC0020220. The work of DM is supported in part by Department of Energy (under DOE grant number DE-SC0020262) and the Julian Schwinger Foundation.

References

1. L. Freidel, R. G. Leigh, and D. Minic, "Metastring Theory and Modular Space-time," https://doi.org/10.1007/JHEP06(2015)006JHEP 06 (2015) 006, arXiv:1502.08005 [hep-th].
2. L. Freidel, R. G. Leigh, and D. Minic, "Modular Spacetime and Metastring Theory," https://doi.org/10.1088/1742-6596/804/1/012032J. Phys. Conf. Ser. 804 no. 1, (2017) 012032.
3. L. Freidel, F. J. Rudolph, and D. Svoboda, "A Unique Connection for Born Geometry," https://doi.org/10.1007/s00220-019-03379-7Comm. Math. Phys. 372 no. 1, (2019) 119–150, arXiv:1806.05992 [hep-th].
4. N. E. Hurt, *Geometric Quantization in Action*. D. Reidel Publishing Company, 1983.
5. N. M. J. Woodhouse, *Geometric Quantization*. Oxford University Press, 2nd ed., 1997.
6. P. Berglund, T. Hübsch, and D. Minic, "On Stringy de Sitter Spacetimes," https://doi.org/10.1007/JHEP12(2019)166JHEP 2019 (2019) 166, arXiv:1902.08617 [hep-th].
7. N. Hitchin, "Generalized calabi-yau manifolds," https://doi.org/10.1093/qmath/hag025The Quarterly Journal of Mathematics 54 no. 3, (Sep, 2003) 281–308. https://doi.org/10.1093/qmath/hag025.
8. P. Berglund, T. Hübsch, and D. Minic, "Stringy bubbles solve de Sitter troubles," https://doi.org/10.3390/universe7100363Universe 7 no. 10, (2021) 363, arXiv:2109.01122 [hep-th]. https://www.mdpi.com/2218-1997/7/10/363.
9. D. J. Gross, J. A. Harvey, E. J. Martinec, and R. Rohm, "The Heterotic String," https://doi.org/10.1103/PhysRevLett.54.502Phys. Rev. Lett. 54 (1985) 502–505.
10. S. R. Coleman, "The Quantum Sine-Gordon Equation as the Massive Thirring Model," DOIurl-http://dx.doi.org/10.1103/PhysRevD.11.2088Phys. Rev. D 11 (1975) 2088.

11. S. Mandelstam, "Soliton Operators for the Quantized Sine-Gordon Equation," https://doi.org/10.1103/PhysRevD.11.3026Phys. Rev. D 11 (1975) 3026.
12. E. Witten, "Non-Abelian bosonization in two dimensions," https://doi.org/10.1007/BF01215276Commun. Math. Phys. 92 (1984) 455–472.
13. K. Harada, "The chiral Schwinger model in terms of chiral bosonization," https://doi.org/10.1103/PhysRevLett.64.139Phys. Rev. Lett. 64 (1990) 139.
14. J. von Delft and H. Schoeller, "Bosonization for beginners: Refermionization for experts," https://doi.org/10.1002/(SICI)1521-3889(199811)7:4<225::AID-ANDP225>3.0.CO;2-LAnn. Phys. 7 (1998) 225–305, arXiv:cond-mat/9805275.
15. D. Sénéchal, https://doi.org/10.1007/0-387-21717-7_4"An Introduction to bosonization," in *Theoretical Methods for Strongly Correlated Electrons*, D. Sénéchal, A. Tremblay, and C. Bourbonnais, eds., CRM Series in Mathematical Physics, pp. 139–186. Springer, 8, 2004. arXiv:cond-mat/9908262.
16. A. Casher, F. Englert, H. Nicolai, and A. Taormina, "Consistent Superstrings as Solutions of the D=26 Bosonic String Theory," https://doi.org/10.1016/0370-2693(85)91072-XPhys. Lett. B 162 (1985) 121–126.
17. W. Siegel, "Superspace duality in low-energy superstrings," https://doi.org/10.1103/PhysRevD.48.2826Phys. Rev. D48 (1993) 2826–2837, arXiv:hep-th/9305073 [hep-th].
18. A. Borel, "Automorphic L-functions," in *Automorphic Forms, Representations, and L-functions*, A. Borel and W. Casselman, eds., vol. 33, Part 2 of *Proceedings of Symposia in Pure Mathematics*, pp. 27–61. American Mathematical Society, Providence, RI, 1979.
19. A. Kapustin and E. Witten, "Electric-Magnetic Duality And The Geometric Langlands Program," https://doi.org/10.4310/CNTP.2007.v1.n1.a1Commun. Num. Theor. Phys. 1 (2007) 1–236, arXiv:hep-th/0604151.
20. E. Frenkel, "Gauge Theory and Langlands Duality," in *Bourbaki Seminar*, 6, 2009. arXiv:0906.2747 [math.RT].
21. P. Berglund, T. Hübsch, and D. Minic, "String theory, the dark sector and the hierarchy problem," https://doi.org/10.31526/lhep.2021.186LHEP 2021 (2021) 186, arXiv:2010.15610 [hep-th]. http://journals.andromedapublisher.com/index.php/LHEP/article/view/186.
22. A. Strominger, S.-T. Yau, and E. Zaslow, "Mirror symmetry is T-duality," https://doi.org/10.1016/0550-3213(96)00434-8Nucl. Phys. B479 (1996) 243–259, arXiv:hep-th/9606040.
23. M. Kontsevich, "Homological algebra of mirror symmetry," in *Proceedings of the International Congress of Mathematicians*, S. D. Chatterji, ed., pp. 120–139. Birkhäuser Basel, Basel, 1995.
24. E. Witten, "Mirror manifolds and topological field theory," in *Mirror Symmetry*, S.-T. Yau, ed., vol. I. AMS, 1991. arXiv:hep-th/9112056 [hep-th]. AMS/IP Stud. Adv. Math. 9 (1998) 121–160.

Double Copying Exceptional Field Theories

David S. Berman, Kwangeon Kim, and Kanghoon Lee

Abstract We examine exceptional field theory through the lens of the generalised double copy formalism. This allows us to construct classical solutions in M-theory using a generalised Kerr-Schild ansatz and along the way indicates hints towards a single copy of M-theory. Based on a talk at the Nankai Symposium given by DSB.

7.1 Introduction

The goal of Exceptional field theory (ExFT) is to provide a reformulation of eleven dimensional supergravity while making manifest the exceptional symmetry acting on supergravity fields. This reformulation uses generalised geometry to combine the metric and p-form potentials into a single geometric object. For a recent review of the topic and a complete set of references see [1]. This reformulation has led to various insights and reinterpretations of M-theory. In particular, M-theory branes such as the membrane and fivebrane are to be viewed as superpositions of waves and gravitational monopoles in the exceptional space as described in [2–4]. One successful approach to studying ExFT is to generalise methods from general relativity and adapt them to ExFT. A very useful technique in general relativity that allows us to find solutions of Einstein's equations is the Kerr-Schild ansatz. The Kerr-Schild ansatz has the extraordinary property that it linearises the Einstein equations and so solutions of the resulting linear equations will give an exact solutions to the non-linear Einstein equation! In fact, the resulting linear equations that come from the KS ansatz are Maxwell's equations. This implies the Kerr-Schild ansatz provides

D. S. Berman (✉)
School of Physical and Chemical Sciences, Queen Mary University of London, Centre for Theoretical Physics, London, England, USA
e-mail: d.s.berman@qmul.ac.uk

K. Kim
Department of Physics, Yonsei University, Seoul, Korea
e-mail: kim64656@yonsei.ac.kr

K. Lee
Asia Pacific Center for Theoretical Physics, Postech, Pohang, Korea

Y.-H. He et al. (eds.), *Nankai Symposium on Mathematical Dialogues*,
https://doi.org/10.1007/978-981-19-2328-9_7

us with a map between solutions of Maxwell's equations and solutions in general relativity. This map has been used to formulate a classical solution form of the so called "double copy" that maps between gravity and Yang-Mills theory [5–11]. Heuristically, the double copy construction formulates gravity as coming from two copies of Yang-Mills theory as originated in a study of the scattering amplitudes of the two theories [12, 13].

We can now ask, instead of just gravity, what if we have a supergravity with additional p-form potentials. What is the single copy of that theory? Very concretely we can ask how we can generalise the Kerr-Schild ansatz in general relativity to include these p-form fields. Answering this is quite hard given the nonlinear nature of Einstein's equations. The key to solving this is to use the reformulation of supergravity in terms of ExFT and then one can use a simple generalisation of the Kerr-Schild ansatz for the generalised metric. This produces again a set of linear equations but this time it is a combination of a Maxwell field and a two form potential. This hints that the single copy of M-theory would be a some product of these one form and two form theories.

7.2 $SL(5)$ Exceptional Field Theory

We willl work with the exceptional geometry associated to 4 dimensions whose coordinates we label x^μ where $\mu = 0, 1, 2, 3$. We will make manifest the U-duality group $SL(5)$. To do this we will extend the dimensionality of the spacetime by including new coordinates that describe membrane wrapping modes. There will be six such coordinates which we call $y_{\rho\nu}$ giving ten in total. A generalised tangent vector V^M with $M = 1..10$ in this extended space lies in the **10** representation of $SL(5)$ and is parametrised by both the 4-vector v^μ and the two-form $\lambda_{\rho\nu}$ in 4d as follows:

$$V^M = \begin{pmatrix} v^\mu \\ \lambda_{\rho\nu} \end{pmatrix}, \tag{7.1}$$

For a more in depth review see [1, 14, 15]. We may also represent the **10** of $SL(5)$ $(M, N = 1 \cdots 10)$ as antisymmetric pairs of **5** indices, $[mn]$, where $m, n = 0, 1, 2, 3$ and 5, so that

$$V^{[mn]} = \begin{cases} V^{\mu 5} = -V^{5\mu} = v^\mu \\ V^{\alpha\beta} = \tilde{\lambda}^{\alpha\beta} = \frac{1}{2}\epsilon^{\alpha\beta\rho\sigma}\lambda_{\rho\sigma} \end{cases} \tag{7.2}$$

A key goal of the construction is to combine the metric and 3-form potential, $C_{\mu\nu\rho}$ into a single geometric object called the generalised metric. For the $SL(5)$ theory, the generalized metric $\mathcal{M}_{MN}$ is given as follows:

$$\mathcal{M}_{MN} = |g|^{\frac{1}{5}} \begin{pmatrix} g_{\mu\nu} + \frac{1}{4}C_{\mu\rho_1\rho_2}g^{\rho_1\rho_2,\sigma_1\sigma_2}C_{\sigma_1\sigma_2\nu} & \frac{1}{2}C_{\mu\rho_1\rho_2}g^{\rho_1\rho_2,\nu_1\nu_2} \\ \frac{1}{2}g^{\mu_1\mu_2,\rho_1\rho_2}C_{\rho_1\rho_2\nu} & g^{\mu_1\mu_2,\nu_1\nu_2} \end{pmatrix}. \tag{7.3}$$

Along with this generalised metric there will be an ordinary metric G_{ij} with $i, j = 1..7$ corresponding to the remaining seven Riemannian dimensions of eleven dimensional supergravity. There will also be additional mixed objects, metric components with legs in the four dimensional space and the seven dimensional space and components of the three form with legs in the seven dimensional space. We will ignore these in what follows. The action that will describe the dynamics for the fields in question is given by:

$$S = \int_\Sigma \mathrm{d}^7 z\, \mathrm{d}^{10} X \sqrt{|G|} \Big[\, R[G_{ij}] + \frac{1}{12} G^{ij} \partial_i \mathcal{M}_{MN} \partial_j \mathcal{M}^{MN} - V[\mathcal{M}, G] \,\Big] \tag{7.4}$$

where $R[G_{ij}]$ is the Ricci scalar with respect to G_{ij}, and the scalar potential $V[\mathcal{M}, G]$ is given by

$$\begin{aligned} V[\mathcal{M}, G] = &-\frac{1}{12}\mathcal{M}^{MN}\partial_M \mathcal{M}^{KL}\partial_N \mathcal{M}_{KL} + \frac{1}{2}\mathcal{M}^{MN}\partial_M \mathcal{M}^{KL}\partial_K \mathcal{M}_{LN} \\ &-\frac{1}{2}\partial_M \mathcal{M}^{MN}\partial_N \ln|G| \\ &-\frac{1}{4}\mathcal{M}^{MN}\big(\partial_M G_{ij}\partial_N G^{ij} + (\partial_M \ln|G|)(\partial_N \ln|G|)\big)\,. \end{aligned} \tag{7.5}$$

The variation of the action with respect to the generalized metric $\mathcal{M}$ gives

$$\delta_{\mathcal{M}} S = \int_\Sigma \mathrm{d}^7 z\, \mathrm{d}^{10} X \sqrt{|G|}\, \delta\mathcal{M}^{MN} \mathcal{K}_{MN}\,, \tag{7.6}$$

where

$$\begin{aligned} \mathcal{K}_{MN} = &-\frac{1}{6\sqrt{|G|}}\partial_P\big(\sqrt{|G|}\mathcal{M}^{PQ}\partial_Q \mathcal{M}_{MN}\big) + \frac{1}{\sqrt{|G|}}\partial_P\big(\sqrt{|G|}\mathcal{M}^{PQ}\partial_{(M}\mathcal{M}_{N)Q}\big) \\ &+\frac{1}{12}\partial_M \mathcal{M}_{PQ}\partial_N \mathcal{M}^{PQ} + \frac{1}{6}\mathcal{M}^{PQ}\mathcal{M}^{RS}\partial_P \mathcal{M}_{RM}\partial_Q \mathcal{M}_{SN} \\ &-\frac{1}{2}\mathcal{M}^{PQ}\mathcal{M}^{RS}\partial_P \mathcal{M}_{RM}\partial_S \mathcal{M}_{QN} - \frac{1}{2}\partial_M \partial_N \ln|G| + \frac{1}{4}\partial_M G_{ij}\partial_N G^{ij} \\ &-\frac{1}{6\sqrt{|G|}}\partial_i\big(\sqrt{|G|}G^{ij}\partial_j \mathcal{M}_{MN}\big)\,. \end{aligned} \tag{7.7}$$

To get a consistent variation of $\mathcal{M}$ that is compatible with the coset structure of the generalised metric we need a projection operator $P_{MN}{}^{PQ}$. The equations of motion are then

$$\hat{\mathcal{R}}_{MN} = P_{MN}{}^{PQ}\mathcal{K}_{PQ} = 0\,, \tag{7.8}$$

Table 7.1 List of duality groups G and their maximal compact subgroups H (Euclidean case) the generalised metric parameterises G/H

G	H	α	ω
$GL(d)$	$SO(d)$	1	0
$O(D, D)$	$O(D) \times O(D)$	2	0
$SL(5)$	$SO(5)$	3	$-\frac{1}{5}$
$Spin(5, 5)$	$Spin(5) \times Spin(5)$	4	$-\frac{1}{4}$
$E_{6(6)}$	$USp(8)$	6	$-\frac{1}{3}$
$E_{7(7)}$	$SU(8)$	12	$-\frac{1}{2}$
$E_{8(8)}$	$SO(16)$	60	-1

with

$$P_{MN}{}^{KL} = \frac{1}{\alpha}\left(\delta_M^{(K}\delta_N^{L)} - \omega \mathcal{M}_{MN}\mathcal{M}^{KL} - \mathcal{M}_{MQ}Y^{Q(K}{}_{RN}\mathcal{M}^{L)R}\right). \tag{7.9}$$

The constants α and ω depend on the theory and are described in Table 7.1.

7.3 Generalised Kerr-Schild

We now present the generalised Kerr-Schild ansatz for the generalised metric that we will use to solve the equations of motion [16].

$$\begin{aligned} \mathcal{M}_{MN} &= \mathcal{M}_{0MN} + \kappa\varphi P_{0MN}{}^{PQ}K_P K_Q\,, \\ \left(\mathcal{M}^{-1}\right)^{MN} &= \left(\mathcal{M}_0^{-1}\right)^{MN} - \kappa\varphi P_0{}^{MN}{}_{PQ}K^P K^Q\,. \end{aligned} \tag{7.1}$$

$\mathcal{M}_{0MN}$ is the background generalised metric which we take to be flat. K_M is a null vector with respect to $\mathcal{M}_0$, which means:

$$K_M\left(\mathcal{M}_0^{-1}\right)^{MN}K_N = 0\,. \tag{7.2}$$

Finally, φ is a scalar field that must be determined.

All the indices are raised and lowered by the background generalised metric $\mathcal{M}_0$ and $P_{0MN}{}^{PQ}$ is constructed with the trivial background generalised metric. The null condition on K_M is not sufficient to linearise the inverse metric. To do this so we need an additional constraint, defining:

$$Q_{MN} = \varphi P_{0MN}{}^{PQ}K_P K_Q \tag{7.3}$$

we need

$$Q_{MP}Q^{PQ} = 0\,. \tag{7.4}$$

In general relativity the projection operator is trivial,

$$\left(P^{\mathrm{GR}}\right)_{\mu\nu}{}^{\rho\sigma} = \delta_\mu{}^{(\rho}\delta_\nu{}^{\sigma)} \,. \tag{7.5}$$

and so from the form (7.1), the Kerr-Schild ansatz for relativity is given by

$$\begin{aligned} g_{\mu\nu} &= \tilde{g}_{\mu\nu} + \kappa\varphi K_\mu K_\nu \,, \\ \left(g^{-1}\right)^{\mu\nu} &= \left(\tilde{g}^{-1}\right)^{\mu\nu} - \kappa\varphi K^\mu K^\nu \,, \end{aligned} \tag{7.6}$$

where $\tilde{g}_{\mu\nu}$ is the background metric which the vector field K^μ is null with respect to:

$$K^\mu \tilde{g}_{\mu\nu} K^\nu = 0 \,. \tag{7.7}$$

The Einstein equations then reduce to Maxwell's equations for φK_μ and so this allows us to identify φK_μ as the single copy Maxwell field, A_μ.

Before moving on to the $SL(5)$ exceptional case, let us do the Kerr Schild ansatz for Double Field Theory as first reported in [17]. For Double Field Theory, the generalised metric $\mathcal{H}_{MN}$ on a $2D$ dimensional doubled space is given by the coset $O(D,D)/O(1,D-1)\times O(1,D-1)$. In terms of the usual D dimensional metric g and the Kalb-Ramond two-form B it is written as follows:

$$\mathcal{H}_{MN} = \begin{pmatrix} g_{\mu\nu} - B_{\mu\rho}g^{\rho\sigma}B_{\sigma\nu} & B_{\mu\rho}g^{\rho\nu} \\ -g^{\mu\rho}B_{\rho\nu} & g^{\mu\nu} \end{pmatrix} . \tag{7.8}$$

It is a symmetric $O(D,D)$ element and satisfies the $O(D,D)$ compatibility constraint, $\mathcal{H}_{MP}\mathcal{J}^{PQ}\mathcal{H}_{QN} = \mathcal{J}_{MN}$, where $\mathcal{J}_{MN}$ is the $O(D,D)$ metric which maybe chosen to be:

$$\mathcal{J}_{MN} = \begin{pmatrix} \mathbf{0} & \delta^a{}_b \\ \delta_a{}^b & \mathbf{0} \end{pmatrix} . \tag{7.9}$$

The projector uses the so called Y-tensor [18] which is given in terms of the $O(D,D)$ metric by

$$Y^{PQ}{}_{MN} = \mathcal{J}^{PQ}\mathcal{J}_{MN} \,. \tag{7.10}$$

The projection operator then factorises as follows:

$$\left(P^{\mathrm{DFT}}\right)_{MN}{}^{PQ} = 2P_{(M}{}^{P}\bar{P}_{N)}{}^{Q} \,, \tag{7.11}$$

where $P_M{}^N$ and $\bar{P}_M{}^N$ are projectors defined by the generalized metric of DFT, $\mathcal{H}_{MN}$, and the $O(D,D)$ metric $\mathcal{J}_{MN}$:

$$P_{MN} = \frac{1}{2}\left(\mathcal{J}_{MN} + \mathcal{H}_{MN}\right), \qquad \bar{P}_{MN} = \frac{1}{2}\left(\mathcal{J}_{MN} - \mathcal{H}_{MN}\right). \tag{7.12}$$

The factorisation of the projector, should not in fact be a surprise, it is in fact a consequence of the left and right decomposition of the closed string modes.

Applying to the general KS ansatz (7.1), we have the KS ansatz for $\mathcal{H}_{MN}$

$$\mathcal{H}_{MN} = \mathcal{H}_{0MN} + \kappa\varphi\big(P_0^{\text{DFT}}\big)_{MN}{}^{PQ} K_P K_Q\,, \tag{7.13}$$

where K_M is a null vector, $K_M \mathcal{J}^{MN} K_N = 0$ and $\big(P_0^{\text{DFT}}\big)_{MN}{}^{PQ}$ is a background projection operator. It is useful to write the projected null vectors as

$$L_M = P_{0M}{}^N K_N\,, \qquad \bar{L}_M = \bar{P}_{0M}{}^N K_N \tag{7.14}$$

and from the completeness relation, $\delta_M{}^N = P_M{}^N + \bar{P}_M{}^N$, K_M is decomposed to L_M and $\bar{L}_M$.

Q_{MN} then may written in terms of $L_M, \bar{L}_N$ as follows:

$$Q_{MN} = \varphi\big(L_M \bar{L}_N + L_N \bar{L}_M\big)\,. \tag{7.15}$$

We require the nilpotency condition on Q_{MN} (7.4) for the linearity of the KS ansatz. This implies that L_M and $\bar{L}_M$ have to be mutually orthogonal null vectors of definite chirality and so,

$$L_M = P_{0M}{}^N L_N\,, \qquad \bar{L}_M = \bar{P}_{0M}{}^N \bar{L}_N \tag{7.16}$$

and

$$L_N L^N = 0\,, \qquad \bar{L}_N \bar{L}^N = 0\,. \tag{7.17}$$

Then the Kerr Schild ansatz is then rewritten as:

$$\mathcal{H}_{MN} = \mathcal{H}_{0MN} + \kappa\varphi\big(L_M \bar{L}_N + L_N \bar{L}_M\big)\,. \tag{7.18}$$

From this the equations of motion become that of two Maxwell fields from which the B-field and metric solutions may be reconstructed [17].

But what about exceptional field theory?

We will now assume the background generalised metric $\mathcal{M}_0$ is:

$$\mathcal{M}_{0MN} = \begin{pmatrix} \eta_{\mu\nu} & 0 \\ 0 & \eta^{\mu\mu',\nu\nu'} \end{pmatrix}, \quad \text{and} \quad \mathcal{M}_{0mm',nn'} = \begin{pmatrix} \eta_{\mu\nu} & 0 \\ 0 & -\eta_{\mu\mu',\nu\nu'} \end{pmatrix}, \tag{7.19}$$

and the associated null vector K_M is parametrized as

$$K_M = \begin{pmatrix} l_\mu \\ k^{\nu\nu'} \end{pmatrix}, \qquad K^M = \begin{pmatrix} l^\mu \\ k_{\nu\nu'} \end{pmatrix}, \tag{7.20}$$

and

$$K_{mm'} = \begin{cases} K_{\mu 5} = l_\mu \\ K_{\nu\nu'} = \tilde{k}_{\nu\nu'} \end{cases} \qquad K^{mm'} = \begin{cases} K^{\mu 5} = l^\mu \\ K^{\nu\nu'} = -\tilde{k}^{\nu\nu'} \end{cases} . \tag{7.21}$$

Here the Greek indices are raised and lowered by the flat background metric $\eta_{\mu\nu}$. $\tilde{k}_{\mu\nu}$ and $\tilde{k}^{\mu\nu}$ are related with $k_{\mu\mu'}$ and $k^{\mu\mu'}$ as

$$\tilde{k}_{\mu\mu'} = \frac{1}{2}\epsilon_{\mu\mu'\nu\nu'}k^{\nu\nu'}, \qquad \tilde{k}^{\mu\mu'} = \frac{1}{2}\eta^{\mu\mu',\nu\nu'}\tilde{k}_{\nu\nu'} = -\frac{1}{2}\epsilon^{\mu\mu'\nu\nu'}k_{\nu\nu'}, \tag{7.22}$$

and their inner product satisfies

$$k_{\mu\mu'}k^{\mu\mu'} = -\tilde{k}_{\mu\mu'}\tilde{k}^{\mu\mu'} . \tag{7.23}$$

Then the null condition for K is then

$$K_M K^M = K_{mm'}K^{mm'} = l_\mu l^\mu + \frac{1}{2}k_{\nu\nu'}k^{\nu\nu'} = l_\mu l^\mu - \frac{1}{2}\tilde{k}_{\nu\nu'}\tilde{k}^{\nu\nu'} = 0 . \tag{7.24}$$

The projection operator $P_{MN}{}^{PQ}$ for the $SL(5)$ ExFT is:

$$P_{MN}{}^{PQ} = \frac{1}{3}\Big(\delta_{(M}{}^P\delta_{N)}{}^Q + \frac{1}{5}\mathcal{M}_{MN}\mathcal{M}^{PQ} - \mathcal{M}_{MR}Y^{R(P}{}_{SN}\mathcal{M}^{Q)S}\Big), \tag{7.25}$$

where the Y-tensor is

$$Y^{MN}{}_{PQ} = Y^{mm'nn'}{}_{pp'qq'} = \epsilon^{mm'nn'r}\epsilon_{pp'qq'r} . \tag{7.26}$$

Combining the above results allows us to write the Kerr-Schild ansatz for $\mathcal{M}_{MN}$ as described in equation (7.1). One may then express the ordinary metric and three form potential in terms of the components of K^M appearing in the Kerr-Schild ansatz as follows:

$$\begin{aligned} |g| &= \left|1 - \frac{\kappa\varphi}{3}l\cdot l\right|^{-\frac{5}{3}}, \\ g_{\mu\nu} &= \left|1 - \frac{\kappa\varphi}{3}l\cdot l\right|^{-\frac{2}{3}}\Big(\eta_{\mu\nu} + \frac{\kappa\varphi}{3}\big(l_\mu l_\nu - \tilde{k}_{\mu\rho}\tilde{k}_\nu{}^\rho\big)\Big), \\ C_{\mu\nu\rho} &= \frac{2\kappa\varphi}{3|1-\frac{\kappa\varphi}{3}l\cdot l|}l_{[\mu}k_{\nu\rho]} . \end{aligned} \tag{7.27}$$

where $l\cdot l = l_\mu\eta^{\mu\nu}l_\nu$. Remarkably, the KS ansatz for the generalized metric is linear in κ, but component fields are highly nonlinear. If we set $k_{\mu\nu} = 0$, then l^μ becomes a null vector and the KS ansatz reduces to the conventional KS ansatz in GR, $g_{\mu\nu} = \eta_{\mu\nu} + \kappa\varphi l_\mu l_\nu$ and $C_{\mu\nu\rho} = 0$.

When we insert this ansatz into the equations of motion we find:

$$\partial^\sigma \partial_\sigma(\varphi l_\mu) - \partial^\sigma \partial_\mu(\varphi l_\sigma) = 0 \,,$$
$$\partial^\rho \partial_{[\rho}(\varphi k_{\mu\nu]}) = 0 \,. \tag{7.28}$$

From which we can identify:

$$A_\mu = \varphi l_\mu \,, \qquad B_{\mu\nu} \;= \varphi k_{\mu\nu} \,. \tag{7.29}$$

as the single copy fields.

This gives the Maxwell field and two form potential as the single copy for gravity with a three form potential.

As a test of this ansatz lets construct the membrane solution in terms of the generalised Kerr-Schild Ansatz and interpret these solutions from the point of view of the single copy fields.

Take the worldvolume directions of the M2-brane to be t, x^1, x^2 and denote them as $x^\alpha = \{t, x^1, x^2\}$ and choose the M-theory circle direction as x^2. The transverse directions are $\mathbf{x}_8 = \{x^3, z^i\}$, where $z^i, z^j, \cdots$ are the coordinates of the 7 extra directions. The M2-brane geometry is given by

$$\mathrm{d}s_{11}^2 = H^{-\frac{2}{3}}\eta_{\alpha\beta}\mathrm{d}x^\alpha \mathrm{d}x^\beta + H^{\frac{1}{3}}\mathrm{d}x^3\mathrm{d}x^3 + H^{\frac{1}{3}}\delta_{ij}\mathrm{d}z^i\mathrm{d}z^j \,,$$
$$C_{t12} = -\left(1 - H^{-1}\right), \qquad H = 1 + \frac{h}{|\vec{x}_8|^6} \,. \tag{7.30}$$

To embed the 11-dimensional supergravity into the $SL(5)$ ExFT, we need to use the following Kaluza-Klein ansatz for the 11-dimensional metric $\hat{g}$

$$\hat{g}_{\hat{\mu}\hat{\nu}} = \begin{pmatrix} |g|^{-\frac{1}{5}}G_{ij} + A_i^\rho A_j^\sigma g_{\rho\sigma} & A_i^\rho g_{\rho\nu} \\ g_{\mu k}A_j^k & g_{\mu\nu} \end{pmatrix} , \tag{7.31}$$

The internal metric and C-field are

$$g_{\mu\nu} = H^{-\frac{2}{3}}\left(\eta_{\alpha\beta}\mathrm{d}x^\alpha \mathrm{d}x^\beta + H\mathrm{d}x^3\mathrm{d}x^3\right), \qquad C_{012} = H^{-1} - 1 \,. \tag{7.32}$$

The single copy of this solution is then given by:

$$A_0 = \varphi \,, \qquad B_{12} = \varphi \,. \tag{7.33}$$

A quite remarkable result. Interpreting in terms of the single copy fields the membrane is described by a two dimensional plane of electric charges. That is the Maxwell field is given by solving the Harmonic equation for an electric source smeared over two spatial dimensions just like the classical electrostatic problem for a plane of charge. The two form solution is like that of a magnetic string that is smeared over one additional dimension to give a plane of string charge.

7.4 Conclusions and Discussion

Whether one appreciates the need for reformulating eleven dimensional supergravity in terms of expectional field theory or not, the generalised Kerr-Schild ansatz allows one to vastly simplify Einstein's equations sourced by a three form potential. It is doubtful that without ExFT one would have the imagination to construct the generalised Kerr-Schild ansatz. This of course can be extended to other fields such as the six form in eleven dimensional supergravity but to do this requires a $E7$ Kerr-Schild ansatz. It is also curious that the single copy produces a Maxwell and a two-form field. Thus indicating that the single copy for M-theory is more complicated than a product of Yang-Mills fields and instead must involve some form of Gerbe.

References

1. D. S. Berman and C. D. A. Blair, *The Geometry, Branes and Applications of Exceptional Field Theory*, *Int. J. Mod. Phys. A***35** (2020), no. 30 2030014, [arXiv:2006.0977].
2. D. S. Berman and F. J. Rudolph, *Strings, Branes and the Self-dual Solutions of Exceptional Field Theory*, *JHEP***05** (2015) 130, [arXiv:1412.2768].
3. D. S. Berman and F. J. Rudolph, *Branes are Waves and Monopoles*, *JHEP***05** (2015) 015, [arXiv:1409.6314].
4. J. Berkeley, D. S. Berman, and F. J. Rudolph, *Strings and Branes are Waves*, *JHEP***06** (2014) 006, [arXiv:1403.7198].
5. R. Monteiro, D. O'Connell, and C. D. White, *Black holes and the double copy*, *JHEP***1412** (2014) 056, [arXiv:1410.0239].
6. A. Luna, R. Monteiro, D. O'Connell, and C. D. White, *The classical double copy for Taub–NUT spacetime*, *Phys. Lett.***B750** (2015) 272–277, [arXiv:1507.0186].
7. N. Bahjat-Abbas, A. Luna, and C. D. White, *The Kerr-Schild double copy in curved spacetime*, *JHEP***12** (2017) 004, [arXiv:1710.0195].
8. D. S. Berman, E. Chacón, A. Luna, and C. D. White, *The self-dual classical double copy, and the Eguchi-Hanson instanton*, *JHEP***01** (2019) 107, [arXiv:1809.0406].
9. Y.-T. Huang, U. Kol, and D. O'Connell, *Double copy of electric-magnetic duality*, *Phys. Rev. D***102** (2020), no. 4 046005, [arXiv:1911.0631].
10. R. Alawadhi, D. S. Berman, B. Spence, and D. Peinador Veiga, *S-duality and the double copy*, *JHEP***03** (2020) 059, [arXiv:1911.0679].
11. K. Kim, K. Lee, R. Monteiro, I. Nicholson, and D. Peinador Veiga, *The Classical Double Copy of a Point Charge*, *JHEP***02** (2020) 046, [arXiv:1912.0217].
12. Z. Bern, J. J. M. Carrasco, and H. Johansson, *Perturbative Quantum Gravity as a Double Copy of Gauge Theory*, *Phys.Rev.Lett.***105** (2010) 061602, [arXiv:1004.0476].
13. Z. Bern, T. Dennen, Y.-t. Huang, and M. Kiermaier, *Gravity as the Square of Gauge Theory*, *Phys.Rev.***D82** (2010) 065003, [arXiv:1004.0693].
14. D. S. Berman and M. J. Perry, *Generalized Geometry and M theory*, *JHEP***06** (2011) 074, [arXiv:1008.1763].
15. D. S. Berman, *A Kaluza–Klein Approach to Double and Exceptional Field Theory*, *Fortsch. Phys.***67** (2019), no. 8-9 1910002, [arXiv:1903.0286].

16. D. S. Berman, K. Kim, and K. Lee, *The classical double copy for M-theory from a Kerr-Schild ansatz for exceptional field theory*, *JHEP* **04** (2021) 071, [arXiv:2010.0825].
17. K. Lee, *Kerr-Schild Double Field Theory and Classical Double Copy*, *JHEP* **10** (2018) 027, [arXiv:1807.0844].
18. D. S. Berman, M. Cederwall, A. Kleinschmidt, and D. C. Thompson, *The gauge structure of generalised diffeomorphisms*, *JHEP* **01** (2013) 064, [arXiv:1208.5884].

Affine Grassmannians and Brane Systems

Antoine Bourget

Abstract This is a contribution to the proceedings for the Nankai Symposium on Mathematical Dialogues, based on my talk given on August 4th, 2021. It is heavily inspired on the paper [1], written with J. F. Grimminger, A. Hanany, M. Sperling and Z. Zhong. I review the connection between the Coulomb branch of framed unitary quivers based on finite Dynkin diagrams, their brane incarnations, and the affine Grassmannian.

8.1 Introduction

In this talk, in the spirit of the mathematical dialogues that are the focus of this Nankai symposium, I would like to describe a particularly beautiful connection between physics and geometry, as illustrated in Fig. 8.1. On the physics side, we are looking at a class of supersymmetric quantum field theories (SQFTs) in three dimensions, which due to the large amount of supersymmetry, possess infinitely many degenerate vacua, i.e. field configurations minimizing the energy. This *moduli space of vacua* is parametrized by the vacuum expectation value of the scalar fields in the theory, and can be seen as an algebraic variety [2, 3]. This space is of fundamental importance for various reasons: it parametrizes the phases the theory can be in, at no energy cost; from a computational point of view, any correlation function obtained perturbatively depends on the vacuum around which the expansion is taken. More abstractly, the moduli space of vacua plays the role of a fingerprint for the theory, and as such it can be used to detect dualities, or as a classification tool. Our first natural question is then:

Given a SQFT, how to compute its moduli space of vacua?

A. Bourget (✉)
CNRS, CEA, Institut de physique théorique, Université Paris-Saclay, Gif-sur-Yvette, France

Laboratoire de Physique, École normale supérieure, ENS, Université PSL, CNRS, Sorbonne Université, Université de Paris, Paris, France
e-mail: antoine.bourget@polytechnique.org

Y.-H. He et al. (eds.), *Nankai Symposium on Mathematical Dialogues*,
https://doi.org/10.1007/978-981-19-2328-9_8

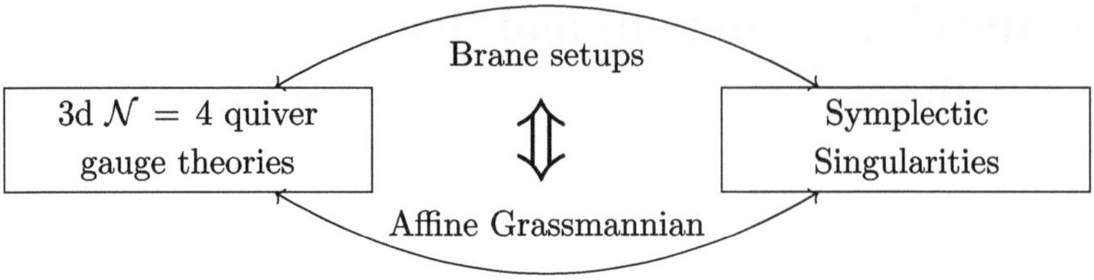

Fig. 8.1 The Coulomb branch part of the moduli space of certain quiver gauge theories (see text) are symplectic singularities, whose geometry can be understood both in terms of brane moves in string theory, and in terms of the Affine Grassmannian. The main message of the present note is that these two points of view are equivalent

This being said, it should be of no surprise that a companion question is of great interest: which spaces can be moduli spaces of vacua of SQFTs? More precisely, one can ask

Given a variety, which SQFT, if any, has it as its moduli space of vacua?

The two questions above are outstanding problems connecting physics and geometry, and are way to broad to be addressed in full generality at the moment. This talk focuses instead on a particular instance of the problem, in which two equivalent solutions are offered, one by physics and the other by mathematics, as shown in Fig. 8.1.

8.2 Quiver Gauge Theories and Their Moduli Spaces

First let's restrict the scope of the SQFTs under consideration. I will be talking today about 3d $\mathcal{N} = 4$ *quiver gauge theories*. The "3d" part means that there is one time and two space dimensions, and the "$\mathcal{N} = 4$" part indicates the amount of supersymmetry, namely 4 Majorana spinors, i.e. 8 real supercharges. With this amount of supersymmetry, it is enough to specify the gauge groups, the matter content, and a few numerical constants, to fully define a theory. This is what is provided by the quiver, which is an unoriented graph in which nodes correspond to unitary groups (gauge groups for round nodes and flavor groups for square nodes) and links are supersymmetric hypermultiplets ("hypers") transforming in the bifundamental representation of the nodes they connect. For instance, the quiver

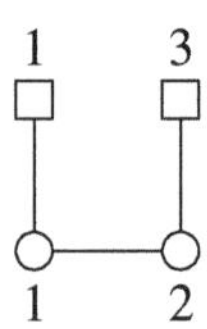

(8.1)

represents a $U(1) \times U(2)$ gauge theory with one $U(1)$ fundamental, three $U(2)$ fundamental hypers and one bifundamental hyper. Note that 3d $\mathcal{N} = 4$ supersymmetry imposes the presence of superpartners for the gauge fields A_μ, including a triplet of real scalars ϕ_i, $i = 1, 2, 3$, transforming in the adjoint representation of the gauge group. As for the numerical constants, these are the gauge couplings—which we set at a fixed (dimensionful) non-zero value, the value itself does not affect the geometry of the moduli space—, and masses and Fayet-Iliopoulos parameters, which we set to zero for simplicity. In this talk, the gauge nodes of the quiver form the Dynkin diagram of a simple finite-dimensional Lie algebra $\mathfrak{g}$.

Turning to the right hand side of Fig. 8.1, we look at a the moduli space of vacua of the quiver gauge theories defined above. One can define a metric on it, which turns out to be of (singular) hyperKähler type. In addition, the spectrum of massless states at a given point in the moduli space jumps at singularities. In the present case, where we recall that masses and FI parameters are assumed to vanish, the space is actually conical, and the structure of nested singularities of various codimensions can be described by a *Hasse diagram*, which can be seen as a phase diagram for the theory [4–12]. For simplicity, we consider here the *Coulomb branch* of the moduli space, which is the part that is classically parametrized by fields from the vector multiplets.[1] For the example (8.1) above, the diagram reads

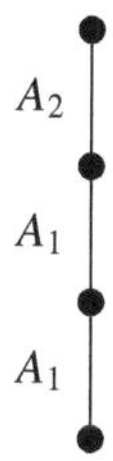

(8.2)

The labels A_n next to the lines correspond to the singularity $\mathbb{C}^2/\mathbb{Z}_{n+1}$ transverse to the singular locus represented by the dot at the bottom of the line. This can be derived using a brane construction [13] of the theory. The brane construction experiences a succession of *Kraft-Procesi transitions* [14–17], which can be turned to an algorithm on the quiver itself, called *quiver subtraction* [1, 8, 16–20]. In the next section, we introduce a geometric object which encodes this algorithm in an algebraico-geometric way.

[1] This includes the ϕ_i and also the gauge field, which can be dualized to a scalar, at least on a generic point of the Coulomb branch where the gauge group is broken to an Abelian subgroup.

8.3 The Affine Grassmannian for $\mathrm{SL}(n, \mathbb{C})$

For simplicity of this presentation, we look in this Section at the affine Grassmannian [21–26] for the group $G = \mathrm{SL}(n, \mathbb{C})$, where the definitions simplify, referring the reader to [1, 21, 27–31] for the general case. Anticipating on what follows, G is connected to the quiver theories by the fact that its Lie algebra is $\mathfrak{g}$ defined below equation (8.1).

The Affine Grassmannian Gr_G for $G = \mathrm{SL}(n, \mathbb{C})$ can be seen as the quotient

$$\mathrm{Gr}_G = G((t))/G[[t]] \tag{8.3}$$

where $G[[t]]$ (respectively $G((t))$) is the ring (resp. the field) of invertible matrices with coefficients in formal series of the form $\sum_{k=0}^{\infty} a_k t^k$ for $a_k \in \mathbb{C}$ (resp. of the form $\sum_{k=-N}^{\infty} a_k t^k$ for $a_k \in \mathbb{C}$ and $N \in \mathbb{N}$). The quotient (8.3) is taken with respect to right multiplication. Elements in Gr_G can be seen as exactly the rank n lattices in $\mathbb{C}((t))^n$: a matrix in $G((t))$ encodes the lattice generated by $G[[t]]$-linear combinations of the column vectors, and the quotient removes the ambiguity in the choice of that basis. Multiplication on the left instead generates $G[[t]]$-orbits in Gr_G. In particular, for $\lambda = (\lambda_1, \ldots, \lambda_n) \in \mathbb{Z}^n$ such that $\sum \lambda_i = 0$ (interpreted as a coweight of G), the orbit of the matrix $\mathrm{Diag}(\lambda_1, \ldots, \lambda_n)$ is denoted $[\mathrm{Gr}_G]^{\lambda}$.

The spaces we are interested in are *transverse slices* to these orbits. We recall that given two coweights λ and μ, one says that $\lambda \leq \mu$ if $\mu - \lambda$ is a $\mathbb{N}$-linear combination of simple coroots of G. For two such coweights, with μ dominant, one can define the transverse slice $\overline{[\mathcal{W}_G]_{\lambda}^{\mu}}$. This is a symplectic singularity, with leaves given by the dominant coweights ν satisfying $\lambda \leq \nu \leq \mu$ and elementary transitions corresponding to pairs of coweights (ν, ν') such that $\nu' - \nu$ is a simple coroot. It turns out slices in the affine Grassmannian [32–34] correspond to moduli space of vacua of interacting theories, with or without a Lagrangian. More precisely, Coulomb branches can be seen as spaces of dressed monopole operators [13, 35–39] and this was identified with slices in the affine Grassmannian in [34, 40–42], building upon the works [43–48]. The correspondence can be made very explicit (see Sect. 3.1 in [1]).

In essence, the geometry of coweights and the partial order defined above are in one-to-one correspondence with the brane transitions and the quiver subtraction algorithm mentioned in the last section. This is illustrated in Fig. 8.2 on the case of $G = \mathrm{SL}(3, \mathbb{C})$. The geometry of the elementary transverse slices is given in term of the pair of adjacent coweights as shown in [32, 49, 50]. This is in perfect agreement with the string theory predictions. In particular, for type A the only elementary slices that occur are Klein/Du Val singularitues of type A along with closures of minimal orbits in $\mathfrak{sl}(m, \mathbb{C})$.

It should be noted that if we consider instead the affine Grassmannian for the centerless group $\mathrm{PSL}(n, \mathbb{C})$, it is made of n distinct connected components—in general, $\pi_0(\mathrm{Gr}_G) = \pi_1(G)$. The Coulomb branch of the quiver (8.1) is a transverse slice in

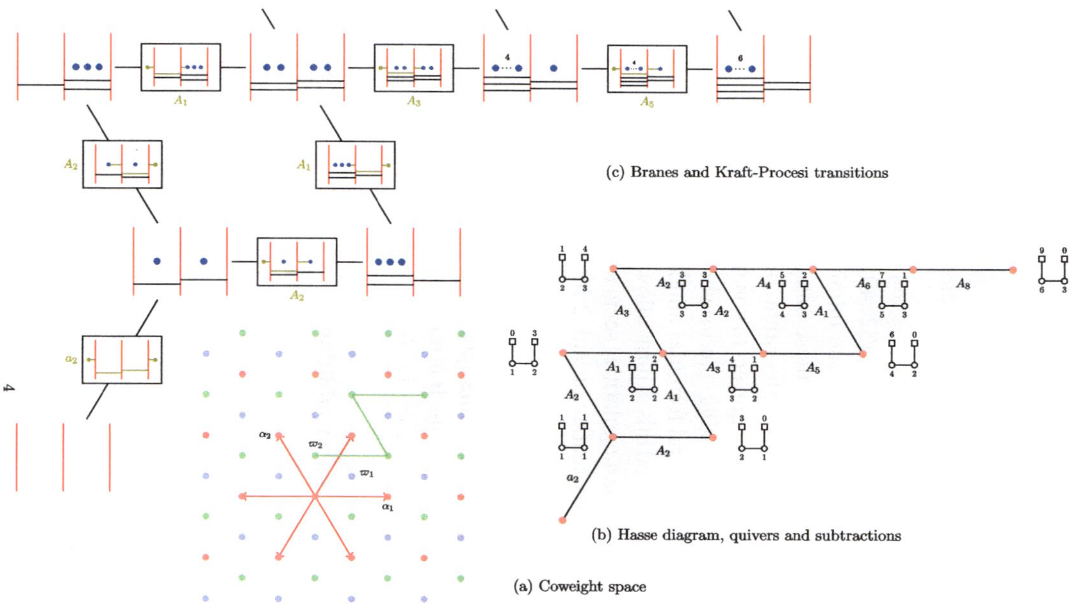

(c) Branes and Kraft-Procesi transitions

(b) Hasse diagram, quivers and subtractions

(a) Coweight space

Fig. 8.2 Three depictions of the coweight space for $G = \mathrm{SL}(3, \mathbb{C})$, taken from [1]. **a** The simple coroots are $\alpha_{1,2}$ and the coroot system is shown as red arrows. The red dots correspond to the coroot (=coweight) lattice for G. The red, blue and green dots correspond to coweights of $G' = \mathrm{SL}(3, \mathbb{C})/\mathbb{Z}_3$—and the affine Grassmannian for G' has three connected components, one for each color. The green line is the Hasse diagram (8.2). **b** The bottom of the Hasse diagram for G, whose dots are the dominant coweights of G, contained in the shaded Weyl chamber. The Coulomb branch of the quiver next to each red dot coincides with the transverse slice between that coweight and 0. The elementary transverse slices are labeled with the Kraft-Procesi notations ($A_n = \mathbb{C}^2/\mathbb{Z}_{n+1}$ and $a_n = \overline{\mathcal{O}}^{\min}_{\mathfrak{sl}(n+1,\mathbb{C})}$). **c** The type IIB string theory translation, with red lines denoting NS5 branes, black lines D3 branes, and blue dots D5 branes. Branes involved in a Kraft-Procesi transition are shown in brown

the component $1 \in \pi_1(\mathrm{PSL}(3, \mathbb{C}) = \mathbb{Z}_3)$, and the Hasse diagram (8.2) is found in the "green" component in Fig. 8.2a.

8.4 Applications in Physics and Open Questions

In this short report I have only discussed the affine Grassmannian for Lie groupe of type A, but the definitions extend straightforwardly to other types. Several insights can be obtained for a careful analysis of slices in these cases.

Quasi-minimal singularities In those cases the geometry of the Weyl chamber gives rise to other types of elementary transitions [32, 49, 50]. These have been dubbed *quasi-minimal singularities*, and come in one infinite family ac_n $(n \geq 2)$ and two outliers ag_2 and cg_2. The knowledge that these transitions exist is crucial to understand the details of the quiver subtraction algorithm. But the affine Grasmannian gives more: provided we enlarge the class of quivers to non simply laced graphs—for which a Lagrangian description is still missing, but to which one can still formally associate a Coulomb branch–, the general formulas alluded to at the end of the previous section provide explicit quivers, and this in turns allows to compute the corresponding Hilbert series, see Fig. 8.3. These singularities seem to appear generically in physical moduli spaces: for instance they feature at rank 2 in the Higgs branch of 4d $\mathcal{N} = 2$ theories [20, 51].

Brane movements on orientifolds It turns out these quivers can be associated to brane systems with orientifolds, and the quasi-minimal slices of Fig. 8.3 in turn correspond to allowed brane transitions. It is remarkable to see that the affine Grassmannian seems to be a universal geometric object which provides a precise analog of a class of brane systems.

Balanced Finite quivers and quiver addition The Coulomb branch of any quiver which has its gauge nodes in the shape of a Dynkin diagram (simply laced or not, but for a finite-dimensional Lie algebra $\mathfrak{g}$), with flavor nodes in such a way that every gauge node is balanced or overbalanced, is a slice $\overline{[\mathcal{W}_G]_\lambda^\mu}$ for certain coweights λ, μ of G, the centerless simple complex group with algebra $\mathfrak{g}$. This allows in turn to define, for that class of quivers, a quiver *addition* algorithm [1, 52], which can be seen as the first step in understanding how elementary symplectic singularities can be combined to form more complex ones.

More general quivers There are several natural extensions of these ideas. For quivers which are not overbalanced, the notion of *generalized affine Grassmannian slices* [34] can be put forward. For quivers beyond the finite Dynkin type—notably those encoding moduli spaces of instantons—one might need to examine slices in the *double affine Grassmannian* [43]. This promises a lot of exciting discoveries at the intersection between geometry and physics in the coming years.

Quasi-minimal singularity	Quiver	Global Symmetry	Plog of HWG
ac_n	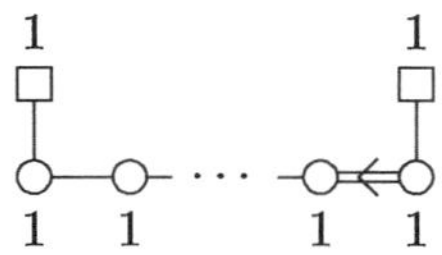	$\mathfrak{su}(n)_\mu \oplus \mathfrak{u}(1)_q$	$t^2 + \mu_1\mu_{n-1}t^2 + (q^{-1}\mu_1^2 + q\mu_{n-1}^2)t^3 - \mu_1^2\mu_{n-1}^2 t^6$
ag_2	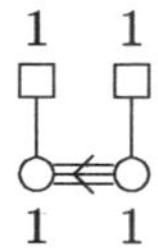	$\mathfrak{su}(2)_\mu \oplus \mathfrak{u}(1)_q$	$t^2 + \mu^2t^2 + (q + q^{-1})\mu^3t^4 - \mu^6t^8$
cg_2	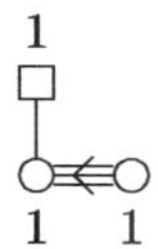	$\mathfrak{su}(2)_\mu \oplus \mathfrak{u}(1)_q$	$t^2 + \mu^2t^2 + (q + q^{-1})\mu^3t^3 - \mu^6t^6$

Fig. 8.3 Quivers for the quasi-minimal singularities. We also give the global symmetry of the Coulomb branch, and the plethystic logarithm of the Highest Weight Generating Function (HWG) [53]

Acknowledgements I would like to thank my long term collaborators Julius Grimminger, Amihay Hanany, Marcus Sperling and Zhenghao Zhong with whom this work was realized. I am also grateful to Daniel Juteau and Alex Weekes for many insights regarding the affine Grassmannian, and to the organizers of the Nankai meeting, in particular Yang-Hui He, for giving me the opportunity to join this exciting conference. I am supported by the ERC Consolidator Grant 772408-Stringlandscape, and by the LabEx ENS-ICFP: ANR-10-LABX-0010/ANR-10-IDEX-0001-02 PSL*.

References

1. A. Bourget, J. F. Grimminger, A. Hanany, M. Sperling and Z. Zhong, *Branes, Quivers, and the Affine Grassmannian*, 2102.06190.
2. N. Seiberg, *IR dynamics on branes and space-time geometry*, https://doi.org/10.1016/0370-2693(96)00819-2Phys. Lett. B 384 (1996) 81–85, [hep-th/9606017].
3. N. Seiberg and E. Witten, *Gauge dynamics and compactification to three-dimensions*, in *The mathematical beauty of physics: A memorial volume for Claude Itzykson. Proceedings, Conference, Saclay, France, June 5–7, 1996*, pp. 333–366, 1996. hep-th/9607163.
4. A. Bourget, S. Cabrera, J. F. Grimminger, A. Hanany, M. Sperling, A. Zajac et al., *The Higgs mechanism — Hasse diagrams for symplectic singularities*, https://doi.org/10.1007/JHEP01(2020)157 JHEP 01 (2020) 157, [1909.00667].
5. A. Bourget, S. Cabrera, J. F. Grimminger, A. Hanany and Z. Zhong, *Brane Webs and Magnetic Quivers for SQCD*, https://doi.org/10.1007/JHEP03(2020)176JHEP 03 (2020) 176, [1909.00667].
6. P. C. Argyres, A. Bourget and M. Martone, *On the moduli spaces of 4d $\mathcal{N} = 3$ SCFTs I: triple special Kähler structure*, 1912.04926.
7. J. F. Grimminger and A. Hanany, *Hasse diagrams for 3d $\mathcal{N} = 4$ quiver gauge theories — Inversion and the full moduli space*, https://doi.org/10.1007/JHEP09(2020)159JHEP 09 (2020) 159, [2004.01675].
8. M. van Beest, A. Bourget, J. Eckhard and S. Schafer-Nameki, *(Symplectic) Leaves and (5d Higgs) Branches in the Poly(go)nesian Tropical Rain Forest*, https://doi.org/10.1007/JHEP11(2020)124JHEP 11 (2020) 124, [2008.05577].
9. J. Eckhard, S. Schäfer-Nameki and Y.-N. Wang, *Trifectas for* T_N *in 5d*, https://doi.org/10.1007/JHEP07(2020)199JHEP 07 (2020) 199, [2004.15007].
10. P. C. Argyres and M. Martone, *Towards a classification of rank r $\mathcal{N} = 2$ SCFTs. Part II. Special Kahler stratification of the Coulomb branch*, https://doi.org/10.1007/JHEP12(2020)022JHEP 12 (2020) 022, [2007.00012].
11. A. Bourget, A. Hanany and D. Miketa, *Quiver origami: discrete gauging and folding*, https://doi.org/10.1007/JHEP01(2021)086JHEP 01 (2021) 086, [2005.05273].
12. G. Arias-Tamargo, A. Bourget and A. Pini, *Discrete gauging and Hasse diagrams*, https://doi.org/10.21468/SciPostPhys.11.2.026SciPost Phys. 11 (2021) 026, [2105.08755].
13. A. Hanany and E. Witten, *Type IIB superstrings, BPS monopoles, and three-dimensional gauge dynamics*, https://doi.org/10.1016/S0550-3213(97)00157-0,10.1016/S0550-3213(97)80030-2Nucl. Phys. B492 (1997) 152–190, [hep-th/9611230].
14. H. Kraft and C. Procesi, *Minimal singularities in GL_n*, *Inventiones mathematicae* 62 (1980) 503–515.
15. H. Kraft and C. Procesi, *On the geometry of conjugacy classes in classical groups*, https://doi.org/10.1007/BF02565876Commentarii Mathematici Helvetici 57 (Dec, 1982) 539–602.
16. S. Cabrera and A. Hanany, *Branes and the Kraft-Procesi Transition*, https://doi.org/10.1007/JHEP11(2016)175JHEP 11 (2016) 175, [1609.07798].
17. S. Cabrera and A. Hanany, *Branes and the Kraft-Procesi transition: classical case*, https://doi.org/10.1007/JHEP04(2018)127, [1711.02378].

18. A. Bourget, S. Giacomelli, J. F. Grimminger, A. Hanany, M. Sperling and Z. Zhong, *S-fold magnetic quivers*, https://doi.org/10.1007/JHEP02(2021)054JHEP 02 (2021) 054, [2010.05889].
19. A. Bourget, J. F. Grimminger, A. Hanany, R. Kalveks, M. Sperling and Z. Zhong, *Folding orthosymplectic quivers*, https://doi.org/10.1007/JHEP12(2021)070JHEP 12 (2021) 070, [2107.00754].
20. A. Bourget, J. F. Grimminger, M. Martone and G. Zafrir, *Magnetic quivers for rank 2 theories*, 2110.11365.
21. A. Beauville and Y. Laszlo, *Conformal blocks and generalized theta functions*, *Communications in mathematical physics* 164 (1994) 385–419.
22. S. Kumar, M. Narasimhan and A. Ramanathan, *Infinite grassmannians and moduli spaces*, *Math. Ann* 300 (1994) 41–75.
23. A. Beauville, Y. Laszio and C. Sorger, *The picard group of the moduli of-bundles on a curve*, *Compositio Mathematica* 112 (1998) 183–216.
24. G. Pappas and M. Rapoport, *Twisted loop groups and their affine flag varieties*, *Advances in Mathematics* 219 (2008) 118–198.
25. C. Sorger, *Lectures on moduli of principal g-bundles over algebraic curves*, *ICTP Lecture Notes* 3.
26. A. Beilinson and V. Drinfeld, "Quantization of hitchin's integrable system and hecke eigensheaves."
27. G. Segal, *Loop groups*, in *Arbeitstagung Bonn 1984*, pp. 155–168. Springer, 1985.
28. U. Görtz, *Affine springer fibers and affine deligne-lusztig varieties*, in *Affine flag manifolds and principal bundles*, pp. 1–50. Springer, 2010.
29. X. Zhu, *An introduction to affine Grassmannians and the geometric Satake equivalence*, arXiv e-prints (Mar., 2016) arXiv:1603.05593, [1603.05593].
30. P. Baumann and S. Riche, *Notes on the geometric satake equivalence*, in *Relative aspects in representation theory, Langlands functoriality and automorphic forms*, pp. 1–134. Springer, 2018.
31. T. Richarz, *Notes: Basics on affine grassmannians*, https://timo-richarz.com/wp-content/uploads/2020/02/BoAG_02.pdf, accessed November 2020.
32. A. Malkin, V. Ostrik and M. Vybornov, *The minimal degeneration singularities in the affine Grassmannians*, arXiv Mathematics e-prints (May, 2003) math/0305095 [math/0305095]
33. J. Kamnitzer, B. Webster, A. Weekes and O. Yacobi, *Yangians and quantizations of slices in the affine grassmannian*, *Algebra & Number Theory* 8 (2014) 857–893.
34. A. Braverman, M. Finkelberg and H. Nakajima, *Coulomb branches of 3d $\mathcal{N} = 4$quiver gauge theories and slices in the affine Grassmannian*, https://doi.org/10.4310/ATMP.2019.v23.n1.a3Adv. Theor. Math. Phys. 23 (2019) 75–166, [1604.03625].
35. G. Chalmers and A. Hanany, *Three-dimensional gauge theories and monopoles*, https://doi.org/10.1016/S0550-3213(97)00036-9Nucl. Phys. B 489 (1997) 223–244, [hep-th/9608105].
36. J. de Boer, K. Hori, H. Ooguri and Y. Oz, *Mirror symmetry in three-dimensional gauge theories, quivers and D-branes*, https://doi.org/10.1016/S0550-3213(97)00125-9Nucl. Phys. B 493 (1997) 101–147, [hep-th/9611063].
37. J. de Boer, K. Hori, H. Ooguri, Y. Oz and Z. Yin, *Mirror symmetry in three-dimensional theories, SL(2,Z) and D-brane moduli spaces*, https://doi.org/10.1016/S0550-3213(97)00115-6Nucl. Phys. B 493 (1997) 148–176, [hep-th/9612131].
38. S. A. Cherkis and A. Kapustin, *Singular monopoles and supersymmetric gauge theories in three-dimensions*, https://doi.org/10.1016/S0550-3213(98)00341-1Nucl. Phys. B 525 (1998) 215–234, [hep-th/9711145].
39. D. Tong, *Three-dimensional gauge theories and ADE monopoles*, https://doi.org/10.1016/S0370-2693(98)01583-4Phys. Lett. B 448 (1999) 33–36, [hep-th/9803148].
40. M. Bullimore, T. Dimofte and D. Gaiotto, *The Coulomb Branch of 3d $\mathcal{N} = 4$ Theories*, https://doi.org/10.1007/s00220-017-2903-0Commun. Math. Phys. 354 (2017) 671–751, [1503.04817].
41. M. Finkelberg, *Double affine grassmannians and coulomb branches of 3d n= 4 quiver gauge theories*, in *Proceedings of the International Congress of Mathematicians*, World Scientific, 2017. https://doi.org/10.1142/9789813272880_0097 DOI.

42. H. Nakajima and A. Weekes, *Coulomb branches of quiver gauge theories with symmetrizers*, 1907.06552.
43. A. Braverman, M. Finkelberg et al., *Pursuing the double affine grassmannian, i: Transversal slices via instantons on a_k-singularities*, https://doi.org/10.1215/00127094-2010-011Duke Mathematical Journal 152 (2010) 175–206.
44. S. Cremonesi, A. Hanany and A. Zaffaroni, *Monopole operators and Hilbert series of Coulomb branches of 3d $\mathcal{N} = 4$ gauge theories*, https://doi.org/10.1007/JHEP01(2014)005JHEP 01 (2014) 005, [1309.2657].
45. H. Nakajima, *Towards a mathematical definition of Coulomb branches of* 3*-dimensional* $\mathcal{N} =$ 4 *gauge theories, I*, https://doi.org/10.4310/ATMP.2016.v20.n3.a4Adv. Theor. Math. Phys. 20 (2016) 595–669, [1503.03676].
46. H. Nakajima, *Questions on provisional Coulomb branches of* 3*-dimensional* $\mathcal{N} = 4$ *gauge theories*, arXiv e-prints (Oct, 2015) arXiv:1510.03908, [1510.03908].
47. A. Braverman, M. Finkelberg and H. Nakajima, *Towards a mathematical definition of Coulomb branches of* 3*-dimensional* $\mathcal{N} = 4$ *gauge theories, II*, https://doi.org/10.4310/ATMP.2018.v22.n5.a1Adv. Theor. Math. Phys. 22 (2018) 1071–1147, [1601.03586].
48. H. Nakajima, *Introduction to a provisional mathematical definition of Coulomb branches of* 3*-dimensional* $\mathcal{N} = 4$ *gauge theories*, 1706.05154.
49. J. R. Stembridge, *The partial order of dominant weights*, *Advances in Mathematics* 136 (1998) 340–364.
50. D. Juteau, *Modular representations of reductive groups and geometry of affine grassmannians*, arXiv preprint arXiv:0804.2041 (2008) .
51. M. Martone, *Testing our understanding of SCFTs: a catalogue of rank-2 $\mathcal{N}$=2 theories in four dimensions*, 2102.02443.
52. K. Gledhill and A. Hanany, *Coulomb Branch Global Symmetry and Quiver Addition*, 2109.07237.
53. A. Hanany and R. Kalveks, *Highest Weight Generating Functions for Hilbert Series*, https://doi.org/10.1007/JHEP10(2014)152JHEP 10 (2014) 152, [1408.4690].

On Lie Algebras in Characteristic p

Lukas Brantner

Differential graded ('d.g.') Lie algebras over fields of characteristic zero have numerous applications, ranging from deformation theory in geometry to rational homotopy theory and the theory of configuration spaces in topology. In finite characteristic, however, they are neither well-behaved nor useful in applications. In this talk, we will survey recently discovered substitutes known as *partition Lie algebras*, following [1, 2].

But before delving into characteristic p, we will review the role of Lie algebras in deformation theory over the complex numbers and introduce some key definitions along the way.

A Reminder on Kodaira–Spencer Theory

Let Z be a smooth and proper complex variety. We fix a first order deformation $\widetilde{Z}$ of Z, i.e., a pullback square

$$\begin{array}{ccc} Z & \longrightarrow & \widetilde{Z} \\ \downarrow & \lrcorner & \downarrow \\ \mathrm{Spec}(\mathbb{C}) & \longrightarrow & \mathrm{Spec}(\mathbb{C}[\epsilon]/\epsilon^2) \end{array}$$

where the right vertical map is smooth and proper. The underlying space of $\widetilde{Z}$ is simply Z, nontrivial information lies in the deformation of the structure sheaf $\mathcal{O}_Z$.

To capture this geometric situation algebraically, we pick a cover Z by open affines $\mathrm{Spec}(B_i)$ with open affine intersections $\mathrm{Spec}(B_i) \cap \mathrm{Spec}(B_j) \cong \mathrm{Spec}(B_{ij})$.

L. Brantner (✉)
University of Oxford, Oxford, UK
e-mail: lukas.brantner@maths.ox.ac.uk

Y.-H. He et al. (eds.), *Nankai Symposium on Mathematical Dialogues*,
https://doi.org/10.1007/978-981-19-2328-9_9

On each of those affines, we can trivialise our first order deformation, and comparing trivialisations on overlaps gives rise to derivations $B_{ij} \to B_{ij}$, i.e. sections in $\Gamma(\mathrm{Spec}(B_{ij}), T_Z)$. These in turn glue to an element $x_{\widetilde{Z}} \in \mathrm{H}^1(Z, T_Z)$, which is in fact independent of the choices we made. Kodaira–Spencer theory [3] asserts that the assignment $\widetilde{Z} \mapsto x_{\widetilde{Z}}$ defines a bijection between first order deformations of Z and elements of $\mathrm{H}^1(Z, T_Z)$. We can now ask:

Question. Given a first order deformation $\widetilde{Z}$, does it extend to a higher-order deformation?

To formulate an algebraic answer, we use the commutator of vector fields to define a graded Lie algebra structure on $\mathrm{H}^*(Z, T_Z)$. Then $\widetilde{Z}$ extends to a second-order deformation precisely if $[x_{\widetilde{Z}}, x_{\widetilde{Z}}] = 0$.

In fact, $\mathrm{H}^*(Z, T_Z)$ is the homology of a *differential graded Lie algebra* $\mathfrak{g}_Z$, i.e., a chain complex with a bilinear bracket satisfying the Jacobi identity, antisymmetry, and the Leibniz rule, all in a graded sense. Concretely, $\mathfrak{g}_Z$ is given by the Dolbeault complex

$$\mathfrak{g}_Z = \left(\mathcal{A}^{0,0}(T_Z) \to \mathcal{A}^{0,1}(T_Z) \to \mathcal{A}^{0,2}(T_Z) \to \dots\right).$$

Miraculously, this d.g. Lie algebra in fact controls *all* infinitesimal deformations of Z. Indeed, if A is a local Artin $\mathbb{C}$-algebra with maximal ideal $\mathfrak{m}_A$, classical deformation theory constructs a bijection

$$\left\{ \begin{matrix} \text{Deformations } \widetilde{Z} \text{ of } Z \\ \text{over Spec}(A) \end{matrix} \right\} \Big/ {\scriptstyle \text{isomorphism} \atop \text{restricting to } \mathrm{id}_Z} \cong \left\{ \begin{matrix} \text{Maurer–Cartan elements} \\ x \in (\mathfrak{g}_Z)_{-1} \otimes \mathfrak{m}_A : dx + \frac{1}{2}[x,x] = 0 \end{matrix} \right\} \Big/ {\scriptstyle \text{gauge} \atop \text{equivalence}}.$$

Here $x, y \in (\mathfrak{g}_Z)_{-1} \otimes \mathfrak{m}_A$ are called *gauge equivalent* if there is some $a \in (\mathfrak{g}_Z)_0 \otimes \mathfrak{m}_A$ satisfying

$$y = x + \sum_{n=0}^{\infty} \frac{[a, -]^{\circ n}}{(n+1)!}([a, x] - da).$$

Towards Derived Algebraic Geometry

There are numerous other algebro-geometric objects Y over $\mathbb{C}$ whose infinitesimal deformations are governed by some d.g. Lie algebra $\mathfrak{g}_Y$, including subschemes, vector bundles, and representations. It is natural to ask:

Question. Given an algebro-geometric object Y over $\mathbb{C}$, how can we construct the d.g. Lie algebra $\mathfrak{g}_Y$ controlling its infinitesimal deformations?

Unfortunately, many inequivalent d.g. Lie algebras can control the same deformation functor, and it is not always possible to pick a preferred one—see [4] for an example. This is a real problem, as it means that obstruction classes do not vary functorially in the deformation functor.

In a visionary letter [5] from 1988, Drinfel'd proposed that one should consider *derived* infinitesimal deformations to resolve this issue, i.e., study deformations over simplicial local Artin $\mathbb{C}$-algebras

$$A = \left(\ldots \substack{\rightarrow\\\rightarrow\\\rightarrow\\\rightarrow} A_1 \substack{\rightarrow\\\rightarrow\\\rightarrow} A_0\right).$$

Formal Moduli Problems

Derived infinitesimal deformation functors are most naturally axiomatised in the language of ∞-categories, which are generalisations of categories where the collection of maps between any two objects forms a space rather than a set. We recall the following definition from [6]:

Definition (*Formal moduli problem*) A *formal moduli problem* over $\mathbb{C}$ is a functor

$$X : \mathrm{SCR}_{\mathbb{C}}^{\mathrm{art}} \to \mathcal{S}$$

from the ∞-category $\mathrm{SCR}_{\mathbb{C}}^{\mathrm{art}}$ of simplicial local Artin $\mathbb{C}$-algebras to the ∞-category $\mathcal{S}$ of spaces such that:

(1) *Normalisation.* $X(\mathbb{C}) \simeq *$ is a contractible space.
(2) *Gluing.* For any pullback square of simplicial local Artin $\mathbb{C}$-algebras

$$\begin{array}{ccc} \widetilde{A} & \longrightarrow & A' \\ \downarrow & & \downarrow \\ A & \longrightarrow & A'' \end{array}$$

with $\pi_0(A') \twoheadrightarrow \pi_0(A'')$ and $\pi_0(A) \twoheadrightarrow \pi_0(A'')$ surjective, applying X gives a pullback square in spaces.

We denote the ∞-category of formal moduli problems by $\mathrm{Moduli}_{\mathbb{C}}$.

Example. Given a smooth and proper complex variety Z as above, we can enhance its classical deformation functor to a formal moduli problem $\mathrm{Def}_Z : \mathrm{SCR}_{\mathbb{C}}^{\mathrm{art}} \to \mathcal{S}$ by sending a simplicial local Artin $\mathbb{C}$-algebra A to the space of pullback diagrams

$$\begin{array}{ccc} Z & \longrightarrow & \widetilde{Z} \\ \downarrow & \lrcorner & \downarrow \\ \mathrm{Spec}(\mathbb{C}) & \longrightarrow & \mathrm{Spec}(A) \end{array},$$

where $\tilde{Z}$ is a smooth and proper *derived* scheme over $\mathrm{Spec}(A)$, cf. [7, Sect. 19.4].

Deformation Theory in Characteristic 0

We will now outline the relation between derived infinitesimal deformation functors, i.e. formal moduli problems, and d.g. Lie algebras over $\mathbb{C}$.

Given a d.g. Lie algebra $\mathfrak{g}$ and a simplicial local Artin $\mathbb{C}$-algebra A with maximal ideal $\mathfrak{m}_A$, Hinich [8] (generalising work of Deligne and Goldman–Millson [9]) constructed a *space* of Maurer–Cartan elements

$$\mathrm{MC}(\mathfrak{m}_A \otimes \mathfrak{g}).$$

Varying A, we obtain a functor $\Psi : \mathrm{dgla}_{\mathbb{C}} \to \mathrm{Moduli}_{\mathbb{C}}$ from d.g. Lie algebras to formal moduli problems. The following result by Lurie [10] and Pridham [11], which generalises earlier work of many others (including [8, 12, 13]) asserts that d.g. Lie algebras control derived infinitesimal deformations:

Theorem (Lurie, Pridham) *The functor* $\Psi : \mathrm{dgla}_{\mathbb{C}} \to \mathrm{Moduli}_{\mathbb{C}}$ *is an equivalence.*

Given a formal moduli problem $X \in \mathrm{Moduli}_{\mathbb{C}}$, it is easy to describe the underlying spectrum of the associated d.g. Lie algebra. Up to a shift, it is given by the *tangent fibre* of X, which is obtained by assembling the sequence of spaces $X(\mathbb{C} \oplus \mathbb{C}[n])$ and the equivalences $\Omega X(\mathbb{C} \oplus \mathbb{C}[n+1]) \simeq X(\mathbb{C} \oplus \mathbb{C}[n])$ (coming from the gluing axiom) into a spectrum. Here $\mathbb{C} \oplus \mathbb{C}[n]$ denotes the trivial square-zero extension of $\mathbb{C}$ by a class in homological degree n.

Remark. The Lurie–Pridham theorem in fact holds over any field of characteristic zero.

Deformation Theory in Characteristic P

Many key players in number theory, arithmetic geometry, and representation theory are not defined over $\mathbb{C}$, but over some field k of characteristic p, e.g. $\mathbb{F}_p$ or $\overline{\mathbb{F}}_p$. It is vital to understand how these objects deform.

Unfortunately, d.g. Lie algebras in characteristic p are ill-behaved, and no longer classify formal moduli problems. In fact, they do not even admit a model structure satisfying the most basic desiderata.

To resolve this problem, we replace d.g. Lie algebras by *partition Lie algebras*; these new algebraic-homotopical structures can be defined in two equivalent ways:

(1) Abstractly via ∞-categories, cf. [1, Construction 1.10].
The (contravariant) tangent fibre functor $A \mapsto \mathrm{cot}(A)^\vee = (k \otimes_A L_{A/k})^\vee$ from augmented simplicial commutative k-algebras to chain complexes over k is part of an adjunction. The associated monad T on Mod_k behaves badly, but we can use Goodwillie's functor calculus to approximate T by a monad

$$\mathrm{Lie}_k^\pi$$

which preserves both filtered colimits and geometric realisations.
If $V^\bullet$ is a cosimplicial k-module with associated chain complex $\mathrm{Tot}(V^\bullet)$, then the monad Lie_k^π satisfies the formula

$$\mathrm{Lie}_k^{\pi}(\mathrm{Tot}(V^{\bullet})) \simeq \bigoplus_n \mathrm{Tot}\left(\widetilde{C}^{\bullet}(\Sigma|\Pi_n|^{\diamond}, k) \otimes (V^{\bullet})^{\otimes n}\right)^{\Sigma_n}.$$

Here $\Sigma|\Pi_n|^{\diamond}$ is a simplicial Σ_n-complex known as the nth (doubly suspended) *partition complex*. For $d > 0$, the non-degenerate d-simplices of $\Sigma|\Pi_n|^{\diamond}$ are given by chains of increasingly coarse partitions $[\hat{0} = x_0 < x_1 < \ldots < x_t = \hat{1}]$ of the set $\{1, \ldots n\}$.

Algebras over the monad Lie_k^{π} are called *partition Lie algebras*, and we write $\mathrm{Alg}_{\mathrm{Lie}_k^{\pi}}$ for the ∞-category of partition Lie algebras.

(2) Concretely via model categories, cf. [2, Construction 5.43].

To model partition Lie algebras more concretely, we construct an ordinary category whose objects are cosimplicial-simplicial k-modules equipped with n-ary operations

$$\{a_1, \ldots, a_r\}_{(\sigma,S)} \qquad \gamma_{(\sigma,S)}(a_1, \ldots, a_r)$$

for any tuple $(a_1, \ldots, a_r) \in \mathfrak{g}^d$ and any pair

$$(\sigma, S) = \left([\hat{0} = x_0 < \cdots < x_t = \hat{1}],\ S_0 \subseteq \cdots \subseteq S_d\right)$$

of chain $\sigma = [\hat{0} = x_0 < x_1 < \ldots < x_t = \hat{1}]$ of increasingly coarse partitions of $\{1, \ldots, n\}$ and chain $S = (S_0 \subseteq \ldots \subseteq S_d)$ of increasingly large subsets of $S_d = \{0, \ldots, t\}$. These operations are then required to satisfy various relations.

The resulting category admits a natural model structure, whose weak equivalences are those maps which become quasi-isomorphisms of chain complexes after applying the Dold–Kan functor. Inverting weak equivalences gives the ∞-category of partition Lie algebras by [2, Theorem 5.42].

The notion of formal moduli problem makes sense over arbitrary fields k, as does the tangent fibre construction outlined above. We write Moduli_k for the resulting ∞-category of formal moduli problems. In [1, Theorem 1.11], we prove that partition Lie algebras control infinitesimal deformations of algebro-geometric objects over general fields:

Theorem (B. Mathew) *Given a field k, the tangent fibre functor refines to an equivalence*

$$\mathrm{Moduli}_k \simeq \mathrm{Alg}_{\mathrm{Lie}_k^{\pi}}.$$

This equivalence singles out partition Lie algebras as the appropriate analogues of d.g. Lie algebras in characteristic p, and an important tool in deformation theory. But partition Lie algebras also have various applications beyond deformation theory—we give two examples:

In algebra, they have been used to construct a Galois correspondence for finite purely inseparable field extensions [14]—these are extensions L/K such that for all $x \in L$, some power x^{p^n} belongs to K, where p is the characteristic prime. For extensions of height one, this correspondence was proven earlier by Jacobson [15].

In geometry, partition Lie algebras facilitate an infinitesimal version of Frobenius' theorem for derived foliations over coherent ground rings [16, 17].

References

1. Lukas Brantner and Akhil Mathew, *Deformation theory and partition lie algebras*, Acta Math. **235**, no. 1 (2025), 1–148
2. Lukas Brantner, Ricardo Campos, and Joost Nuiten, *PD operads and explicit partition lie algebras*, **315**, no. 1597 Mem. Amer. Math. Soc. (2025)
3. K. Kodaira and D. C. Spencer, *On deformations of complex analytic structures. I, II*, Ann. of Math. (2) **67** (1958), 328–466. 0112154
4. Bertrand Toën, *Derived algebraic geometry*, EMS Surv. Math. Sci. **1** (2014), no. 2, 153–240. 3285853
5. V. Drinfeld, *A letter from Kharkov to Moscow*, EMS Surv. Math. Sci. **1**, no. 2, 241–248, Translated from Russian by Keith Conrad. 3285856
6. Jacob Lurie, *Derived algebraic geometry X: Formal moduli problems*, Preprint available from the author's web page (2011).
7. Jacob Lurie, *Spectral algebraic geometry*, Preprint available from the author's web page (2016).
8. Vladimir Hinich, *DG coalgebras as formal stacks*, J. Pure Appl. Algebra **162** (2001), no. 2-3, 209–250. 1843805
9. William M. Goldman and John J. Millson, *The deformation theory of representations of fundamental groups of compact Kähler manifolds*, Publ. Math. Inst. Hautes Études Sci. (1988), no. 67, 43–96. 972343
10. Jacob Lurie, *Moduli problems for ring spectra*, Proceedings of the International Congress of Mathematicians. Volume II, Hindustan Book Agency, New Delhi, 2010, pp. 1099–1125. 2827833
11. Jon P Pridham, *Unifying derived deformation theories*, Adv. Math. **224** (2010), no. 3, 772–826.
12. Maxim Kontsevich and Yan Soibelman, *Deformation theory*, Livre en préparation (2002).
13. Marco Manetti, *Differential graded Lie algebras and formal deformation theory*, Algebraic geometry—Seattle 2005. Part 2, Proc. Sympos. Pure Math., vol. 80, Amer. Math. Soc., Providence, RI, 2009, pp. 785–810. 2483955
14. Lukas Brantner and Joe Waldron, *Purely inseparable Galois theory I: the fundamental theorem*, arXiv preprint arXiv:2010.15707 (2020).
15. Nathan Jacobson, *Galois theory of purely inseparable fields of exponent one*, Amer. J. Math. **66** (1944), no. 4, 645–648.
16. Lukas Brantner, Kirill Magidson, and Joost Nuiten, *Formal integration of derived foliations*, arXiv preprint arXiv:2502.05257 (2025).
17. Jiaqi Fu, *A duality between Lie algebroids and infinitesimal foliations*, arXiv preprint arXiv:2410.04950 (2024).

Recent Developments in Line Bundle Cohomology and Applications to String Phenomenology

Callum Brodie, Andrei Constantin, James Gray, Andre Lukas, and Fabian Ruehle

Abstract Vector bundle cohomology represents a key ingredient for string phenomenology, being associated with the massless spectrum arising in string compactifications on smooth compact manifolds. Although standard algorithmic techniques exist for performing cohomology calculations, they are laborious and ill-suited for scanning over large sets of string compactifications to find those most relevant to particle physics. In this article (based on the second author's lecture at the Nankai Symposium, August 2021) we review some recent progress in deriving closed-form expressions for line bundle cohomology and discuss some applications to string phenomenology.

1 Introduction

Computing cohomology is a crucial and time-consuming step in the derivation of the spectrum of low-energy particles resulting from string compactifications. The state-of-the-art consists of computer implementations of algorithmic methods, based on Čech cohomology and spectral sequences [1–3]. However, these methods are

C. Brodie (✉) · J. Gray
Department of Physics, Robeson Hall, Virginia Tech, Blacksburg, VA, USA
e-mail: callumb@vt.edu

J. Gray
e-mail: jamesgray@vt.edu

A. Constantin · A. Lukas
Rudolf Peierls Centre for Theoretical Physics, University of Oxford, Oxford, UK
e-mail: andrei.constantin@physics.ox.ac.uk

A. Lukas
e-mail: andre.lukas@physics.ox.ac.uk

F. Ruehle
Department of Physics & Department of Mathematics, Northeastern University, Boston, MA, USA
e-mail: f.ruehle@northeastern.edu

The NSF AI Institute for Artificial Intelligence and Fundamental Interactions, Boston, USA

Y.-H. He et al. (eds.), *Nankai Symposium on Mathematical Dialogues*,
https://doi.org/10.1007/978-981-19-2328-9_10

computationally intensive and provide little insight into the origin of the results. Recently, it has been shown that in many cases of interest in string theory topological formulae for line bundle cohomology exist, providing mathematical shortcuts that can reduce the time needed for deciding the physical viability of a string compactification from several months of computer algebra to the split of a second.

Line bundles feature in many string theory contexts and have proven to be a fruitful setting for realistic phenomenology [4–11]. Moreover, they represent building blocks for higher-rank vector bundles, such as monad or extension bundles [12–17]. Topological formulae for line bundle cohomology were initially discovered empirically, through a combination of direct observation [18–22] and machine learning techniques [23–26] and covered examples from several classes of complex manifolds of dimensions two and three. The main observation was that the Picard group decomposes into disjoint regions, called *cohomology chambers*, in each of which the cohomology function is polynomial or very close to polynomial, a pattern observed both for the zeroth and the higher cohomologies, with a different chamber structure emerging for each type of cohomology. The number of regions often increases dramatically with the Picard number.

It is well-known that given a complex manifold X and a holomorphic bundle V, the Euler characteristic $\chi(X, V)$, that is the alternating sum of cohomologies $h^i(X, V)$ can be equated to a purely topological index:

$$\chi(X, V) = \sum_{i=0}^{\dim(X)} (-1)^i h^i(X, V) = \int_X \mathrm{ch}(V) \cdot \mathrm{td}(X),$$

which is the statement of the Hirzebruch-Riemann-Roch theorem. Similar formulae have now been found to hold for each individual cohomology dimension $h^i(X, V)$, for large classes of complex manifolds and abelian vector bundles.

2 Dimension Two

The first advancement in understanding the mathematical origin of these formulae came through the study of line bundles on several classes of complex surfaces widely used in string theory, including compact toric surfaces, weak Fano surfaces (generalised del Pezzo surfaces), and K3 surfaces [27, 28]. Note that for surfaces it suffices to study the zeroth cohomology function $h^0(X, V)$, from which the formulae for the first and second cohomologies then follow by Serre duality and the Riemann-Roch theorem.

Line bundle cohomology formulae on these families of surfaces arise through a combination of Zariski decomposition and vanishing theorems. Zariski decomposition is the statement that on any complex projective surface X, any effective divisor D can be uniquely decomposed as $D = P + N$, where P is nef, N is effective, and P intersects no components in the curve decomposition of N. In general, the divisors

P and N are rational. When D is integral, its class defines an effective line bundle and the following equality holds

$$h^0\big(X, \mathcal{O}_X(D)\big) = h^0\big(X, \mathcal{O}_X(\lfloor P \rfloor)\big) = h^0\big(X, \mathcal{O}_X(\lceil P \rceil)\big), \tag{1}$$

where $\lfloor P \rfloor$ and $\lceil P \rceil$ are the maximal integral subdivisor of P and, respectively, the minimal integral divisor that has P as a subdivisor. The classes $[P]$, $[\lfloor P \rfloor]$ and $[\lceil P \rceil]$ depend only on the class $[D]$, and, crucially, can be computed purely from the intersection properties of D. In particular this computation requires knowledge of the Mori cone and the intersection form on X. If the cohomology on the right-hand side of Eq. (1) can be computed more easily than on the left, the relation becomes practically useful.

2.1 Zariski Chambers

The map $[D] \to [P]$ gives rise to a *Zariski chamber decomposition* of the effective cone [29]. When D is nef it follows that N is trivial and $D = P$, hence the nef cone is a chamber in itself. Furthermore, every face F of the nef cone not contained in the boundary of the Mori cone gives rise to a Zariski chamber Σ_F, by translating the face F along the dual Mori cone generators. Thus, if D is an effective divisor within the chamber $\Sigma_{i_1 \ldots i_n}$ obtained by translating the codimension n face F of the nef cone along the set of Mori cone generators $R = \{\mathcal{M}_{i_1}, \mathcal{M}_{i_2}, \ldots, \mathcal{M}_{i_n}\}$ orthogonal to F with respect to the intersection form, then P lies on F and

$$P = D - \sum_{k=1}^{n} (-D \cdot \mathcal{M}_{i_k,R}^{\vee})\, \mathcal{M}_{i_k}, \tag{2}$$

where the dual $\mathcal{M}_{i_k,R}^{\vee}$ is an effective divisor with support R defined by $\mathcal{M}_{i_k,R}^{\vee} \cdot \mathcal{M}_{i_m} = -\delta_{km}, \forall \mathcal{M}_{i_m} \in R$.

2.2 Cohomology Chambers

While Zariski chambers exist for every smooth complex projective surface, whether these become cohomology chambers or not depends on the presence of an appropriate vanishing theorem that interacts well with the floor/ceiling operations in Eq. (1). If the image of a Zariski chamber under the map $[D] \to [\lfloor P \rfloor]$ or under the map $[D] \to [\lceil P \rceil]$ is covered by a vanishing theorem, the index function can be 'pulled back' to give a single function for zeroth cohomology throughout the Zariski chamber, which is now also a *cohomology chamber*. This is indeed the case for the following classes of surfaces:

a. Generalised del Pezzo surfaces (including del Pezzo surfaces). The Kawamata-Viehweg vanishing theorem is relevant in this case and, in the above notation

$$h^0(X, \mathcal{O}_X(D)) = \chi\left(X, \mathcal{O}_X\left(D - \sum_{k=1}^{n} \lceil -D \cdot \mathcal{M}^\vee_{i_k,R} \rceil \, \mathcal{M}_{i_k}\right)\right). \tag{3}$$

b. Projective toric surfaces. Demazure's vanishing theorem is relevant in this case and

$$h^0(X, \mathcal{O}_X(D)) = \chi\left(X, \mathcal{O}_X\left(D - \sum_{k=1}^{n} \lfloor -D \cdot \mathcal{M}^\vee_{i_k,R} \rfloor \, \mathcal{M}_{i_k}\right)\right). \tag{4}$$

c. K3 surfaces. The Kawamata-Viehweg vanishing theorem can be used in this case too, however, this gives cohomology formulae only in the interior of the effective cone, leading to the same equality as in Eq. (3). On the boundary of the effective cone, for the integral divisors D whose support has negative definite intersection matrix it follows that the positive part is trivial $P = 0$ so that $h^0(X, \mathcal{O}_X(D)) = h^0(X, \mathcal{O}_X) = 1$. In general this determines the cohomology on some but not all the faces of the Mori cone. The cohomologies of the remaining line bundles on the boundary require a separate treatment.

2.3 Example

We illustrate the above discussion with a simple example of a del Pezzo surface of degree 7, obtained by blowing up $\mathbb{P}^2$ at two generic points and denoted dP_2 in the Physics literature. The Picard lattice of dP_2 is spanned by the hyperplane class H of $\mathbb{P}^2$ and the two exceptional divisor classes E_1 and E_2 resulting from the two blow-ups. The effective cone (Mori cone) is generated by $\mathcal{M}_1 = E_1$, $\mathcal{M}_2 = E_2$, and $\mathcal{M}_3 = H - E_1 - E_2$. All three generators are rigid, satisfying $\mathcal{M}_i^2 = -1$.

The Zariski/cohomology chamber structure of the effective cone is shown in Fig. 1. The nef cone has three co-dimension 1 faces and one co-dimension 2 face not shared with the Mori cone. Each of these gives rise to a cohomology chamber, four in total. The zeroth cohomology is given by the locally polynomial function

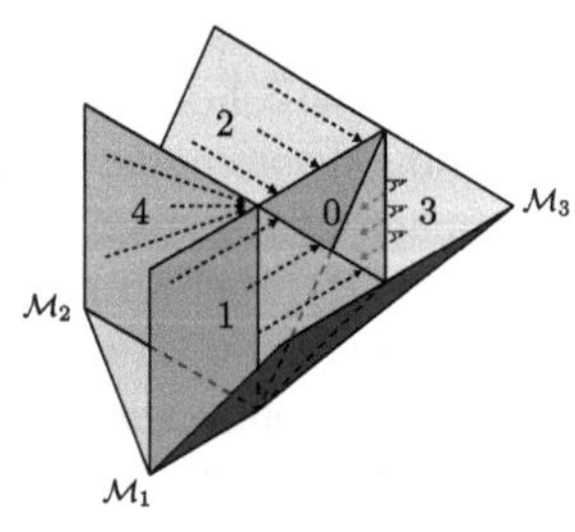

Fig. 1 Zeroth cohomology chamber structure of the effective cone of dP_2. The projection arrows correspond to the map $[D] \to [\lceil P \rceil]$. The nef cone is the chamber labelled as 0

$$h^0\left(\mathrm{dP}_2, \mathcal{O}_{\mathrm{dP}_2}(D)\right) = \chi\left(\mathrm{dP}_2, \mathcal{O}_{\mathrm{dP}_2}\left(D - \sum_{i=1}^{3} \theta(-D \cdot \mathcal{M}_i)\,(-D \cdot \mathcal{M}_i)\mathcal{M}_i\right)\right),$$

where $\theta(x)$ equals one for $x \geq 0$ and zero otherwise.

3 Dimension Three

Calabi-Yau three-folds provide qualitatively new phenomena for line bundle cohomology. Several examples of low Picard number Calabi-Yau threefolds realised as complete intersections in products of projective spaces (CICY threefolds) or hypersurfaces in toric varieties were studied in Refs. [30–32]. As in the case of complex surfaces, in these examples the zeroth cohomology determines a decomposition of the effective cone into polyhedral chambers, in each of which the zeroth cohomology can be expressed as a topological index. The chambers were understood to be either:

(i) Images of Kähler cones of birational models of X under the natural identifications of the Picard groups. Inside such chambers and along their common walls the zeroth cohomology can be equated to the Euler characteristic on the corresponding flopped manifold, due to Kodaira's vanishing theorem, the Kawamata-Viehweg vanishing theorem and the invariance of the number of global sections of a line bundle under a flop. The union of these (images of) Kähler cones is known in the Physics literature as the *extended Kähler cone* and in the Mathematics literature as the *movable cone*.
(ii) Zariski chambers where a cohomology-preserving projection operates in a similar way as in the two-dimensional case. The existence of a Zariski chamber neighbouring a boundary of the (extended) Kähler cone is signalled by the shrinking of an effective divisor to zero volume as computed with respect to the Kähler form on the corresponding flopped manifold.

The appearance of the first type of chambers is novel and is related to the fact that in dimension 3 minimal models are not unique. Calabi-Yau threefolds have trivial and hence nef canonical bundle, therefore they are minimal models. It is well-known that in dimension 3 minimal models are not unique in general, but any two distinct minimal models are birationally related by a sequence of flops, being isomorphic outside subsets of codimension 2.

The zeroth line bundle cohomology encodes the information about the flops connecting the birational models of the manifold, as well as about the Gromov-Witten invariants counting the number of flopping curves. We discovered that the vast majority of known Calabi-Yau threefolds admit flops, while many of these flop to manifolds isomorphic to themselves [33]. In particular, many threefolds admit infinite sequences of flops and have an infinite number of contractible rational curves.

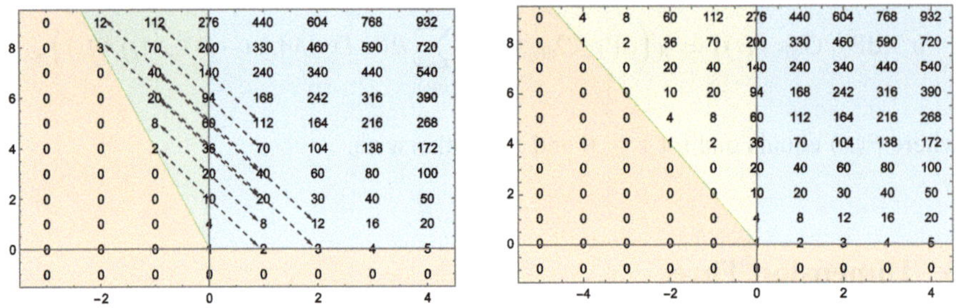

Fig. 2 Zeroth line bundle cohomology for generic (left) and tuned (right) Calabi-Yau hypersurface of bi-degree (2, 4) in $\mathbb{P}^1 \times \mathbb{P}^3$

3.1 Example

For illustration, consider a Calabi-Yau threefold X with two Kähler parameters, defined as the zero locus of a generic homogeneous polynomial of bi-degree $(2, 4)$ in $\mathbb{P}^1 \times \mathbb{P}^3$. This is the manifold with identifier 7887 in the CICY list [34, 35] with configuration matrix

$$\begin{matrix}\mathbb{P}^1\\ \mathbb{P}^3\end{matrix}\left[\begin{matrix}2\\ 4\end{matrix}\right]^{2,86} \tag{5}$$

where the superscripts indicate the Hodge numbers $(h^{1,1}(X), h^{2,1}(X))$. The manifold has been recently studied in Ref. [30] and earlier on in Ref. [36]. The structure of the zeroth line bundle cohomology is shown in the left plot of Fig. 2. The blue cone corresponds to the Kähler cone of X and here the zeroth cohomology is given by $h^0(X, L) = \chi(X, L)$ due to Kodaira's vanishing theorem. The green cone corresponds to the Kähler cone of the flopped manifold X' and here $h^0(X, L) = h^0(X', L') = \chi(X', L')$ where L' is the line bundle on X' corresponding to L on X. The wall separating the two Kähler cones, excluding the origin, is covered by the Kawamata-Viehweg vanishing theorem and here $h^0(X, L) = \chi(X, L) = \chi(X', L')$. The line bundles lying along the two boundaries of the effective cone require a separate discussion. Thus, if $L = \mathcal{O}_X(k_1[D_1] + k_2[D_2])$ is a line bundle with first Chern class written in the basis $\{[D_1], [D_2]\}$ consisting of the restrictions to X of the hyperplane classes on $\mathbb{P}^1$ and $\mathbb{P}^3$ (which are simultaneously generators of the Picard group and of the Kähler cone of X), then $h^0(X, L) = \chi(\mathbb{P}^1, \mathcal{O}_{\mathbb{P}^1}(k_1))$ if $k_1 > 0,\ k_2 = 0$ and $h^0(X, L) = \chi(\mathbb{P}^1, \mathcal{O}_{\mathbb{P}^1}(-k_1))$, if $k_1 < 0$ and $k_2 = -4k_1$. Outside of these regions the zeroth cohomology is trivial. The cohomology formulae are summarised below:

region in eff. cone	$h^0(X, L = \mathcal{O}_X(k_1[D_1] + k_2[D_2]))$
$\mathcal{K}(X)$	$\chi(X, L)$
$\mathcal{K}(X')$	$\chi(X', L')$
$k_1 = 0,\ k_2 > 0$	$\chi(X, L) = \chi(X', L')$
$k_1 < 0,\ k_2 = -4k_1$	$\chi(\mathbb{P}^1, \mathcal{O}_{\mathbb{P}^1}(-k_1))$
$k_1 > 0,\ k_2 = 0$	$\chi(\mathbb{P}^1, \mathcal{O}_{\mathbb{P}^1}(k_1))$
$k_1 = k_2 = 0$	1

The flop $X \to X'$ can be constructed explicitly as follows. Let $[x_0 : x_1]$ denote the homogeneous coordinates on the $\mathbb{P}^1$ and $[y_0 : y_1 : y_2 : y_3]$ the coordinates on the $\mathbb{P}^3$. The hypersurface equation reads

$$x_0^2 P(y) + x_0 x_1 Q(y) + x_1^2 R(y) = 0, \tag{6}$$

where P, Q and R are generic quartic polynomials in the y-coordinates. For a generic $y \in \mathbb{P}^3$ the equation gives two solutions for x. On the other hand, when $P(y) = Q(y) = R(y) = 0$, which happens at 64 points in $\mathbb{P}^3$, the equation admits an entire $\mathbb{P}^1$ as a solution. By contracting these curves one obtains a singular threefold. The flopped threefold X' is obtained by resolving these singularities into new rational curves. Explicitly, X' is given by the hypersurface in $\mathbb{P}^1 \times \mathbb{P}^3$

$$x_0^2 R(y) + x_0 x_1 Q(y) + x_1^2 P(y) = 0, \tag{7}$$

which is just Eq. (6) with the roles of P and R swapped. The manifold X' is isomorphic to the original one, as reflected by the $\mathbb{Z}_2$-symmetry of the cohomology data in the left plot of Fig. 2. The threefolds X and X' share not only the same Hodge numbers, but also the same triple intersection numbers and c_2-form when written in the basis of generators of their respective Kähler cones.

3.2 Complex Structure Dependence

An important aspect of the cohomology formulae is their complex structure dependence. In the previous example, when the quartic polynomials P, Q and R in Eq. (6) are generic, the cohomology data corresponds to the situation described above. On the other hand, when all the coefficients in Q vanish there is a dramatic change in the structure of the zeroth line bundle cohomology, as shown in the plot on the right in Fig. 2. The defining equation is now

$$x_0^2 P(y) + x_1^2 R(y) = 0, \tag{8}$$

where P and R are generic. At a generic point $y \in \mathbb{P}^3$, the equation gives two solutions for x. However, when $P(y) = R(y) = 0$, which corresponds to a curve in $\mathbb{P}^3$, the solution is an entire $\mathbb{P}^1$. The $\mathbb{P}^1$-fibration over this curve gives rise to a rigid divisor

Γ in the class $-2[D_1] + 4[D_2]$. Note that $h^0(X, \mathcal{O}_X(\Gamma)) = 1$ and the volume of Γ shrinks to zero when the Kähler form approaches the relevant boundary of $\mathcal{K}(X)$. Thus at this locus in complex structure moduli space there are no isolated rational curves that can be flopped.

Zeroth cohomology formulae can be obtained as follows. As before, inside $\mathcal{K}(X)$ and along the boundary separating $\mathcal{K}(X)$ from Σ, excluding the origin, we have $h^0(X, L) = \chi(X, L)$, which follows from Kodaira vanishing inside $\mathcal{K}(X)$, and Kawamata-Viehweg vanishing for the big line bundles on the boundary.

The presence of the rigid divisor Γ gives rise to a Zariski chamber. Γ is part of the fixed locus of every linear system in the cone Σ, as indicated by the fact that the cohomology data inside Σ is invariant under Γ-shifts. The presence of Γ and the amount by which it is present can be detected by intersection properties, as follows. Let $[C_1], [C_2]$ denote the curve classes dual to $[D_1], [D_2]$, such that $[C_i] \cdot [D_j] = \delta_{ij}$. Let D be an effective divisor on X and assume it has a Zariski decomposition $D = P + N$, where P and N are $\mathbb{Q}$-divisors, N is effective and P is nef, that is $[P] \in \overline{\mathcal{K}}(X)$. If D itself is nef, the Zariski decomposition is trivial. Thus we assume $[D] \in \Sigma$. In this case, $[P]$ lies on the boundary shared by Σ and $\overline{\mathcal{K}}(X)$. From general properties of Zariski decomposition we know that

$$h^0(X, \mathcal{O}_X([D])) = h^0(X, \mathcal{O}_X([\lfloor P \rfloor])), \tag{9}$$

where $\lfloor P \rfloor$ is the integral divisor obtained by rounding down all the coefficients in the expansion of P. In terms of divisor classes, the round down class $[\lfloor P \rfloor]$ is well-defined when $[D]$ is an integral class. The divisor N is a rational multiple $N = \gamma\Gamma$ of the rigid divisor Γ and $[P]$ is a rational multiple of $[D_2]$. Since $D = P + N$ and $D_2 \cdot C_1 = 0$, it follows that $D \cdot C_1 = N \cdot C_1 = \gamma\,\Gamma \cdot C_1$, from which $\gamma = D \cdot C_1 / \Gamma \cdot C_1$. Consequently,

$$h^0(X,\ \mathcal{O}_X([D])) = h^0(X, \mathcal{O}_X([D] - \lceil\gamma\rceil\,[\Gamma])) = \chi\left(X, \mathcal{O}_X\left([D] - \left\lceil \frac{D \cdot C_1}{\Gamma \cdot C_1} \right\rceil [\Gamma]\right)\right). \tag{10}$$

With $[D] = k_1[D_1] + k_2[D_2]$ it is straightforward to compute $\gamma = -k_1/2$. The zeroth line bundle cohomology formulae are summarised below:

region in eff. cone	$h^0(X, L = \mathcal{O}_X([D] = k_1[D_1] + k_2[D_2]))$
$\mathcal{K}(X) \cup (k_1 = 0, k_2 > 0)$	$\chi(X, L)$
$\overline{\Sigma}$	$\chi\left(X, \mathcal{O}_X\left([D] - \left\lceil \frac{D \cdot C_1}{\Gamma \cdot C_1} \right\rceil [\Gamma]\right)\right)$
$k_1 > 0,\ k_2 = 0$	$\chi(\mathbb{P}^1, \mathcal{O}_{\mathbb{P}^1}(k_1))$
$k_1 = k_2 = 0$	1

3.3 Infinite Flop Chains

Infinite flop chains arise whenever isomorphic flops exist through at least two different boundaries of the Kähler cone. In this case, the extended Kähler cone consists of an infinite number of sub-cones, and when $h^{1,1}(X) > 2$ it has infinitely many edges (provided that the birational automorphism group is infinite). When $h^{1,1}(X) = 2$ the envelope of the extended Kähler cone is an irrational two-dimensional cone. In Refs. [31, 32] infinite flop chains were studied in relation to the swampland distance conjecture and the Kawamata-Morrison conjecture. For illustration, we present the following example. Let X be a three-fold obtained by intersecting two generic linear hypersurfaces in a product of five $\mathbb{P}^1$ spaces. This corresponds to the configuration matrix with identifier 7447 in the CICY list,

$$X = \begin{matrix} \mathbb{P}^1 \\ \mathbb{P}^1 \\ \mathbb{P}^1 \\ \mathbb{P}^1 \\ \mathbb{P}^1 \end{matrix} \begin{bmatrix} 1 & 1 \\ 1 & 1 \\ 1 & 1 \\ 1 & 1 \\ 1 & 1 \end{bmatrix}^{5,45} \tag{11}$$

The manifold admits flops in all five directions and the flopped spaces are isomorphic to X. This can be seen by explicitly constructing the flops. Let $x = [x_0 : x_1]$ denote the coordinates on the first $\mathbb{P}^1$ and let y, z, t, s be the coordinates of the remaining four $\mathbb{P}^1$'s. In this notation, the two equations can be written as

$$\begin{aligned} x_0 P_0(y, z, t, s) + x_1 P_1(y, z, t, s) &= 0, \\ x_0 Q_0(y, z, t, s) + x_1 Q_1(y, z, t, s) &= 0. \end{aligned} \tag{12}$$

Since $(x_0, x_1) = (0, 0)$ does not correspond to a point in $\mathbb{P}^1$, we must have $P_0 Q_1 - P_1 Q_0 = 0$. When this condition is satisfied, the equations determine a unique point in the first $\mathbb{P}^1$, except when $P_0 = P_1 = Q_0 = Q_1 = 0$, which happens at 24 points, when an entire $\mathbb{P}^1$ is obtained as a solution. Collapsing these 24 rational curves produces a singular tetra-quadric three-fold $\{P_0 Q_1 - P_1 Q_0 = 0\} \subset (\mathbb{P}^1)^{\times 4}$. The flopped manifold can be constructed by analogy with the Atiyah flop,

$$\begin{aligned} x_0 P_0(y, z, t, s) + x_1 Q_0(y, z, t, s) &= 0, \\ x_0 P_1(y, z, t, s) + x_1 Q_1(y, z, t, s) &= 0 \end{aligned} \tag{13}$$

and belongs to the same configuration class (11). The same discussion can be carried out in all other four directions, leading to a total of five flops for X. Each of the flopped manifolds being isomorphic to X, admits itself five different flops, one of which takes it back to X. The process can be iterated indefinitely, leading to an infinite number of flops and a fractal-like structure of the extended Kähler cone,

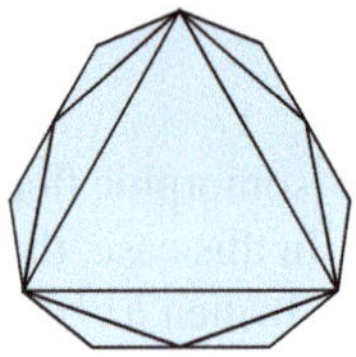

Fig. 3 Schematic representation of the extended Kähler cone of the CICY manifold 7447. There are infinitely many subcones, though only the largest ones are included here

which consists of the union of the Kähler cones of all birational models connected to X through a sequence of flops. A schematic representation of the extended Kähler cone is shown in Fig. 3, which should be interpreted as a cross-section.

4 String Theory Applications

4.1 Heterotic Model Building

Cohomology computations represent the main limiting factor in the analysis of heterotic string compactifications on Calabi-Yau threefolds with holomorphic vector bundles. The new cohomology formulae essentially remove this obstacle, allowing for a quick check of the entire low-energy spectrum. This represents a significant improvement from the standard situation where spectrum checks are usually limited to making sure that the number of chiral families, computed as a topological index, is equal to 3.

More concretely, heterotic compactifications with sums of line bundles have proven to be a rich arena for model building leading to a large number of standard-like models with many phenomenologically attractive features [7, 9, 37]. In this context cohomology constraints can be imposed to ensure: three families of quarks and leptons, the presence of a Higgs field, no exotic matter charged under the SM gauge group, a hierarchy of Yukawa couplings compatible with the observations, the absence of proton decay inducing operators, the existence of right-handed neutrinos, Yukawa unification etc.

While the cohomology formulae could, in principle, be inverted to infer in a bottom-up fashion the topological data of the compactification directly from the low-energy physical requirements, this approach is not feasible in practice. The constraints take the form of a system of linear, quadratic and cubic diophantine equations which are hard to solve explicitly or even undecidable [38–40]. However, the cohomology formulae can be used in conjunction with heuristic methods of search, such as Reinforcement Learning and Genetic Algorithms, to identify models resembling the Standard Model at a detailed level of analysis. These methods have recently been shown to be extremely efficient in identifying phenomenologically rich patches of the string landscape without systematically scanning over the entire range of solutions [41–48].

Going beyond abelian bundles, the cohomology formulae discussed above can also be useful in the context of heterotic model building with monad bundles. The monad sequence involves sums of line bundles and their cohomologies feed into the cohomology computations for the monad bundle. It is conceivable that simple analytic formulae for bundle cohomology exist also in the case of monad bundles, though more work is needed in this direction. However, even at this stage certain necessary conditions required for the consistency of heterotic compactifications with monad bundles, such as bundle stability, can be expressed in terms of the vanishing of line bundle cohomology groups. Such checks are now immediate and have already been used in Refs. [44, 46, 47].

4.2 Moduli Stabilisation

Most of the cohomology formulae studied so far correspond to manifolds defined at generic points in the complex structure moduli space and indicate that line bundle cohomology is related in a simple way to the topological and algebraic properties of the base manifold. As the complex structure is varied, cohomology dimensions can jump. The example presented in Sect. 3.2 indicates that such changes are not isolated and are correlated with something drastic happening to the structure of the cohomology cones. Understanding the manner in which cohomology chambers change with complex structure can be a valuable asset for moduli stabilisation.

The basic idea is that, for certain bundles and certain choices for the complex structure of the base manifold, the latter cannot be altered without ruining the holomorphicity of the bundle, which introduces a potential for the moduli [49, 50]. The presence of such configurations is signalled by certain cohomology jumps. Working out all the loci in complex structure moduli space where a given bundle stabilises some of the moduli is a hard problem, requiring Gröbner base calculations that have double-exponential running time [39, 51] and hence are often completely unfeasible. The complex structure dependence of the new cohomology formulae can lead to a quick identification of jumping loci and this information can be used in the construction of hidden sector bundles that can stabilise a maximal number of complex structure moduli in this way.

Acknowledgements The work of CB and JG is supported by the NSF grant PHY-2014086. The work of AC is supported by the EPSRC grant EP/T016280/1. The work of FR is supported by startup funding from Northeastern University. We are grateful to John Christian Ottem for useful comments on an earlier version of this article.

References

1. L. B. Anderson, J. Gray, Y.-H. He, S.-J. Lee, and A. Lukas, CICY package, based on methods described in arXiv:0911.1569,arXiv:0911.0865, arXiv:0805.2875, hep-th/0703249, hep-th/0702210.
2. cohomcalg package, 2010. High-performance line bundle cohomology computation based on methods described in arXiv:1003.5217, arXiv:1006.2392, arXiv:1006.0780. Download link: http://wwwth.mppmu.mpg.de/members/blumenha/cohomcalg/
3. M Larfors and R. Schneider, pycicy - a python cicy toolkit, https://doi.org/10.5281/zenodo.3243914, [github-link], (2019).
4. Jacques Distler and Brian R. Greene, Aspects of (2,0) String Compact? cations, Nucl. Phys. B 304 (1988), 1-62.
5. Ralph Blumenhagen, Sebastian Moster, and Timo Weigand, Heterotic GUT and standard model vacua from simply connected Calabi-Yau manifolds, Nucl. Phys. B751 (2006), 186–221, available at hep-th/0603015.
6. Ralph Blumenhagen, Sebastian Moster, Rene Reinbacher, and Timo Weigand, Massless Spectra of Three Generation U(N) Heterotic String Vacua, JHEP 0705 (2007), 041, available at hep-th/0612039.
7. Lara B. Anderson, James Gray, Andre Lukas, and Eran Palti, Two Hundred Heterotic Standard Models on Smooth CalabiYau Threefolds, Phys. Rev. D84 (2011), 106005, available at 1106.4804.
8. Heterotic Line Bundle Standard Models, JHEP 06 (2012), 113, available at 1202.1757.
9. Lara B. Anderson, Andrei Constantin, James Gray, Andre Lukas, and Eran Palti, A Comprehensive Scan for Heterotic SU(5) GUT models, JHEP 01 (2014), 047, available at 1307.4787.
10. Evgeny I. Buchbinder, Andrei Constantin, and Andre Lukas, Non-generic Couplings in Supersymmetric Standard Models, Phys. Lett. B 748 (2015), 251-254, available at 1409.2412.
11. A heterotic standard model with B? L symmetry and a stable proton, JHEP 06 (2014), 100, available at 1404.2767.
12. Shamit Kachru, Some three generation (0,2) Calabi-Yau models, Phys. Lett. B 349 (1995), 76-82, available at hep-th/9501131.
13. Ralph Blumenhagen, Gabriele Honecker, and Timo Weigand, Loop-corrected compact? cations of the heterotic string with line bundles, JHEP 0506 (2005), 020, available at hep-th/0504232.
14. Lara Briana Anderson, Heterotic and M-theory Compact? cations for String Phenomenology, Ph.D. Thesis, 2008.
15. Lara B. Anderson, Yang-Hui He, and Andre Lukas, Monad Bundles in Heterotic String Compact? cations, JHEP 07 (2008), 104, available at 0805.2875.
16. Lara B. Anderson, James Gray, Yang-Hui He, and Andre Lukas, Exploring Positive Monad Bundles And A New Heterotic Standard Model, JHEP 02 (2010), 054, available at 0911.1569.
17. Yang-Hui He, Seung-Joo Lee, and Andre Lukas, Heterotic Models from Vector Bundles on Toric Calabi-Yau Manifolds, JHEP 05 (2010), 071, available at 0911.0865.
18. Andrei Constantin, Heterotic String Models on Smooth Calabi-Yau Threefolds, Ph.D. Thesis, 2013.
19. Evgeny I. Buchbinder, Andrei Constantin, and Andre Lukas, The Moduli Space of Heterotic Line Bundle Models: a Case Study for the Tetra-Quadric, JHEP 03 (2014), 025, available at 1311.1941.
20. Andrei Constantin and Andre Lukas, Formulae for Line Bundle Cohomology on Calabi-Yau Threefolds, Fortsch. Phys. 67 (2019), no. 12, 1900084, available at 1808.09992.
21. Magdalena Larfors and Robin Schneider, Line bundle cohomologies on CICYs with Picard number two, Fortsch. Phys. 67 (2019), no. 12, 1900083, available at 1906.00392.
22. Callum R. Brodie, Andrei Constantin, Rehan Deen, and Andre Lukas, Index Formulae for Line Bundle Cohomology on Complex Surfaces, Fortsch. Phys. 68 (2020), no. 2, 1900086, available at 1906.08769.

23. Daniel Klaewer and Lorenz Schlechter, Machine Learning Line Bundle Cohomologies of Hypersurfaces in Toric Varieties, Phys. Lett. B789 (2019), 438-443, available at 1809.02547.
24. Callum R. Brodie, Andrei Constantin, Rehan Deen, and Andre Lukas, Machine Learning Line Bundle Cohomology, Fortsch. Phys. 68 (2020), no. 1, 1900087, available at 1906.08730.
25. Yang-Hui He, Deep-Learning the Landscape (2017), available at 1706.02714.
26. Fabian Ruehle, Evolving neural networks with genetic algorithms to study the String Landscape, JHEP 08 (2017), 038, available at 1706.07024.
27. Callum R. Brodie, Andrei Constantin, Rehan Deen, and Andre Lukas, Topological Formulae for the Zeroth Cohomology of Line Bundles on del Pezzo and Hirzebruch Surfaces, Complex Manifolds 08 (2021), 223-229, available at 1906.08363.
28. Callum R. Brodie and Andrei Constantin, Cohomology Chambers on Complex Surfaces and Elliptically Fibered Calabi-Yau Three-folds (2020), available at 2009.01275.
29. Th. Bauer, A. Kuronya, and T. Szemberg, Zariski chambers, volumes, and stable base loci, J. Reine Angew. Math. 576 (2004), 209-233, available at 0312211v1.
30. Callum R. Brodie, Andrei Constantin, and Andre Lukas, Flops, Gromov-Witten Invariants and Symmetries of Line Bundle Cohomology on Calabi-Yau Three-folds, J. Geom. Phys. 171 (2020), 104398, available at 2010.06597.
31. Callum R. Brodie, Andrei Constantin, Andre Lukas, and Fabian Ruehle, Swampland Conjectures and In? nite Flop Chains, Phys. Rev. D 104 (2021), no. 4, 046008, available at 2104.03325.
32. Geodesics in the extended K ahler cone of Calabi-Yau threefolds, JHEP 03 (2021), 024, available at 2108.10323.
33. Flops for Complete Intersection Calabi-Yau Threefolds (2021), available at 2112.12106.
34. P. Candelas, A.M. Dale, C.A. Lutken, and R. Schimmrigk, Complete Intersection Calabi-Yau Manifolds, Nucl. Phys. B 298 (1988), 493.
35. Paul S. Green, Tristan Hubsch, and Carsten A. Lutken, All Hodge Numbers of All Complete Intersection Calabi-Yau Manifolds, Class. Quant. Grav. 6 (1989), 105–124.
36. John Christian Ottem, Birational geometry of hypersurfaces in products of projective spaces, Mathematische Zeitschrift 280 (2015), 135–148, available at 1305.0537.
37. Andrei Constantin, Yang-Hui He, and Andre Lukas, Counting String Theory Standard Models, Phys. Lett. B 792 (2019), 258–262, available at 1810.00444.
38. Ju. V. Matijasevi?, The Diophantineness of enumerable sets, Dokl. Akad. Nauk SSSR 191 (1970), 279–282.
39. James Halverson and Fabian Ruehle, Computational Complexity of Vacua and Near-Vacua in Field and String Theory, Phys. Rev. D 99 (2019), no. 4, 046015, available at 1809.08279.
40. James Halverson, Michael Plesser, Fabian Ruehle, and Jiahua Tian, K ahler Moduli Stabilization and the Propagation of Decidability, Phys. Rev. D 101 (2020), no. 4, 046010, available at 1911.07835.
41. Alex Cole, Andreas Schachner, and Gary Shiu, Searching the Landscape of Flux Vacua with Genetic Algorithms, JHEP 11 (2019), 045, available at 1907.10072.
42. James Halverson, Brent Nelson, and Fabian Ruehle, Branes with Brains: Exploring String Vacua with Deep Reinforcement Learning, JHEP 06 (2019), 003, available at 1903.11616.
43. Magdalena Larfors and Robin Schneider, Explore and Exploit with Heterotic Line Bundle Models, Fortsch. Phys. 68 (2020), no. 5, 2000034, available at 2003.04817.
44. Andrei Constantin, Thomas R. Harvey, and Andre Lukas, Heterotic String Model Building with Monad Bundles and Reinforcement Learning, Fortschr. Phys. 70 (2022), 2100186, available at 2108.07316.
45. Sven Krippendorf, Rene Kroepsch, and Marc Syvaeri, Revealing systematics in phenomenologically viable ?ux vacua with reinforcement learning (2021), available at 2107.04039.
46. Steven Abel, Andrei Constantin, Thomas R. Harvey, and Andre Lukas, Evolving Heterotic Gauge Backgrounds: Genetic Algorithms versus Reinforcement Learning, Fortschr. Phys. (2022), 2200034, available at 2110.14029.
47. String Model Building, Reinforcement Learning and Genetic Algorithms, Nankai Symposium on Mathematical Dialogues: In celebration of S.S. Chern's 110th anniversary, 2021.

48. Alex Cole, Sven Krippendorf, Andreas Schachner, and Gary Shiu, Probing the Structure of String Theory Vacua with Genetic Algorithms and Reinforcement Learning, 2021.
49. Lara B. Anderson, James Gray, Andre Lukas, and Burt Ovrut, Stabilizing the Complex Structure in Heterotic Calabi-Yau Vacua, JHEP 02 (2011), 088, available at 1010.0255.
50. Stabilizing All Geometric Moduli in Heterotic Calabi-Yau Vacua, Phys.Rev. D83 (2011), 106011, available at 1102.0011.
51. H. Michael M oller and Ferdinando Mora, Upper and lower bounds for the degree of groebner bases, Proceedings of the international symposium on symbolic and algebraic computation, 1984, pp. 172–183.

Computation and Data in the Classification of Fano Varieties

Gavin Brown, Tom Coates, Alessio Corti, Tom Ducat, Liana Heuberger, and Alexander M. Kasprzyk

Abstract Fano varieties are 'atomic pieces' of algebraic varieties, the shapes that can be defined by polynomial equations. We describe the role of computation and database methods in the construction and classification of Fano varieties, with an emphasis on three-dimensional Fano varieties with mild singularities called $\mathbb{Q}$-Fano threefolds. The classification of $\mathbb{Q}$-Fano threefolds has been open for several decades, but there has been significant recent progress. These advances combine computational algebraic geometry and large-scale data analysis with new ideas that originated in theoretical physics.

Keywords Fano variety · Mirror symmetry · Classification

G. Brown (✉) · A. M. Kasprzyk
Mathematics Institute, University of Warwick, Coventry, UK
e-mail: G.Brown@warwick.ac.uk

A. M. Kasprzyk
e-mail: alexander.kasprzyk@warwick.ac.uk

T. Coates · A. Corti
Department of Mathematics, Imperial College London, London, UK
e-mail: t.coates@imperial.ac.uk

A. Corti
e-mail: a.corti@imperial.ac.uk

T. Ducat
Department of Mathematical Sciences, Durham University, Durham, UK
e-mail: thomas.ducat@durham.ac.uk

L. Heuberger
Institut de Mathématiques de Marseille, Marseille, France
e-mail: liana.heuberger@univ-amu.fr

Y.-H. He et al. (eds.), *Nankai Symposium on Mathematical Dialogues*,
https://doi.org/10.1007/978-981-19-2328-9_11

1 Introduction

Fano varieties are an important class of geometric shapes. They are basic building blocks in algebraic geometry, both via the Minimal Model Program [11, 45, 65, 66] and as a source of explicit constructions. For example, one can construct three-dimensional Calabi–Yau manifolds (smooth varieties with zero curvature) as anticanonical sections in smooth four-dimensional Fano varieties; these are of particular importance in constructing models of spacetime in Type II string theory [6, 71]. The classification of Fano varieties is a fundamental problem in geometry yet, despite almost a century of study, this classification is still far from understood. In this paper we discuss the classification of Fano varieties over $\mathbb{C}$. In particular, the n-dimensional Fano varieties that we consider can also be thought of, at least when they are smooth, as $2n$-dimensional real manifolds with positive curvature.

There are finitely many deformation families of Fano manifolds (that is, smooth Fano varieties) in each dimension [60]. There is exactly 1 one-dimensional Fano manifold: the Riemann sphere $\mathbb{P}^1$. In two dimensions there are 10 deformation families, the del Pezzo surfaces: $\mathbb{P}^1 \times \mathbb{P}^1$, $\mathbb{P}^2$, and $\mathbb{P}^2$ blown-up in at most eight points [37]. In dimension three the classification is due to Fano [41], Iskovskikh [49–51], and Mori–Mukai [67, 68]: there are 105 deformation families.[1] Very little is known about the classification of Fano manifolds in dimensions four or more.

When we allow mild singularities even less is known. There are finitely many deformation families in each dimension with singularities of bounded complexity [10, 61], but there are few explicit classification results. In dimension three it is natural to focus on the $\mathbb{Q}$-Fano threefolds: Fano varieties with $\mathbb{Q}$-factorial terminal singularities—this is the natural setting for the Minimal Model Program. Even here there is as yet no classification, although several hundred deformation families have been described; see e.g. [17, 19, 20, 34, 40, 48, 55, 74, 79–81]. What is known, however, are the possible Hilbert series of $\mathbb{Q}$-Fano threefolds[2] [5, 16]. The Hilbert series of a Fano variety X is the generating series for the dimensions of the graded pieces of the anticanonical ring

$$R(X, -K_X) = \bigoplus_{n=0}^{\infty} H^0(X, -nK_X). \tag{1.1}$$

Hilbert series remain invariant under $\mathbb{Q}$-Gorenstein deformation of X. The possible Hilbert series are recorded in the Graded Ring Database [13, 15]. One can think of this database as a first sketch, giving numerical invariants only, of the possible 'geography' of $\mathbb{Q}$-Fano threefolds.

---

[1] The three-dimensional classification can be explored online via the excellent Fanography website [9].

[2] Strictly speaking, we mean here the possible Hilbert series of Mori–Fano threefolds only—see [16, Sect. 5] for a precise discussion—but as every known $\mathbb{Q}$-Fano threefold has a Hilbert series that is in this list, we will blur the distinction in the rest of the paper. In any case, this is a large and important part of the landscape of $\mathbb{Q}$-Fano threefolds.

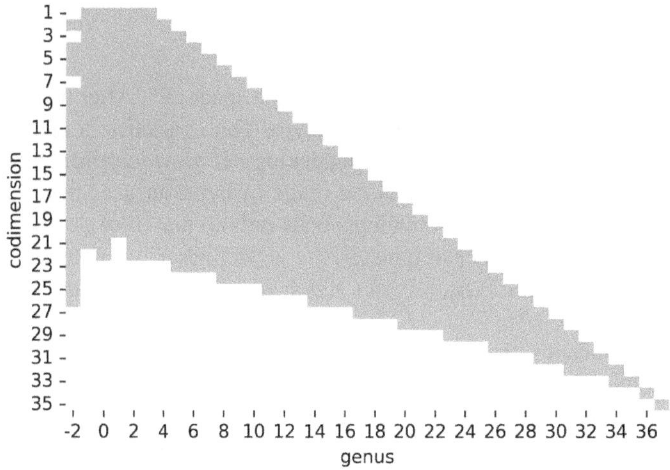

Fig. 1 A sketch of the landscape of $\mathbb{Q}$-Fano threefolds, from the Graded Ring Database [15]

A Fano variety X can be recovered as Proj of its anticanonical ring (1.1). Choosing homogeneous generators for the anticanonical ring defines an embedding of X into a weighted projective space $w\mathbb{P}$, and we can read off various measures of the complexity of this embedding (or of the complexity of $R(X, -K_X)$ as a ring) directly from the Hilbert series. In particular, the Hilbert series can be used to estimate a lower bound for the *codimension*

$$\operatorname{codim} X := \dim w\mathbb{P} - \dim X,$$

that is, the number of independent equations required to define X (locally) inside $w\mathbb{P}$. The Hilbert series also determines the *genus* of the embedding

$$g(X) := \dim H^0(X, -K_X) - 2.$$

If X is smooth then $g(X)$ is the genus of the curve defined by intersecting X with two generic hyperplanes in $w\mathbb{P}$; for the general case, including an interpretation of the case where $g(X) < 0$, see [5, Sect. 4]. The genus increases with the complexity of the embedding. Figure 1 shows the range of pairs $(g(X), \operatorname{codim} X)$ for entries in the database of possible Hilbert series of $\mathbb{Q}$-Fano threefolds. See [16] for precise details: in particular, codimension here means estimated codimension as there.

The database [13, 15] is produced using combinatorial methods that do not guarantee existence of a $\mathbb{Q}$-Fano threefold that realises any given entry: there can be zero, one, or many deformation families of $\mathbb{Q}$-Fano threefolds with a given candidate Hilbert series. In the remainder of the paper, we discuss techniques for constructing and classifying $\mathbb{Q}$-Fano threefolds, guided by the geography in Fig. 1.

2 Classifications in Low Codimension

The first examples of Fano varieties are projective spaces $\mathbb{P}^n$. After these, projective hypersurfaces of low degree give the archetypal Fano varieties: a cubic surface, a cubic or quartic threefold, and so on. Broadening our view to orbifolds with cyclic quotient singularities greatly extends the range of hypersurfaces that we can construct. For example, a weighted homogeneous polynomial f of degree d in $n+1$ variables with positive integer weights $a_0, \dots, a_n$ defines a hypersurface $(f=0)$ in weighted projective space $\mathbb{P}(a_0, \dots, a_n)$. If $d < a_0 + \cdots + a_n$ then this hypersurface is Fano. See [38, 48, 64].

In three dimensions, there are exactly 95 families of $\mathbb{Q}$-Fano threefolds that arise as quasismooth weighted projective hypersurfaces [14, 53, 77]. Here *quasismoothness* ensures that all singularities of the hypersurface arise from singularities of the ambient weighted projective space [48]. The 95 hypersurfaces are all of the form $X_d \subset \mathbb{P}(1, a_1, a_2, a_3, a_4)$, where $d = a_1 + a_2 + a_3 + a_4$, and correspond to the 95 families of K3 surfaces $X_d \subset \mathbb{P}(a_1, a_2, a_3, a_4)$. Quasismooth members X of each family are birationally rigid [22, 32], meaning if X is birational to another $\mathbb{Q}$-Fano threefold or Mori fiber space Y then X is in fact biregular isomorphic to Y. One may consider this to be a strong form of non-rationality [52] since, in particular, they do not admit birational maps to $\mathbb{P}^3$. The 95 families have been studied from many points of view, and serve as a testing ground for conjectures and computations. For example, Kim–Okada–Won compute their log canonical thresholds, and deduce K-stability results for 50 of the 95 families [59]. In four dimensions, $\mathbb{Q}$-Fano quasismooth weighted projective hypersurfaces have also been classified [7]: there are 11 618 families.

Hypersurfaces $(f=0)$ have codimension one, because they are defined (globally) by a single equation. One can construct Fano varieties with higher codimension by considering complete intersections

$$(f_1 = \cdots = f_c = 0)$$

in weighted projective spaces, where the f_i are homogeneous polynomials of sufficiently low degree. Iano-Fletcher describes 85 families of $\mathbb{Q}$-Fano threefolds that arise as codimension-two complete intersections in weighted projective space [48]; this list is in fact complete [23]. In higher codimension most Fano varieties are not weighted projective complete intersections. In codimension three, for example, sections of weighted Grassmannians $w\mathrm{Gr}(2,5)$, defined by five equations of Pfaffian type, are typical [33]. Codimension four embeddings are also tractable [4, 19] but beyond codimension four, a comprehensive understanding is restricted to special classes such as smooth Fano threefolds.

3 Birational Geometry and Steps Into Higher Codimension

Whilst the low codimension studies discussed in Sect. 2 are very successful and comprehensive, they break down in higher codimension. There are several approaches to this problem. One is to find simple examples whose anticanonical embedding happens to be in high codimension: for example, $\mathbb{P}^3$ itself has $-K_X = 4H$, for a linear hyperplane $H \subset \mathbb{P}^3$, so that the anticanonical embedding is the 4th Veronese embedding $\mathbb{P}^3 \hookrightarrow \mathbb{P}^{34}$ in codimension 31. Another approach is to find varieties that naturally live in high codimension whose sections are $\mathbb{Q}$-Fano threefolds [18, 76]. A more sophisticated approach, the *Takeuchi program* [82], analyses possible birational links between Fano threefolds with the aim of constructing complicated examples from simpler ones; cf. [3, 31]. This was exploited by Takagi [81] to classify all $\mathbb{Q}$-Fano threefolds X having only $\mathbb{Z}/2$ hyperquotient singularities, Picard rank $\rho_X = 1$, and dim $H^0(X, -K_X) \geq 4$; and by Prokhorov [73] to analyse $\mathbb{Q}$-Fano threefolds with $-K_X$ divisible.

We discuss a fourth approach here, applying *unprojection* [19, 70, 78]. This is close in spirit to the Takeuchi program, and is rooted in Fano's original approach to the classification problem via projection. The idea is as follows. If $X_4 \subset \mathbb{P}^4$ is a terminal Fano quartic threefold that contains a linear plane $\mathbb{P}^2 \cong D \subset X_4$, then there is a birational map $X_4 \dashrightarrow Y = Y_{3,3} \subset \mathbb{P}(1, 1, 1, 1, 1, 2)$. In coordinates $x_0, \dots x_4$ on $\mathbb{P}^4$, this is

$$D = (x_0 = x_1 = 0) \subset X_4 = (x_0 A - x_1 B = 0) \subset \mathbb{P}^4$$

where A and B are cubic forms, and setting $y = A/x_1 = B/x_0$ defines Y and a morphism

$$\begin{aligned} X_4 &\dashrightarrow Y = (yx_1 = A, yx_0 = B) \subset \mathbb{P}(1, 1, 1, 1, 1, 2) \\ (x_0, \dots, x_4) &\mapsto (x_0, \dots, x_4, y(x_0, \dots, x_4)). \end{aligned} \tag{3.1}$$

Here $x_0, \dots, x_4$ and y are the coordinates on $\mathbb{P}(1, 1, 1, 1, 1, 2)$ of weights $1, \dots, 1$ and 2. The map (3.1) is called the *unprojection* of $D \subset X_4$, and conversely, elimination of the variable y from the equations of Y recovers X_4 as a projection of Y.

Multiplying up the denominator in (3.1) of either expression for y shows that general points of D are mapped to $[0:0:0:0:0:1] \in Y$, and so D is indeed contracted. However, the map (3.1) is not a morphism. The geometry of the map depends on the singularities that X_4 acquires at the points $(A = B = 0)$ on D. Typically these are ordinary double points, and the map consists of the D-ample small resolution of each such point followed by the contraction of D, but this requires detailed analysis in any given case. If one can repeat this with two planes $D_1, D_2 \subset X_4$, unprojecting them one after the other, the result would be a Fano threefold embedded in codimension three. It is easy to imagine continuing this process, but imposing successive planes becomes an increasingly difficult geometrical and combinatorial problem.

This approach works well into codimension four, with several hundred families constructed by unprojecting weighted planes $D = \mathbb{P}(1, a, b)$ [19]. In the example above, Brown–Kerber–Reid construct three different deformation families of $\mathbb{Q}$-Fano threefolds, using different configurations of planes. (Takagi's classification describes two of these families; the third has higher Picard rank.) Unprojection techniques have been extended into codimension ≥ 5 by Brown and Ducat [12]. The real power of this technique becomes apparent when combined with ideas from mirror symmetry: see Sect. 9 below.

4 Polytopes, Fans, and Toric Varieties

Toric varieties are highly symmetrical algebraic varieties. An n-dimensional normal projective variety X is a *toric variety* if it contains a dense algebraic torus $(\mathbb{C}^\times)^n$ and the action of that torus on itself extends to an action of the torus on all of X: see [35, 42]. For example, projective spaces and weighted projective spaces are toric varieties. In our context, toric varieties will play two different roles. Three-dimensional toric varieties will occur as degenerations of $\mathbb{Q}$-Fano threefolds; in general these toric varieties will be highly singular. Higher-dimensional toric varieties will occur as ambient spaces when we construct $\mathbb{Q}$-Fano threefolds as hypersurfaces or complete intersections; we have already seen this in Sect. 2.

The study of toric varieties is underpinned by a rich 'dictionary' that translates geometry into combinatorics. Let $M \cong \mathbb{Z}^n$ denote the lattice of characters of $T = (\mathbb{C}^\times)^n$, with dual lattice $N := \mathrm{Hom}(M, \mathbb{Z})$. A toric variety X with dense torus T has a combinatorial description in terms of a fan Σ in $N_{\mathbb{Q}}$, and many geometric properties of X can be rephrased in terms of combinatorial statements about Σ. For example, let $\rho_1, \ldots, \rho_k$ be the one-dimensional cones of Σ. Each ρ_i is generated by a (unique) primitive lattice element $v_i \in N$, and X is Fano if and only if $v_1, \ldots, v_k$ correspond to the vertices of a convex lattice polytope in $N_{\mathbb{Q}}$ containing the origin in its interior. Such a polytope is called a *Fano polytope*. Conversely, given a Fano polytope P the fan Σ (and hence the toric Fano variety X) can be recovered by taking the collection of cones spanned by the faces of P; we call this the *spanning fan* of P. Under this dictionary there is a correspondence between the dual polytope

$$P^* := \{u \in M_{\mathbb{Q}} \mid u(v) \geq -1\}$$

of P and the anticanonical divisor $-K_X$ of X. Since P^* always contains the origin, it follows that $\dim H^0(X, -K_X) = |P^* \cap M|$ is strictly positive.

5 Mirror Symmetry

A new approach to the classification of Fano varieties centres around a set of ideas from string theory called *mirror symmetry* [21, 44, 47]. Mirror symmetry gives a correspondence between Fano manifolds and certain Laurent polynomials [24]. From this perspective, the key invariant of a Fano variety X is its *regularised quantum period*

$$\widehat{G}_X(t) = \sum_{d=0}^{\infty} c_d t^d. \tag{5.1}$$

This is a power series with coefficients $c_0 = 1$, $c_1 = 0$, and $c_d = r_d d!$, where r_d is a certain Gromov–Witten invariant of X. Intuitively speaking, r_d is the number of rational curves in X of degree $d \geq 2$ that pass through a fixed generic point and have a certain constraint on their complex structure. Under mirror symmetry, an n-dimensional Fano manifold X corresponds to a Laurent polynomial $f \in \mathbb{C}[x_1^{\pm 1}, \ldots, x_n^{\pm 1}]$ if the regularised quantum period $\widehat{G}_X$ agrees with the *classical period*

$$\pi_f(t) = \sum_{d=0}^{\infty} c'_d t^d. \tag{5.2}$$

Here c'_d is the coefficient of the unit monomial in f^d.

It is expected that if a Fano manifold X corresponds under mirror symmetry to a Laurent polynomial f, then there is a $\mathbb{Q}$-Gorenstein degeneration of X to a (singular) toric variety X_f. This toric variety is defined by the spanning fan of the Newton polytope of f. One can hope, therefore, to recover X directly from its mirror partner f: by regarding f as encoding a log structure on X_f that determines the smoothing from X_f to X, and then constructing X as this smoothing.

Gromov–Witten invariants are deformation invariant, so the regularised quantum period $\widehat{G}_X$ is a deformation invariant of X. The classical period π_f, on the other hand, is invariant under an equivalence relation called *mutation* [2]. It is conjectured that mirror symmetry gives a one-to-one correspondence between deformation-equivalence classes of $\mathbb{Q}$-Fano varieties[3] that admit a toric degeneration, and mutation-equivalence classes of *rigid maximally mutable Laurent polynomials* (rigid MMLPs) [28]. For the classification of two- and three-dimensional Fano manifolds from this perspective, see [1, 25, 39, 58, 72].

If X is a Fano projective hypersurface then the Givental/Hori–Vafa mirror construction [43, 47] gives rise to a Laurent polynomial f that corresponds to X under mirror symmetry. For example, if X is a cubic surface in $\mathbb{P}^3$ then the Givental/Hori–Vafa mirror is given by

[3] That is, Fano varieties with $\mathbb{Q}$-factorial terminal locally toric singularities.

$$W = x_0 + x_1 + x_2 + x_3 \qquad \text{where} \qquad \begin{aligned} x_0 x_1 x_2 x_3 &= 1 \\ x_1 + x_2 + x_3 &= 1. \end{aligned}$$

Eliminating the second equation by setting

$$x_1 = \frac{1}{1+x+y} \qquad x_2 = \frac{x}{1+x+y} \qquad x_3 = \frac{y}{1+x+y}$$

and solving for x_0 yields the Laurent polynomial

$$f = W - 7 = \frac{(1+x+y)^3}{xy} - 6.$$

This is the Przyjalkowski method. It applies in greater generality to give Laurent polynomials that correspond under mirror symmetry to complete intersections in weighted projective spaces [75] and toric varieties [29, 39] that satisfy mild combinatorial conditions.

The Givental/Hori–Vafa construction above can be reversed to recover a presentation of X as a toric complete intersection from the corresponding Laurent polynomial f: this is *Laurent inversion* [30]. This process cannot always work—for example, only 89 of the 105 smooth Fano threefolds are known to be toric complete intersections—but nonetheless Laurent inversion gives a powerful tool for constructing $\mathbb{Q}$-Fano varieties. In outline: one starts with a Fano polytope P, searches for rigid MMLPs f with Newton polytope P, and then applies Laurent inversion to construct a corresponding Fano variety. Heuberger has used this approach to construct many new $\mathbb{Q}$-Fano threefolds [46].

6 Fano Polytopes

The classification of Fano polytopes is an attractive combinatorial problem: see [57] for an overview. Smooth Fano polytopes—corresponding to toric Fano manifolds—can be classified efficiently using an algorithm of Øbro [69]. The number grows slowly with dimension: just 18 out of the 105 three-dimensional Fano manifolds are toric, and even by dimension eight there are only 749 892 smooth Fano polytopes. *Reflexive* Fano polytopes—corresponding to Gorenstein toric Fano varieties—are of particular importance in mirror symmetry: topologically mirror-symmetric pairs of Calabi–Yau varieties can be constructed as hypersurfaces in Gorenstein toric Fano varieties, and this duality is reflected in the combinatorics of the polytopes P and P^* [8]. There are 4319 reflexive polytopes in dimension three, and 473 800 776 in dimension four [62, 63]. Toric Fano varieties with only terminal singularities correspond to Fano polytopes $P \subset N_{\mathbb{Q}}$ such that the only lattice points in P are the origin and the vertices. These have been classified in dimension three [55]: there are 634 cases. Toric Fano varieties with canonical singularities correspond

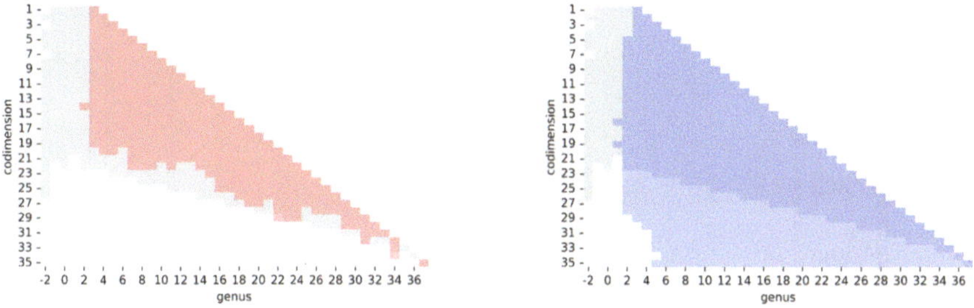

(a) Canonical polytopes (1 interior point).

(b) Fano polytopes with 2 interior points.

Fig. 2 The portions of the landscape of $\mathbb{Q}$-Fano threefolds that might be realised by three-dimensional Fano polytopes with one and two interior lattice points, overlaid on Fig. 1

to Fano polytopes $P \subset N_{\mathbb{Q}}$ such that the only interior lattice point of P is the origin. These *canonical Fano polytopes* have also been classified in dimension three [56]: there are 674 688 cases.

These polytope classifications provide a natural place to search for rigid MMLPs. For example, an analysis of the 4319 reflexive three-dimensional polytopes was used in [2, 25] to recover the 98 deformation families of three-dimensional Fano manifolds with very ample anticanonical bundle. Extending this idea, by constructing rigid MMLPs supported on Fano polytopes more generally, we can begin to populate the geography of $\mathbb{Q}$-Fano threefolds. This is developed further in [26, 46], but let us first consider how much of the geography in Fig. 1 we might hope to realise this way.

Genus and codimension are deformation invariants. We expect, as discussed above, that a $\mathbb{Q}$-Fano threefold X corresponding to a rigid MMLP f admits a degeneration to the toric variety X_f defined by the spanning fan of the Newton polytope $P \subset N_{\mathbb{Q}}$ of f. In this case we can read off the genus and codimension of X_f (and hence of X) from the combinatorics of the dual polytope P^* of P. Namely, the genus is given by $|P^* \cap M| - 2$, and the codimension is computed from the Hilbert basis of the four-dimensional cone $\mathbb{R}_{\geq 0} \cdot (P^* \times \{1\})$. Figure 2a shows the pairs $(g(Y), \operatorname{codim} Y)$ realised by toric Fano varieties Y with at worst canonical singularities[4]; here Y is playing the role of X_f. It should be emphasised that not every canonical Fano polytope will support a rigid MMLP: initial experiments indicate $\sim$15% of the canonical Fano polytopes support at least one rigid MMLP, resulting in approximately 8300 mutation-equivalence classes of rigid MMLPs. The codimensions of the hypothetical $\mathbb{Q}$-Fano threefolds discovered (or rather, suggested) in this way are summarised in Fig. 3: note that the majority of these have quite high codimension.

We expect that increasing the number of interior lattice points allowed in the Newton polytope (equivalently, allowing log-terminal singularities in the corresponding toric Fano varieties) will reveal new mutation-equivalence classes of rigid MMLPs. For example, Fano polytopes with two interior lattice points are classified in [7]. The pairs $(g(Y), \operatorname{codim} Y)$ for the resulting toric Fano varieties Y are shown in Fig. 2b.

[4] See [16, Figure 6] for a more detailed analysis.

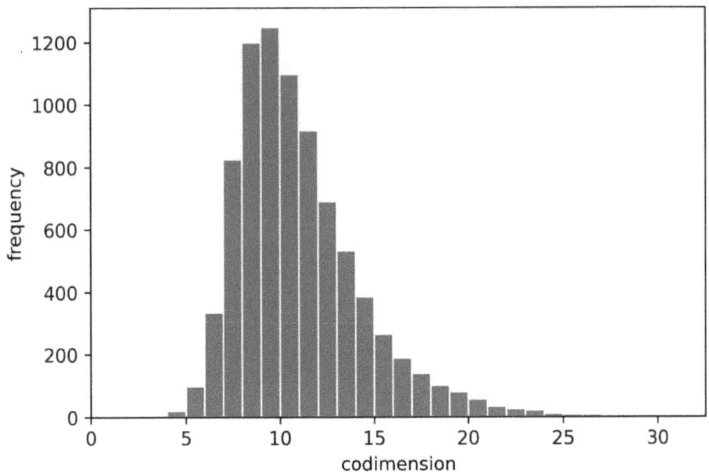

Fig. 3 Known mutation-equivalence classes of rigid MMLPs supported on 3-dimensional canonical Fano polytopes, by codimension

Once again we expect only a fraction of these Fano polytopes to support a rigid MMLP, and hence degenerate to a $\mathbb{Q}$-Fano threefold. In particular, we expect that none of the two-point polytopes that fall outside the gray shaded region in Fig. 2b—or equivalently the region shown in Fig. 1—support a rigid MMLP. Figure 2 suggests that increasing the number of interior lattice points might allow us to find $\mathbb{Q}$-Fano threefolds in both higher codimension and lower genus. The latter point is intuitively plausible: as the Newton polytope P grows larger, so the dual polytope P^* grows smaller and so we expect $|P^* \cap M|$ and hence the corresponding genus to decrease.

7 The 95 Hypersurfaces

We return to the 95 $\mathbb{Q}$-Fano threefold hypersurfaces X_d in weighted projective space discussed in Sect. 2. As in Sect. 5, we can apply the Givental/Hori–Vafa construction to obtain a Laurent polynomial mirror for X_d. Write

$$W = x_0 + x_1 + x_2 + x_3 + x_4 \qquad \text{where} \qquad \begin{aligned} x_0 x_1^{a_1} x_2^{a_2} x_3^{a_3} x_4^{a_4} &= 1 \\ x_1 + x_2 + x_3 + x_4 &= 1. \end{aligned}$$

Eliminating the second equation by setting

$$x_1 = \frac{1}{1+x+y+z} \qquad x_2 = \frac{x}{1+x+y+z} \qquad x_3 = \frac{y}{1+x+y+z} \qquad x_4 = \frac{z}{1+x+y+z}$$

and solving for x_0 yields the Laurent polynomial

$$f = W - 1 - c = \frac{(1 + x + y + z)^d}{x^{a_2} y^{a_3} z^{a_4}} - c \qquad \text{where} \qquad c = \frac{d!}{a_1! a_2! a_3! a_4!}$$

Here the constant c is chosen to ensure that the coefficient c_1' of the classical period (5.2) is zero. The corresponding Newton polytope P is a simplex; the number of interior lattice points is given by the tetrahedral number $(d-3)(d-2)(d-1)/6$.

The hypersurface with the largest value for d is $X_{66} \subset \mathbb{P}(1, 5, 6, 22, 33)$. It has genus -1, and gives a simplex with 43 680 interior lattice points. Experiments mutating the Laurent polynomial show that we cannot expect to substantially decrease the number of interior lattice points. This agrees with our intuition in Sect. 6 that constructing low-genus $\mathbb{Q}$-Fano threefolds requires Fano polytopes with large numbers of interior points. Classifying Fano polytopes with tens of interior lattice points, let alone tens of thousands, is well beyond the reach of current methods, so it is implausible that we could find examples such as the 95 hypersurfaces using mirror symmetry methods alone.

8 An Example in Codimension Two

As discussed, we expect that if a $\mathbb{Q}$-Fano threefold X corresponds under mirror symmetry to a Laurent polynomial f, then X admits a degeneration to the toric variety X_f defined by the spanning fan of the Newton polytope $P \subset N_{\mathbb{Q}}$ of f. If this is the case, then only those $\mathbb{Q}$-Fano threefolds that admit toric degenerations can be detected using mirror symmetry methods. For example, consider the codimension two $\mathbb{Q}$-Fano threefold $X_{12,14} \subset \mathbb{P}(2, 3, 4, 5, 6, 7)$, which is part of the Iano-Fletcher classification [48]. This has $\dim H^0(X, -K_X) = 0$, and so can never deform to a toric variety.[5] So mirror symmetry methods alone will not be enough to complete the classification.

9 A Two-Pronged Approach

Classical construction methods (Sects. 2, 7–8) and the mirror symmetry approach (Sects. 5–6) complement each other well, with the classical methods most effective for $\mathbb{Q}$-Fano threefolds of low codimension and mirror symmetry giving remarkable coverage in high codimension. Mirror symmetry also gives new insight at one of the most difficult points of the analysis: determining whether a given set of possible numerical invariants is realised by a $\mathbb{Q}$-Fano variety. Conjecturally this problem is

[5] It would be interesting to know if in general the vanishing of $H^0(X, -K_X)$ is the only obstruction to X admitting a toric degeneration.

Table 1 The predicted number of deformation families of $\mathbb{Q}$-Fano threefolds $Y^{(i)}$ from Sect. 9

i	0	1	2	3	4	5	6	7	8
GRDB ID	20521	20522	20543	20652	20938	21436	22093	22801	23386
Codimension	1	2	3	4	5	6	7	8	9
Number of families	1	1	1	3	7	11	11	4	0

equivalent to determining the existence (or non-existence) of rigid MMLPs with appropriate Newton polytopes: a challenging but tractable combinatorial problem, which is amenable to machine computation. Traditionally there are no systematic methods available here.

This suggests a new approach—a hybrid of classical and mirror symmetry methods—to populate and analyse the landscape of $\mathbb{Q}$-Fano threefolds. First, use graded ring methods [5, 13, 16] to determine an explicit list of possible Hilbert series. Then, for each candidate Hilbert series $H(t)$ on that list, classify rigid MMLPs f supported on Fano polytopes P with dual Ehrhart series $\mathrm{Ehr}(P^*, t)$ equal to $H(t)$, where P lies in an appropriate combinatorial class. This predicts the number of deformation families, as well as specific $\mathbb{Q}$-Gorenstein toric degenerations X_f of these families. Finally, use deformation, Laurent inversion, or unprojection methods to construct each family. Meaningful analysis beyond particular cases will require cluster-scale computation and database methods, because of the numbers of cases involved (thousands of deformation families, hundreds of thousands of polytopes), but this is well within what is now possible. For a substantial step in this direction, see [26, 27, 46].

We close by discussing a classical example from this new point of view. Consider a sequence of $\mathbb{Q}$-Fano threefolds $Y^{(i)} \subset \mathbb{P}(1, 1, 1, 1, 1, 2, \ldots, 2)$ in codimension $i + 1$ that arise by successively unprojecting i planes inside a quartic hypersurface $X_4 \subset \mathbb{P}^3$. The case when $i = 1$ is discussed in Sect. 3. When $i \geq 3$, there may be more than one configuration of planes: see [12]. By determining mutation-equivalence classes of rigid MMLPs, we predict numbers of distinct deformation families as shown in Table 1. There remains much detailed analysis still to do, but at a coarse level this prediction matches an observation of Brown–Ducat [12]: that a quartic threefold can contain at most seven planes meeting only pairwise in points. Indeed a quartic X that contains a plane Π is birational to a cubic del Pezzo fibration, obtained as the projection away from Π. Any other plane in X that meets Π only in points corresponds to a section of (-1)-curves in this fibration, and thus there are at most six such planes, corresponding to a maximal set of six disjoint lines in the general fibre of this cubic del Pezzo fibration; cf. [54]. Moreover a set of seven such planes can be explicitly realised inside the Burkhardt quartic $X_4^{\mathrm{Bur}} \subset \mathbb{P}^4$, which is the unique quartic having the maximal number of 45 ordinary double points [36] and contains a configuration of 40 planes. Therefore, if we constrain the unprojection plane configurations to meet as simply as possible, there is no configuration of eight planes to unproject into codimension nine, in accordance with the conjecture.

FundingTC is supported by ERC Consolidator Grant 682603 and EPSRC Programme Grant EP/N03189X/1. LH is supported by Projet Étoiles Montantes GeBi de la Région Pays de la Loire. AK is supported by EPSRC Fellowship EP/N022513/1.

References

1. Mohammad Akhtar, Tom Coates, Alessio Corti, Liana Heuberger, Alexander M. Kasprzyk, Alessandro Oneto, Andrea Petracci, Thomas Prince, and Ketil Tveiten. Mirror symmetry and the classification of orbifold del Pezzo surfaces. *Proc. Amer. Math. Soc.*, 144(2):513–527, 2016.
2. Mohammad Akhtar, Tom Coates, Sergey Galkin, and Alexander M. Kasprzyk. Minkowski polynomials and mutations. *SIGMA Symmetry Integrability Geom. Methods Appl.*, 8:Paper 094, 17, 2012.
3. Valery Alexeev. General elephants of **Q**-Fano 3-folds. *Compositio Math.*, 91(1):91–116, 1994.
4. Selma Altınok. *Graded Rings Corresponding to Polarised K3 Surfaces and $\mathbb{Q}$-Fano 3-folds.* PhD thesis, University of Warwick, 1998.
5. Selma Altınok, Gavin Brown, and Miles Reid. Fano 3-folds, $K3$ surfaces and graded rings. In *Topology and geometry: commemorating SISTAG*, volume 314 of *Contemp. Math.*, pages 25–53. Amer. Math. Soc., Providence, RI, 2002.
6. Paul S. Aspinwall, Brian R. Greene, and David R. Morrison. Calabi-Yau moduli space, mirror manifolds and spacetime topology change in string theory. *Nuclear Phys. B*, 416(2):414–480, 1994.
7. Gabriele Balletti and Alexander M. Kasprzyk. Three-dimensional lattice polytopes with two interior lattice points. arXiv:1612.08918 [math.CO], 2016.
8. Victor V. Batyrev. Dual polyhedra and mirror symmetry for Calabi-Yau hypersurfaces in toric varieties. *J. Algebraic Geom.*, 3(3):493–535, 1994.
9. Pieter Belmans. Fanography: a tool to visually study the geography of Fano 3-folds. https://www.fanography.info.
10. Caucher Birkar. Singularities of linear systems and boundedness of Fano varieties. *Ann. of Math. (2)*, 193(2):347–405, 2021.
11. Caucher Birkar, Paolo Cascini, Christopher D. Hacon, and James McKernan. Existence of minimal models for varieties of log general type. *J. Amer. Math. Soc.*, 23(2):405–468, 2010.
12. Gavin Brown and Tom Ducat. Double unprojection of Fano 3-folds. In preparation.
13. Gavin Brown and Alexander M. Kasprzyk. The Graded Ring Database. http://www.grdb.co.uk.
14. Gavin Brown and Alexander M. Kasprzyk. Four-dimensional projective orbifold hypersurfaces. *Exp. Math.*, 25(2):176–193, 2016.
15. Gavin Brown and Alexander M. Kasprzyk. The Fano 3-fold database. *Zenodo* https://doi.org/10.5281/zenodo.5820338, 2022.
16. Gavin Brown and Alexander M. Kasprzyk. Kawamata boundedness for Fano threefolds and the Graded Ring Database. arXiv:2201.07178 [math.AG], 2022.
17. Gavin Brown, Alexander M. Kasprzyk, and Muhammad Imran Qureshi. Fano 3-folds in $\mathbb{P}^2 \times \mathbb{P}^2$ format, Tom and Jerry. *Eur. J. Math.*, 4(1):51–72, 2018.
18. Gavin Brown, Alexander M. Kasprzyk, and Lei Zhu. Gorenstein formats, canonical and Calabi–Yau threefolds. *Experimental Mathematics*, pages 1–19, 2019.
19. Gavin Brown, Michael Kerber, and Miles Reid. Fano 3-folds in codimension 4, Tom and Jerry. Part I. *Compos. Math.*, 148(4):1171–1194, 2012.
20. F. Campana and H. Flenner. Projective threefolds containing a smooth rational surface with ample normal bundle. *J. Reine Angew. Math.*, 440:77–98, 1993.
21. Philip Candelas, Xenia C. de la Ossa, Paul S. Green, and Linda Parkes. A pair of Calabi-Yau manifolds as an exactly soluble superconformal theory. *Nuclear Phys. B*, 359(1):21–74, 1991.

22. Ivan Cheltsov and Jihun Park. Birationally rigid Fano threefold hypersurfaces. *Mem. Amer. Math. Soc.*, 246(1167):v+117, 2017.
23. Jheng-Jie Chen, Jungkai A. Chen, and Meng Chen. On quasismooth weighted complete intersections. *J. Algebraic Geom.*, 20(2):239–262, 2011.
24. Tom Coates, Alessio Corti, Sergey Galkin, Vasily Golyshev, and Alexander M. Kasprzyk. Mirror symmetry and Fano manifolds. In *European Congress of Mathematics*, pages 285–300. Eur. Math. Soc., Zürich, 2013.
25. Tom Coates, Alessio Corti, Sergey Galkin, and Alexander M. Kasprzyk. Quantum periods for 3-dimensional Fano manifolds. *Geom. Topol.*, 20(1):103–256, 2016.
26. Tom Coates, Liana Heuberger, Alexander M. Kasprzyk, and Giuseppe Pitton. Mirror symmetry and the classification of $\mathbb{Q}$-Fano threefolds. In preparation.
27. Tom Coates, Alexander M. Kasprzyk, and Giuseppe Pitton. Computing maximally mutable Laurent polynomials. In preparation.
28. Tom Coates, Alexander M. Kasprzyk, Giuseppe Pitton, and Ketil Tveiten. Maximally mutable Laurent polynomials. *Proc. of the Royal Society A.*, 477(2254):Paper No. 20210584, 21, 2021.
29. Tom Coates, Alexander M. Kasprzyk, and Thomas Prince. Four-dimensional Fano toric complete intersections. *Proc. of the Royal Society A.*, 471(2175):Paper No. 20140704, 14, 2015.
30. Tom Coates, Alexander M. Kasprzyk, and Thomas Prince. Laurent inversion. *Pure Appl. Math. Q.*, 15(4):1135–1179, 2019.
31. Alessio Corti. Factoring birational maps of threefolds after Sarkisov. *J. Algebraic Geom.*, 4(2):223–254, 1995.
32. Alessio Corti, Aleksandr Pukhlikov, and Miles Reid. Fano 3-fold hypersurfaces. In *Explicit birational geometry of 3-folds*, volume 281 of *London Math. Soc. Lecture Note Ser.*, pages 175–258. Cambridge Univ. Press, Cambridge, 2000.
33. Alessio Corti and Miles Reid. Weighted Grassmannians. In *Algebraic geometry*, pages 141–163. de Gruyter, Berlin, 2002.
34. Stephen Coughlan and Tom Ducat. Constructing Fano 3-folds from cluster varieties of rank 2. *Compos. Math.*, 156(9):1873–1914, 2020.
35. V. I. Danilov. The geometry of toric varieties. *Uspekhi Mat. Nauk*, 33(2(200)):85–134, 247, 1978.
36. A. J. de Jong, N. I. Shepherd-Barron, and A. Van de Ven. On the Burkhardt quartic. *Math. Ann.*, 286(1-3):309–328, 1990.
37. Pasquale Del Pezzo. Sulle superficie dell'n^{mo} ordine immerse nello spazio ad n dimensioni. *Rend. del Circolo Mat. di Palermo*, 1:241–255, 1887.
38. Igor Dolgachev. Weighted projective varieties. In *Group actions and vector fields (Vancouver, B.C., 1981)*, volume 956 of *Lecture Notes in Math.*, pages 34–71. Springer, Berlin, 1982.
39. Charles F. Doran and Andrew Harder. Toric degenerations and Laurent polynomials related to Givental's Landau-Ginzburg models. *Canad. J. Math.*, 68(4):784–815, 2016.
40. Tom Ducat. Constructing $\mathbb{Q}$-Fano 3-folds à la Prokhorov & Reid. *Bull. Lond. Math. Soc.*, 50(3):420–434, 2018.
41. Gino Fano. Nuove ricerche sulle varietà algebriche a tre dimensioni a curve-sezioni canoniche. *Pont. Acad. Sci. Comment.*, 11:635–720, 1947.
42. William Fulton. *Introduction to toric varieties*, volume 131 of *Annals of Mathematics Studies*. Princeton University Press, Princeton, NJ, 1993. The William H. Roever Lectures in Geometry.
43. Alexander Givental. A mirror theorem for toric complete intersections. In *Topological field theory, primitive forms and related topics (Kyoto, 1996)*, volume 160 of *Progr. Math.*, pages 141–175. Birkhäuser Boston, Boston, MA, 1998.
44. B. R. Greene and M. R. Plesser. Duality in Calabi-Yau moduli space. *Nuclear Phys. B*, 338(1):15–37, 1990.
45. Christopher D. Hacon and James McKernan. Existence of minimal models for varieties of log general type. II. *J. Amer. Math. Soc.*, 23(2):469–490, 2010.
46. Liana Heuberger. $\mathbb{Q}$-Fano threefolds and Laurent inversion. Preprint, 2022.
47. Kentaro Hori and Cumrun Vafa. Mirror symmetry. arXiv:hep-th/0002222, 2000.

48. A. R. Iano-Fletcher. Working with weighted complete intersections. In *Explicit birational geometry of 3-folds*, volume 281 of *London Math. Soc. Lecture Note Ser.*, pages 101–173. Cambridge Univ. Press, Cambridge, 2000.
49. V. A. Iskovskih. Fano threefolds. I. *Izv. Akad. Nauk SSSR Ser. Mat.*, 41(3):516–562, 717, 1977.
50. V. A. Iskovskih. Fano threefolds. II. *Izv. Akad. Nauk SSSR Ser. Mat.*, 42(3):506–549, 1978.
51. V. A. Iskovskih. Anticanonical models of three-dimensional algebraic varieties. In *Current problems in mathematics, Vol. 12 (Russian)*, pages 59–157, 239 (loose errata). VINITI, Moscow, 1979.
52. V. A. Iskovskih and Ju. I. Manin. Three-dimensional quartics and counterexamples to the Lüroth problem. *Mat. Sb. (N.S.)*, 86(128):140–166, 1971.
53. Jennifer M. Johnson and János Kollár. Fano hypersurfaces in weighted projective 4-spaces. *Experiment. Math.*, 10(1):151–158, 2001.
54. Anne-Sophie Kaloghiros. *The topology of terminal quartic 3-folds*. PhD thesis, University of Cambridge, 2007.
55. Alexander M. Kasprzyk. Toric Fano three-folds with terminal singularities. *Tohoku Math. J. (2)*, 58(1):101–121, 2006.
56. Alexander M. Kasprzyk. Canonical toric Fano threefolds. *Canad. J. Math.*, 62(6):1293–1309, 2010.
57. Alexander M. Kasprzyk and Benjamin Nill. Fano polytopes. In *Strings, gauge fields, and the geometry behind*, pages 349–364. World Sci. Publ., Hackensack, NJ, 2013.
58. Alexander M. Kasprzyk, Benjamin Nill, and Thomas Prince. Minimality and mutation-equivalence of polygons. *Forum Math. Sigma*, 5:Paper No. e18, 48, 2017.
59. In-Kyun Kim, Takuzo Okada, and Joonyeong Won. K-stability of birationally superrigid Fano 3-fold weighted hypersurfaces. arXiv:2011.07512 [math.AG], 2020.
60. János Kollár, Yoichi Miyaoka, and Shigefumi Mori. Rational connectedness and boundedness of Fano manifolds. *J. Differential Geom.*, 36(3):765–779, 1992.
61. János Kollár, Yoichi Miyaoka, Shigefumi Mori, and Hiromichi Takagi. Boundedness of canonical **Q**-Fano 3-folds. *Proc. Japan Acad. Ser. A Math. Sci.*, 76(5):73–77, 2000.
62. Maximilian Kreuzer and Harald Skarke. Classification of reflexive polyhedra in three dimensions. *Adv. Theor. Math. Phys.*, 2(4):853–871, 1998.
63. Maximilian Kreuzer and Harald Skarke. Complete classification of reflexive polyhedra in four dimensions. *Adv. Theor. Math. Phys.*, 4(6):1209–1230, 2000.
64. Shigefumi Mori. On a generalization of complete intersections. *J. Math. Kyoto Univ.*, 15(3):619–646, 1975.
65. Shigefumi Mori. Threefolds whose canonical bundles are not numerically effective. *Ann. of Math. (2)*, 116(1):133–176, 1982.
66. Shigefumi Mori. Flip theorem and the existence of minimal models for 3-folds. *J. Amer. Math. Soc.*, 1(1):117–253, 1988.
67. Shigefumi Mori and Shigeru Mukai. Classification of Fano 3-folds with $B_2 \geq 2$. *Manuscripta Math.*, 36(2):147–162, 1981/82.
68. Shigefumi Mori and Shigeru Mukai. Erratum: "Classification of Fano 3-folds with $B_2 \geq 2$". *Manuscripta Math.*, 110(3):407, 2003.
69. Mikkel Øbro. An algorithm for the classification of smooth Fano polytopes. arXiv:0704.0049 [math.CO], 2007.
70. Stavros Argyrios Papadakis and Miles Reid. Kustin-Miller unprojection without complexes. *J. Algebraic Geom.*, 13(3):563–577, 2004.
71. Joseph Polchinski. *String theory. Vol. II*. Cambridge Monographs on Mathematical Physics. Cambridge University Press, Cambridge, 2005. Superstring theory and beyond, reprint of 2003 edition.
72. Thomas Prince. Smoothing toric Fano surfaces using the Gross-Siebert algorithm. *Proc. Lond. Math. Soc. (3)*, 117(3):617–660, 2018.
73. Yuri Prokhorov. $\mathbb{Q}$-Fano threefolds of large Fano index, I. *Doc. Math.*, 15:843–872, 2010.
74. Yuri Prokhorov and Miles Reid. On $\mathbb{Q}$-Fano 3-folds of Fano index 2. In *Minimal models and extremal rays (Kyoto, 2011)*, volume 70 of *Adv. Stud. Pure Math.*, pages 397–420. Math. Soc. Japan, [Tokyo], 2016.

75. Victor Przyjalkowski. Hori-Vafa mirror models for complete intersections in weighted projective spaces and weak Landau-Ginzburg models. *Cent. Eur. J. Math.*, 9(5):972–977, 2011.
76. Muhammad Imran Qureshi and Balázs Szendrői. Constructing projective varieties in weighted flag varieties. *Bull. Lond. Math. Soc.*, 43(4):786–798, 2011.
77. Miles Reid. Canonical 3-folds. In *Journées de Géometrie Algébrique d'Angers, Juillet 1979/Algebraic Geometry, Angers, 1979*, pages 273–310. Sijthoff & Noordhoff, Alphen aan den Rijn—Germantown, Md., 1980.
78. Miles Reid. Graded rings and birational geometry. In *Proceedings of algebraic geometry symposium (Kinosaki, Oct 2000), K. Ohno (Ed.)*, pages 1–72. 2002. Available from http://www.warwick.ac.uk/~masda/3folds/.
79. Takeshi Sano. On classifications of non-Gorenstein **Q**-Fano 3-folds of Fano index 1. *J. Math. Soc. Japan*, 47(2):369–380, 1995.
80. Takeshi Sano. Classification of non-Gorenstein **Q**-Fano d-folds of Fano index greater than $d-2$. *Nagoya Math. J.*, 142:133–143, 1996.
81. Hiromichi Takagi. On classification of $\mathbb{Q}$-Fano 3-folds of Gorenstein index 2. I, II. *Nagoya Math. J.*, 167:117–155, 157–216, 2002.
82. Kiyohiko Takeuchi. Some birational maps of Fano 3-folds. *Compositio Math.*, 71(3):265–283, 1989.

Quantum Cluster Varieties, Scattering Diagrams, and the Deformed Poisson Families

Man Wai Mandy Cheung

Abstract In this survey article, we discuss the quantum $\mathcal{A}$ and $\mathcal{X}$ cluster varieties and indicate how to relate the two settings. We further state the counter-example to quantum positivity of quantum theta functions. At the end, we illustrate that the quantization of cluster $\mathcal{X}$ varieties can be extended to the deformation families of the $\mathcal{X}$ varieties and their fibers have Poisson structures.

1 Introduction

Cluster algebras, which were introduced by Fomin and Zelevinsky [1] as a tool for studying total positivity and dual canonical bases in Lie theory. On the geometry side, one can define cluster varieties as schemes such that the rings of regular functions carry cluster structures. There are two versions of mutation which gives the two types of cluster varieties $\mathcal{A}$ and $\mathcal{X}$. From there, Gross-Hacking-Keel-Konstevich [2] made use of the machinery, scattering diagrams and theta functions, from mirror symmetry to construct bases for the cluster algebras.

From the classical setting, Fock-Goncharov [3] constructed the quantum $\mathcal{X}$ cluster varieties while Berenstein-Zelevinsky [4] gave the quantum version of the $\mathcal{A}$ cluster algebras. One may wonder the relation between the two quantum constructions and if one can generalize the results in the classical setting to the quantum version.

This article is a survey for [5]. We will first outline the classical setting. Then we will lay out the quantum $\mathcal{A}$ and the $\mathcal{X}$ cluster algebras and show how to relate them. Then we will illustrate the construction of the quantum scattering diagrams and indicate the failure of quantum positivity in an example. At last, we will discuss the deformation families for the cluster $\mathcal{X}$ varieties in [6] and give a natural quantization of the deformed families. As a consequence, the families carry Poisson structures, as do the fibers.

M. W. Mandy Cheung (✉)
Harvard University, Cambridge, MA, USA
e-mail: mandywai24@gmail.com

Y.-H. He et al. (eds.), *Nankai Symposium on Mathematical Dialogues*,
https://doi.org/10.1007/978-981-19-2328-9_12

2 Background for the Classical Cluster Algebras

Cluster Algebras
Cluster algebras are subalgebras of rational functions in m variables. The generators of cluster algebras are called the cluster variables. Instead of being given the complete sets of generators and relations, a cluster algebra is defined from an (initial) seed $\mathbf{s}$, which includes a subset of the generators $\{A_1, \ldots, A_m\}$ and an exchange matrix B. The matrix B is a $m \times m$ matrix with the first $n \times n$ part is skew-symmetrizable [1, Definition 4.1], for $n \leq m$. This means that the cluster algebra is with n unfrozen indices and $(m - n)$ many frozen indices.

To obtain other generators, there is a procedure called mutation to obtain a new seed $\mu_k(\mathbf{s})$ from a given seed $\mathbf{s}$, where $k = 1, \ldots n$. The new seed $\mu_k(\mathbf{s})$ would be a new subset of generators $\{A_1, A_2, \ldots, A'_k, A_{k+1}, \ldots, A_m\}$, and another matrix B'. The new variable A'_k is obtained by the exchange relation

$$A_k A'_k = \prod_{B_{ik}>0} A_i^{B_{ik}} + \prod_{B_{ik}<0} A_i^{-B_{ik}}. \tag{1}$$

and

$$B'_{i,j} = \begin{cases} -B_{ij} & if\ k = i\ or\ k = j; \\ B_{ij} + \frac{|B_{ik}|B_{kj} + B_{ik}|B_{kj}|}{2} & otherwise. \end{cases} \tag{2}$$

Note that $\mu_k \circ \mu_k$ is an identity. Starting from an initial seed, one can apply all possible sequences of compositions of mutations (possibly infinite). This gives a set (possibly infinite) of all cluster variables generated under mutations. The $\mathcal{A}$ cluster algebra is then defined to be the ring generated by all cluster variables.

There is another type of cluster algebras which is called the $\mathcal{X}$ cluster algebras. The $\mathcal{X}$ cluster algebras arise from $\mathcal{A}$ cluster algebras with principal coefficients in [7]. We can see the $\mathcal{X}$ cluster algebra is defined from another type of initial seed with the variables $X_1, \ldots, X_n$ and the same exchange matrix B. Mutation at $k = 1, \ldots, n$ will give another seed with new $\mathcal{X}$ variables by the exchange relation

$$\mu_k^* X'_i = \begin{cases} X_k^{-1} & \text{for } i = k, \\ X_i (1 + X_k^{-\operatorname{sgn}(B_{ik})})^{-B_{ik}} & \text{for } i \neq k. \end{cases} \tag{3}$$

Then the $\mathcal{X}$ cluster algebras would be the algebras generated by all the $\mathcal{X}$ variables from the mutation process.

Cluster Varieties
Cluster varieties are schemes such that the ring of regular functions carries cluster structures. One can reinterpret the Fomin-Zelevinsky algebraic definition into geometric by translating seed to algebraic tori, and exchange relation to birational map between the torus. Let us first set up the notations for this reinterpretation.

A fixed data consists of a lattice N with a skew-symmetric bilinear form $\{\cdot,\cdot\}: N \times N \to \mathbb{Q}$, an index set I with $|I| = \operatorname{rank} N$, positive integers d_i for $i \in I$, a sublattice $N^\circ \subseteq N$ of finite index with some integral properties, the dual lattice $M = \operatorname{Hom}(N, \mathbb{Z})$ and the corresponding $M^\circ = \operatorname{Hom}(N^\circ, \mathbb{Z})$. One can refer to [8] for the full definition of fixed data. Consider $N_{\mathbb{R}} = N \otimes \mathbb{R}$ and $M_{\mathbb{R}} = M \otimes \mathbb{R}$.

Given this fixed data, a seed data for this fixed data is $\mathbf{s} := (e_i \in N \mid i \in I)$, where $\{e_i\}$ is a basis for N. The basis for M° would then be $f_i = \frac{1}{d_i} e_i^*$. Then one can define the exchange matrix ϵ, where $\epsilon_{ij} = \{e_i, e_j\} d_j$, which is the same as the matrix B in earlier section. We will keep the two notations for readers from different perspectives. One can then associate the seed tori

$$\mathcal{A}_{\mathbf{s}} = T_{N^\circ} = \operatorname{Spec} \Bbbk[M^\circ], \qquad \mathcal{X}_{\mathbf{s}} = T_M = \operatorname{Spec} \Bbbk[N].$$

We will denote the coordinates as $X_i = z^{e_i}$ and $A_i = z^{f_i}$ and they are called the *cluster variables*. Similar to the definition of cluster algebras, there is a procedure, called *mutation*, to produce a new seed data $\mu(\mathbf{s})$ from a given seed $\mathbf{s}$. The mutation formula is stated in [8, Equation 2.3] which we will skip here. The essence is that we will obtain new seed tori $\mathcal{A}_{\mu(\mathbf{s})}$, $\mathcal{X}_{\mu(\mathbf{s})}$ from the mutated seed. Between the tori, there are birational maps $\mu_{\mathcal{X}} : \mathcal{X}_{\mathbf{s}} \dashrightarrow \mathcal{X}_{\mu(\mathbf{s})}$, $\mu_{\mathcal{A}} : \mathcal{A}_{\mathbf{s}} \dashrightarrow \mathcal{A}_{\mu(\mathbf{s})}$ which are stated in [8, Equations 2.5, 2.6]. Note that those birational maps are basically the mutations of cluster variables as in Fomin and Zelevinsky stated in (3), (1).

Let $\mathcal{A}$ be an union of tori glued by $\mathcal{A}$-mutation $\mu_{\mathcal{A}}$. A smooth scheme V is a cluster variety of type $\mathcal{A}$ if there is a birational map $\mu : V \dashrightarrow \mathcal{A}$ which is an isomorphism outside codimension two subsets of the domain and range. The cluster variety of type $\mathcal{X}$ is defined analogously.

Cluster Varieties with Principal Coefficients

The $\mathcal{A}$ and $\mathcal{X}$ cluster varieties can be fit into the formalism of the cluster varieties with principal coefficients $\mathcal{A}_{\text{prin}}$. The scheme $\mathcal{A}_{\text{prin}}$ is defined similarly to the $\mathcal{A}$ by 'doubling' the fixed data, i.e. considering $\widetilde{N} = N \oplus M^\circ$ as fixed data as in [8, Construction 2.11]. Then there are two natural inclusions. The first one is

$$\begin{aligned} \widetilde{p}^* : N &\to \widetilde{M}^\circ = M^\circ \oplus N, \\ n &\mapsto (p^*(n), n), \end{aligned}$$

where $p^*(n) = \{n, \cdot\} \in M^\circ$ in the case of no frozen variable [8]. Then for any seed $\mathbf{s}$, note that $\mathcal{A}_{\text{prin},\mathbf{s}} = T_{\widetilde{N}^\circ}$, and $\mathcal{X}_{\mathbf{s}} = T_M$, the there is the exact sequence of tori

$$1 \to T_{N^\circ} \to \mathcal{A}_{\text{prin},\mathbf{s}} \xrightarrow{\widetilde{p}} \mathcal{X}_{\mathbf{s}} \to 1.$$

The map $\widetilde{p}$ commutes with the mutation maps and thus we get the morphism $\widetilde{p}\colon \mathcal{A}_{\text{prin}} \to \mathcal{X}$. Further the T_{N° action on $\mathcal{A}_{\text{prin},\mathbf{s}}$ extends to $\mathcal{A}_{\text{prin}}$ which makes $\widetilde{p}$ a quotient map. Thus, the $\mathcal{X}$ variety can be seen as $\mathcal{A}_{\text{prin}}/T_{N^\circ}$.

Scattering Diagrams and Theta Functions
Next, we will recall the framework of scattering diagrams and theta functions in the Gross-Hacking-Keel-Kontsevich [2] basis construction for cluster algebras.

First, a wall in $M_{\mathbb{R}}$ is a pair $(\mathfrak{d}, f_{\mathfrak{d}})$ where $\mathfrak{d} \subseteq M_{\mathbb{R}}$ is a convex rational polyhedral cone of codimension one, contained in $n^{\perp}$ for some $n \in N$, and $f_{\mathfrak{d}} = 1 + \sum_{k \geq 1} c_k z^{kp^*(n)}$, where $c_k \in \mathbb{C}$. A scattering diagram $\mathfrak{D}$ is then a collection of walls with certain finiteness conditions.

Given a seed, an $\mathcal{A}_{\text{prin}}$-cluster scattering diagram can be constructed [2] and canonically determined by this given seed data. The $\mathcal{A}$ scattering diagram can be obtained by the projection $\widetilde{M}_{\mathbb{R}} \to M_{\mathbb{R}}$ while the $\mathcal{X}$ scattering diagrams can be defined as slicing the $\mathcal{A}_{\text{prin}}$ scattering diagrams by considering $\{(m, n) \mid m = p^*(n)\}$.

Theta functions are defined by combinatorial object broken lines which are piecewise linear paths on the scattering diagrams together with decorating monomials at each linear segment. Each lattice point on scattering diagrams is associated to a theta function. One can see theta functions as extending how toric monomials associated to fans in toric geometry. With a convexity condition [2], the free module generated by theta functions is endowed with an algebra structure which would agree with the Fomin-Zelevinsky case.

3 Quantum Cluster Algebras

Quantum ($\mathcal{A}$-)cluster algebras were introduced by Berenstein and Zelevinsky in [4] with additional data, a *compatible pair*. On the other hand, Fock and Goncharov [9] discuss the quantization of the $\mathcal{X}$ varieties.

Quantum $\mathcal{X}$-varieties

We again begin with fixed data Γ and a seed $\mathbf{s}$ in previous section. Consider the noncommutative torus

$$\mathcal{T}_N := \mathbb{C}[q^{\pm \frac{1}{d}}]\langle X^n \ : n \in N,\ X^n \cdot X^{n'} = q^{\{n,n'\}} X^{n+n'} \rangle.$$

Thus, we have

$$q^{\{n',n\}} X^n \cdot X^{n'} = q^{\{n,n'\}} X^{n'} \cdot X^n. \tag{4}$$

Quantum Dilogarithm
To construct a non-commutative deformation of the space $\mathcal{X}$, the quantum space $\mathcal{X}_q$, Fock and Goncharov [3] consider the *quantum dilogarithm* defined by

$$\Psi_q(x) = \prod_{\ell=1}^{\infty} \left(1 + q^{2\ell-1}x\right)^{-1}.$$

Let $\mathrm{Li}_2\,(x; q) := \sum_{\ell \geq 1} \frac{x^\ell}{\ell(q^\ell - q^{-\ell})}$. One can check that [3, Section 3.2]

$$\Psi_q\,(x) = \exp\left(-\mathrm{Li}_2\,(-x; q)\right). \tag{5}$$

For each seed $\mathbf{s}$ we will have a copy of $\mathcal{T}_N$, which we denote $\mathcal{T}_{N;\mathbf{s}}$. Since $\mathcal{T}_{N;\mathbf{s}}$ is an Ore domain, we can take its noncommutative fraction field $\mathbb{T}_{N;\mathbf{s}}$. The *quantum mutation map* μ_k^q, $k \in I_{\mathrm{uf}}$, is the isomorphism $\mu_k^q : \mathbb{T}_{N;\mathbf{s}'} \to \mathbb{T}_{N;\mathbf{s}}$ given by conjugation by $\Psi_{q^{1/d_k}}\left(X_{k;\mathbf{s}}\right)$,

$$\mu_k^q := \mathbf{Ad}_{\Psi_{q^{1/d_k}}(X_{k;\mathbf{s}})}.$$

Explicitly, in terms of cluster coordinates,

$$\mu_k^q\left(X_{i;\mathbf{s}'}\right) = \begin{cases} X_{i;\mathbf{s}}^{-1} & \text{if } i = k \\ X_{i;\mathbf{s}} \left(\prod_{\ell=1}^{|\epsilon_{ik}|}\left(1 + q^{(2\ell-1)/d_k} X_{k;\mathbf{s}}^{-\operatorname{sgn}\epsilon_{ik}}\right)\right)^{-\operatorname{sgn}(\epsilon_{ik})} & \text{if } i \neq k. \end{cases} \tag{6}$$

Quantum $\mathcal{A}$-cluster Algebras

Now we recall Berenstein and Zelevinsky's treatment of quantum $\mathcal{A}$-cluster algebras from [4]. Let Λ be an $|I| \times |I|$ skew-symmetric integer matrix, and let $\widetilde{B}$ be an $|I| \times |I_{\mathrm{uf}}|$ integer matrix. The pair $\left(\Lambda, \widetilde{B}\right)$ is called *compatible* if $\widetilde{B}^T \Lambda = \left(D\ 0\right)$, for a diagonal matrix D. Berenstein and Zelevinsky show [4, Proposition 3.3] that the pair $(\Lambda, \widetilde{B})$ being compatible implies $\widetilde{B}$ is of full rank and the principal $|I_{\mathrm{uf}}| \times |I_{\mathrm{uf}}|$ submatrix B is skew-symmetrizable by noting $DB = \widetilde{B}^T \Lambda \widetilde{B}$. A compatible pair is the input for the Berenstein-Zelevinsky quantum $\mathcal{A}$-cluster algebra construction. Note in particular that while we can always quantize an $\mathcal{X}$-variety, if the exchange matrix $\widetilde{B}$ (which is $\epsilon^t_{I_{\mathrm{uf}} \times I}$ in the Fock-Goncharov formalism) is not full rank, then $\mathcal{A}$ does not admit a quantization by this procedure.

We view Λ as a skew form on M° and define $\mathcal{T}_{M^\circ}$ as

$$\mathcal{T}_{M^\circ} := \mathbb{C}[q^{\pm\frac{1}{2}}]\langle A^m \ :\ m \in M^\circ,\ A^m \cdot A^{m'} = q^{\frac{1}{2}\Lambda(m,m')} A^{m+m'}\rangle.$$

Denote the noncommutative fraction field of $\mathcal{T}_{M^\circ;\mathbf{s}}$ by $\mathbb{T}_{M^\circ;\mathbf{s}}$. We denote $A^{f_{i;\mathbf{s}}} \in \mathcal{T}_{M^\circ;\mathbf{s}}$ by $A_{i;\mathbf{s}}$. The quantum $\mathcal{A}$-mutation defining $\mathcal{A}_q$ is the $*$-algebra isomorphism $\mathbb{T}_{M^\circ;\mathbf{s}'} \to \mathbb{T}_{M^\circ;\mathbf{s}}$ given by [4, Proposition 4.9]

$$\mu_k^q\left(A_{i;\mathbf{s}'}\right) = \begin{cases} A^{-f_{k;\mathbf{s}} + \sum_{j:\epsilon_{kj}>0} \epsilon_{kj} f_{j;\mathbf{s}}} + A^{-f_{k;\mathbf{s}} - \sum_{j:\epsilon_{kj}<0} \epsilon_{kj} f_{j;\mathbf{s}}} & \text{for } i = k \\ A_{i;\mathbf{s}} & \text{for } i \neq k \end{cases}.$$

3.1 *Relating the $\mathcal{A}$ and $\mathcal{X}$ Quantum Cluster Algebras*

After recalling the quantum $\mathcal{A}$ and $\mathcal{X}$ cluster algebras, one may wonder, if there is a p^*-map in the quantum setting to put the quantum $\mathcal{A}$ and $\mathcal{X}$ cluster algebras in a unifying way. The answer is affirmative:

Proposition 3.1 ([5]) *After making the identification $q_{\mathrm{FG}}^{\frac{1}{d}} = q_{\mathrm{BZ}}^{-\frac{1}{2}}$, a p^*-map on the level of lattices induces a $*$-algebra homomorphism from the quantum $\mathcal{X}$-torus algebra over $R_{q_{\mathrm{FG}}} := \mathbb{C}[t_i : i \in I]\left[q_{\mathrm{FG}}^{\pm\frac{1}{d}}\right]$ to the quantum $\mathcal{A}$-torus algebra over $R_{q_{\mathrm{BZ}}} := \mathbb{C}[t_i : i \in I]\left[q_{\mathrm{BZ}}^{\pm\frac{1}{2}}\right]$ that commutes with mutation:*

$$p^*\left(\mu^q_{k,\mathbf{t};\mathcal{G}}\left(\widetilde{X}_{i;\mathcal{G}'}\right)\right) = \mu^q_{k,\mathbf{t};\mathcal{G}}\left(p^*\left(\widetilde{X}_{i;\mathcal{G}'}\right)\right),$$

where $\mathcal{G}$ denotes the maximal cone (chamber) of the $\mathcal{A}$ scattering diagram. That is, there is a map of quantum cluster varieties $p : \mathscr{A}_{\mathrm{prin},q} \to \mathscr{X}_q$ from the Berenstein-Zelevinsky quantum $\mathcal{A}$-variety with principal coefficients to the Fock-Goncharov quantum $\mathcal{X}$-variety with principal coefficients in the sense of [6].

The statement above is stated with variables t_i for later discussion. We can assume all the $t_i = 1$ for the classical situation.

4 Quantum Scattering Diagrams

The next question is that if we can extend the scattering diagrams and theta bases construction to the quantum case. The key lies on (5) by noting the quantum mutations are coming from the quantum dilogarithm. Then one can repeat almost the same setup as the non-quantum case but only considering the (initial) quantum $\mathcal{X}$ scattering diagram

$$\widehat{\mathfrak{D}}_{\mathrm{in,s}} := \left\{\left(v_i^{\perp}, \Psi_{q^{1/d_i}}\left(X^{e_i}\right)\right) \mid i \in I_{\mathrm{uf}}\right\},$$

where $v_i = p^*(e_i)$. Then by [2, 10], there is a unique (up to equivalence) consistent scattering diagram $\widehat{\mathfrak{D}}_{\mathbf{s}}$ containing $\widehat{\mathfrak{D}}_{\mathrm{in,s}}$, with only outgoing walls in $\widehat{\mathfrak{D}}_{\mathbf{s}} \setminus \widehat{\mathfrak{D}}_{\mathrm{in,s}}$. The resulting consistent scattering diagrams would be the quantum $\mathcal{X}$ scattering diagrams where one can define the quantum $\mathcal{X}$ theta functions.

For the $\mathcal{A}$-cluster algebra, if the compatible pair exists, one can put Λ as the skew-symmetric form in the construction. Then one can similarly define the quantum $\mathcal{A}$ scattering diagram. Instead of repeating the whole construction, we will demonstrate the quantum $\mathcal{A}$ scattering diagram as an example in the next subsection.

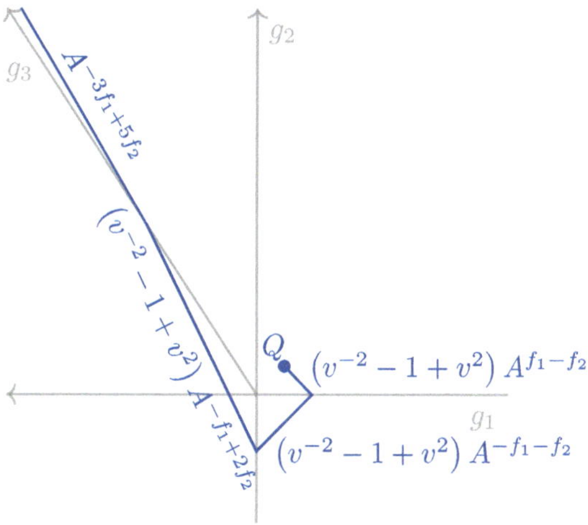

Fig. 1 The broken line giving the $A^{f_1-f_2}$ summand of $\vartheta_{-3f_1+5f_2}$. Further $q_{\mathrm{BZ}} = v^{-2}$ to adopt the notation of [14]

4.1 Non-positivity for Theta Functions

Davison-Mandel [11] showed the strong positivity conjecture for quantum theta bases for skew-symmetric type quantum cluster algebras. On the other hand, there are hints that the positivity fails for the general skew-symmetrizable case by combining the ideas from [12, 13]. The failure can be verified explicitly in the quantum rank two cluster algebra $\mathcal{A}(2,3)$ and the theta function $\vartheta_{(-3,5)}$ of $\mathcal{A}(2,3)$. The broken line with negative quantum coefficient is stated in the following figure in [5] (Fig. 1).

Hence, we have

Proposition 4.1 ([5]) *Quantum positivity fails for the quantum theta basis. That is, for a quantum cluster algebra of skew-symmetrizable type, there may be quantum theta functions which, when expressed in terms of initial quantum cluster variables, have coefficients which are in* $\mathbb{Z}[q^{\pm\frac{1}{2}}]$ *but not* $\mathbb{Z}_{\geq 0}[q^{\pm\frac{1}{2}}]$.

5 Quantization of Deformed $\mathcal{X}$ Varieties

While the schemes $\mathcal{A}_{\mathrm{prin}}$ give the deformations of the $\mathcal{A}$ varieties with suitable convexity condition in [2], one cannot perform analogously in the $\mathcal{X}$ case. The $\mathcal{X}$-mutation with principal coefficients defining the corresponding family of deformations $\mathscr{X}$ of $\mathcal{X}$ from [6] is given by

$$\mu_k^*\left(\widetilde{X}_{i;\mathbf{s}'}\right) = \begin{cases} \widetilde{X}_{i;\mathbf{s}}^{-1} & \text{for } i = k \\ \widetilde{X}_{i;\mathbf{s}}\left(\mathbf{t}^{[\operatorname{sgn}(\epsilon_{ik})\mathbf{c}_{k;\mathbf{s}}]_+} + \mathbf{t}^{[-\operatorname{sgn}(\epsilon_{ik})\mathbf{c}_{k;\mathbf{s}}]_+}\widetilde{X}_{k;\mathbf{s}}^{-\operatorname{sgn}(\epsilon_{ik})}\right)^{-\epsilon_{ik}} & \text{for } i \neq k \end{cases}, \tag{7}$$

where $\mathbf{c}_{k;\mathbf{s}}$ is the $\mathbf{c}$-vector in [7]. Note that $\mathscr{X}$ is *not* the scheme $\mathcal{X}_{\text{prin}}$ of [2]. The difference stems from treatment of coefficients and frozen variables. In the skew-symmetric case, $\mathscr{X}$ is dual to $\mathcal{A}_{\text{prin}}$ as cluster varieties over $\mathbb{C}\left[t_i^{\pm 1} : i \in I\right]$, while $\mathcal{X}_{\text{prin}}$ is dual to $\mathcal{A}_{\text{prin}}$ as cluster varieties over $\mathbb{C}$.

A natural question at this moment would be if we can similarly give a quantum version of this deformation family:

Proposition 5.1 ([5]) *The quantum $\mathcal{X}$-mutation with coefficients $\mu^q_{k,\mathbf{t};\mathcal{G}} : \mathcal{K}\left(\mathbb{A}^{|I|}_{M;\mathcal{G}',q}(R)\right) \to \mathcal{K}\left(\mathbb{A}^{|I|}_{M;\mathcal{G},q}(R)\right)$ is given in cluster coordinates by*

$$\mu^q_{k,\mathbf{t};\mathcal{G}}\left(\widetilde{X}_{i;\mathcal{G}'}\right) = \begin{cases} \widetilde{X}^{-1}_{i;\mathcal{G}} & \text{if } i = k, \\ \widetilde{X}_{i;\mathcal{G}} \left(\prod_{\ell=1}^{|\epsilon_{ik}|} \left(\mathbf{t}^{[\operatorname{sgn}(\epsilon_{ik})\mathbf{c}_{k;\mathcal{G}}]_+} + \mathbf{t}^{[-\operatorname{sgn}(\epsilon_{ik})\mathbf{c}_{k;\mathcal{G}}]_+} q_k^{2\ell-1} \widetilde{X}^{-\operatorname{sgn}(\epsilon_{ik})}_{k;\mathcal{G}} \right) \right)^{-\operatorname{sgn}(\epsilon_{ik})} & \text{if } i \neq k. \end{cases} \tag{8}$$

Corollary 5.2 ([5]) *Setting $\mathbf{t} = \mathbf{1}$ we recover the coefficient free quantum mutation formula* (6)*. Setting $q = 1$, we recover the degeneration formula* (7)*.*

As a result, one can see that the fibers of the families have Poisson structures.

Proposition 5.3 ([5]) *$\widehat{\mathscr{X}}$ is a Poisson scheme over R, with the Poisson structure inherited from the Poisson structure on affine patches $\mathbb{A}^{|I|}_{M;\mathcal{G},q}(R) := \operatorname{Spec}\left(R\left[\widetilde{X}_{i;\mathcal{G}} : i \in I\right]\right)$:*

$$\left\{\widetilde{X}^{v_1}_{\mathcal{G}}, \widetilde{X}^{v_2}_{\mathcal{G}}\right\}_{\mathcal{G}} := 2\left\{v_1, v_2\right\} \widetilde{X}^{v_1+v_2}_{\mathcal{G}}.$$

Acknowledgements The author would like to thank the Chern Institute of Mathematics for organizing the exciting conference 'Nankai Symposium on Mathematical Dialogues'.

References

1. S. Fomin and A. Zelevinsky. Cluster algebras I: Foundations. *J. Amer. Math. Soc.*, 15:497–529, 2002.
2. Mark Gross, Paul Hacking, Sean Keel, and Maxim Kontsevich. Canonical bases for cluster algebras. *J. Amer. Math. Soc.*, 31(2):497–608, 2018.
3. Vladimir V. Fock and Alexander B. Goncharov. Cluster ensembles, quantization and the dilogarithm. *Ann. Sci. Éc. Norm. Supér. (4)*, 42(6):865–930, 2009.
4. Arkady Berenstein and Andrei Zelevinsky. Quantum cluster algebras. *Advances in Mathematics*, 195(2):405–455, 2005.
5. Man-Wai Mandy Cheung, Juan Bosco Frías-Medina, and Timothy Magee. Quantization of deformed cluster poisson varieties. arXiv preprint arXiv:2007.02479, 2020.
6. Lara Bossinger, Bosco Frías-Medina, Timothy Magee, and Alfredo Nájera Chávez. Toric degenerations of cluster varieties and cluster duality. *Compos. Math.*, 156(10):2149– 2206, 2020.

7. Sergey Fomin and Andrei Zelevinsky. Cluster algebras IV: Coefficients. *Compositio Mathematica*, 143(1):112–164, 2007.
8. Mark Gross, Paul Hacking, and Sean Keel. Birational geometry of cluster algebras. *Algebr. Geom.*, 2(2):137–175, 2015.
9. Vladimir V. Fock and Alexander B. Goncharov. The quantum dilogarithm and representations of quantum cluster varieties. *Invent. Math.*, 175(2):223–286, 2009.
10. Maxim Kontsevich and Yan Soibelman. Wall-crossing structures in Donaldson-Thomas invariants, integrable systems and mirror symmetry. In *Homological mirror symmetry and tropical geometry*, volume 15 of *Lect. Notes Unione Mat. Ital.*, pages 197–308. Springer, Cham, 2014.
11. Ben Davison and Travis Mandel. Strong positivity for quantum theta bases of quantum cluster algebras. arXiv preprint arXiv:1910.12915, 2019.
12. Man Wai Cheung, Mark Gross, Greg Muller, Gregg Musiker, Dylan Rupel, Salvatore Stella, and Harold Williams. The greedy basis equals the theta basis: a rank two haiku. *Journal of Combinatorial Theory, Series A*, 145:150–171, 2017.
13. Kyungyong Lee, Li Li, Dylan Rupel, and Andrei Zelevinsky. The existence of greedy bases in rank 2 quantum cluster algebras. *Advances in Mathematics*, 300:360–389, 2016.
14. Kyungyong Lee, Li Li, Dylan Rupel, and Andrei Zelevinsky. Greedy bases in rank 2 quantum cluster algebras. *Proceedings of the National Academy of Sciences*, 111(27):9712–9716, 2014.

(K)not Machine Learning

Jessica Craven, Mark Hughes, Vishnu Jejjala, and Arjun Kar

Abstract We review recent efforts to machine learn relations between knot invariants. Because these knot invariants have meaning in physics, we explore aspects of Chern–Simons theory and higher dimensional gauge theories. The goal of this work is to translate numerical experiments with Big Data to new analytic results.

1 Introduction

In numerous settings, machine learning successfully identifies associations between features in large data sets. Neural networks, which are often used to model these relationships, are universal approximators [1, 2]. They are also black boxes. Even when the predictions are highly accurate, we typically do not know how the machine learns. In certain examples in algebraic geometry, the way the training time scales with the size of the input is sufficient to surmise that the machine makes its prediction in a different manner than how a human would perform the calculation [3, 4]. This suggests that there are better ways to calculate [5, 6]. As theoretical physicists and mathematicians, we aim to obtain simple analytic expressions that are descriptive of a system of interest and to refine our methods of computation. We seek to promote machine learning to a discovery tool.

J. Craven · V. Jejjala (✉)
Mandelstam Institute for Theoretical Physics, School of Physics, NITheCS, and CoE-MaSS, University of the Witwatersrand, Johannesburg, South Africa
e-mail: vishnu@neo.phys.wits.ac.za

J. Craven
e-mail: jessica.craven1@students.wits.ac.za

M. Hughes
Department of Mathematics, Brigham Young University, Provo, UT, USA
e-mail: hughes@mathematics.byu.edu

A. Kar
Department of Physics and Astronomy, University of British Columbia, Vancouver, BC, Canada
e-mail: arjunkar@phas.ubc.ca

Y.-H. He et al. (eds.), *Nankai Symposium on Mathematical Dialogues*,
https://doi.org/10.1007/978-981-19-2328-9_13

Fig. 1 Trefoil and figure-eight knots. These figures are from [15]

One advantage of the large data sets in mathematics and formal theory is that they are clean. There is no measurement error or noise infecting the data. Looking at such exact data from a phenomenological angle has historically yielded new insights. Mirror symmetry in Calabi–Yau manifolds was an experimental observation about string compactification and conformal field theories [7–10] before it became a mathematical fact [11–13]. There are half a billion four dimensional reflexive polytopes with which to explore this relationship further [14]. Similarly large data sets exist in knot theory. As with Calabi–Yau manifolds, the knot theory data sets can be approached on purely mathematical grounds or in terms of theoretical physics.

A knot is a circle S^1 embedded in a three manifold $\mathcal{M}$, which for simplicity, we may take to be S^3. Figure 1 depicts the trefoil and figure-eight knots, with three and four crossings, respectively. Because the representation of a knot is not unique, we must use topological data to distinguish one knot from another. These topological invariants are tabulated in several databases [16, 17] or can be calculated from code [18, 19]. They can also be constructed in quantum field theory. The existing knot invariants are not independent. Some relations between them are known, while others can be experimentally found and mathematically proved, sometimes using artificial intelligence [20, 21].

In this article, which summarizes the results of [22–24], we explore new relations between knot invariants. In Sect. 2, we introduce the relevant knot invariants. In Sect. 3, we apply machine learning to the data. In Sect. 4, we outline targets for future work in this area.

2 Dramatis Personæ

The *Jones polynomial* is a Laurent polynomial computed using the Kauffman bracket [25]. While the mathematical definition is intrinsically two dimensional, Witten established that this polynomial knot invariant has a physical meaning as the unknot normalized vacuum expectation value of a Wilson loop operator in three dimensional $SU(2)$ Chern–Simons theory [26]:

$$J_n(K;q) = \frac{\int_{\mathcal{U}} [DA]\, U_n(K) e^{iS_{\text{CS}}}}{\int_{\mathcal{U}} [DA]\, U_n(\bigcirc) e^{iS_{\text{CS}}}} = \frac{\langle U_n(K)\rangle}{\langle U_n(\bigcirc)\rangle} , \tag{2.1}$$

$$S_{\text{CS}} = \frac{k}{4\pi} \int_{\mathcal{M}} \text{Tr} \left(A \wedge dA + \frac{2}{3} A \wedge A \wedge A \right) . \tag{2.2}$$

The path integral is taken over the space of $SU(2)$ connections modulo gauge transformations; this is denoted $\mathcal{U}$. To ensure that the action is gauge invariant, the coupling k, or Chern–Simons level, is integer quantized. The Wilson loops are

$$U_R(\gamma) = \text{Tr}_R\, \mathcal{P} \exp \left(i \oint_\gamma A \right) \tag{2.3}$$

and denote the trace in the representation R of the path ordered exponential of the holonomy of the gauge connection along some curve $\gamma \subset \mathcal{M}$, which in (2.1) we take to be the knot K and the unknot $\bigcirc$. The ordinary Jones polynomial $J_2(K;q)$ corresponds to an evaluation of (2.1) at $q = e^{\frac{2\pi i}{k+2}}$, with $n = 2$ denoting the fundamental representation of $SU(2)$. Because it is a topological invariant, the expression is independent of the metric on $\mathcal{M}$. The Jones polynomial does not uniquely identify a knot. In fact, it is not known whether the Jones polynomial even uniquely identifies the unknot.

The coefficients in the Jones polynomial are integer valued. This is clear from the mathematical construction but surprising from the definition using Chern–Simons theory. This is also evident through the work of Khovanov, who defined bigraded groups $\mathcal{H}(K)$ for any knot K [27, 28]. The information in Khovanov's homology theory can be captured by the *Khovanov polynomial*:

$$\text{Kh}(K;q,t) = \sum_{m,n} \dim \mathcal{H}^{m,n}(K) t^m q^n , \tag{2.4}$$

where m and n are, respectively, the homological grading and the quantum grading. Because the coefficients are the dimensions of certain homology groups, they are integer valued. A specialization of the Khovanov polynomial recovers the Jones polynomial:

$$J_2(K;q^2) = \frac{\text{Kh}(K;q,-1)}{q+q^{-1}} . \tag{2.5}$$

As the denominator is the Khovanov polynomial of the unknot, this expression for the Jones polynomial is appropriately normalized. While Khovanov's definition of $\mathcal{H}^{m,n}(K)$ in (2.4) is also two dimensional, it is conjectured to be related to quantum field theories in higher dimensions. From a four dimensional perspective, there is a super-Yang–Mills path integral that counts classical supersymmetric solutions weighted by $(-1)^F q^I$, where the homological grading is identified with the fermion number and the quantum grading with the instanton number [29]. Going up a dimension allows for a Hilbert space interpretation of $\mathcal{H}(K)$. In particular, the space $\mathcal{H}(K)$ is given by the Q-cohomology in the Hilbert space of a five dimensional super-Yang–

Mills theory, with a bigrading given by a $U(1) \times U(1)$ symmetry that is generated by fermion and instanton number operators [30]. This five dimensional theory is not ultraviolet complete, and there is as well a six dimensional description with $\mathcal{H}(K)$ the cohomology of a supercharge Q in the (0, 2) M5-brane theory on a Cauchy slice with surface operator Σ_K whose topology is set by the knot [30].

Many knots—and nearly every knot at small crossing number—are hyperbolic. For such knots, when we excise an ϵ-size tubular neighborhood around the knot, the knot complement $S^3 \setminus K$ admits a unique constant negative curvature metric [31]. Thurston showed that the *hyperbolic volume* computed using this metric is a knot invariant that can be conveniently calculated from a tetrahedral decomposition [32]. Complexifying the gauge group of the Chern–Simons theory from $SU(2)$ to $SL(2, \mathbb{C})$, the volume appears as the saddle point in the partition function [33]

$$\mathcal{Z}(\mathcal{M}) = \int_{\mathcal{U}_{\mathbb{C}}} [D\mathcal{A}][D\bar{\mathcal{A}}] \exp\left(\frac{it}{2}\mathcal{W}(\mathcal{A}) + \frac{i\tilde{t}}{2}W(\bar{\mathcal{A}})\right) . \tag{2.6}$$

The critical point corresponds to a flat $SL(2, \mathbb{C})$ connection $\mathcal{A}_+$ known as the geometric conjugate connection:

$$W(\mathcal{A}_+) = -\frac{i}{2\pi}\,\mathrm{Vol}(S^3 \setminus K) + \pi \mathrm{CS}(S^3 \setminus K) . \tag{2.7}$$

(The second term in (2.7) is proportional to the Chern–Simons invariant of the knot.) The complexification serves as a physical motivation for the volume conjecture [34–37], which states

$$\mathrm{Vol}(S^3 \setminus K) = \lim_{n\to\infty} \frac{2\pi \log |J_n(K; \omega_n)|}{n} , \qquad \omega_n = e^{\frac{2\pi i}{n}} . \tag{2.8}$$

This is implicitly at large-k as well.

Two other mathematical invariants are the *Rasmussen s-invariant* and the *smooth slice genus g*. The s-invariant is defined using the Lee spectral sequence [38] which collapses $\mathcal{H}(K)$ to $\mathbb{Q} \oplus \mathbb{Q}$ with instanton grading $s \pm 1$ and $s \in 2\mathbb{Z}$ [39]. The slice genus of a knot K is the least integer g for which there is a smooth orientable surface $\Sigma \subset B^4$ with genus g such that $K = \partial\Sigma \subset S^3$. The slice genus constrains the s-invariant [39]:

$$|s| \leq 2g . \tag{2.9}$$

These are four dimensional invariants. To date, despite notable efforts [40, 41], there is no simple gauge theoretic interpretation of these quantities.

3 Learning from Polynomial Invariants

Fully connected neural networks will be our principal machine learning tool. For our purposes, two hidden layer networks with $\mathcal{O}(100)$ neurons per layer are sufficient to extract relationships between knot invariants. As the results are not especially sensitive to the architecture of the network or the precise non-linearities introduced, the results could be interpretable. Indeed, this is our objective.

There are 313, 209 hyperbolic knots up to fifteen crossings. The Jones polynomials have up to sixteen coefficients. Including the minimum and maximum degree, there are eighteen input features. Using only 10% of the data for training, a neural network accomplishes $2.45 \pm 0.10\%$ mean relative error in its predictions of the volume. Figure 2 plots the true volume vs. the neural network prediction. When knots with the same Jones polynomial have knot complements with different volumes, the volumes differ by about 2.83%. Thus, the neural network predicts the volume near the theoretical upper limit of performance.

The Jones polynomial is defined through Chern–Simons theory. It is a polynomial invariant that is intrinsically quantum in character. In contrast, the volume of the knot complement is a classical knot invariant. Its connection to the quantum field theory arises through the volume conjecture only in the large-n limit. Nevertheless, one predicts the other. Why?

It turns out that the degrees of the polynomials are superfluous data for an accurate prediction. This suggests that the evaluation of the Jones polynomial at particular phases can also predict the volume. By feeding a neural network a set of these evaluations, we use layer-wise relevance propagation [42] to assign a relevance score to each neuron and to each input feature. The evaluation of the Jones polynomial at the phase $e^{\frac{3\pi i}{4}}$ has the largest relevance score. From regression, we find that the formula

$$\mathrm{Vol}(S^3 \setminus K) = 6.20 \log(|J_2(K; e^{\frac{3\pi i}{4}})| + 6.77) - 0.94 \tag{3.1}$$

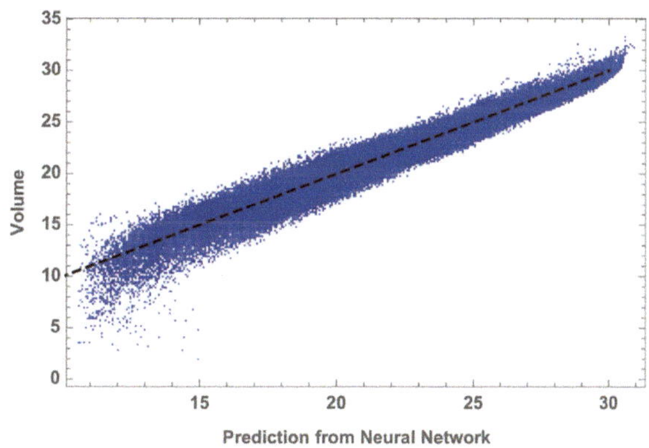

Fig. 2 Prediction of hyperbolic volume from Jones polynomial using knots up to 15 crossings with training fraction 0.1. This figure appears in [22]

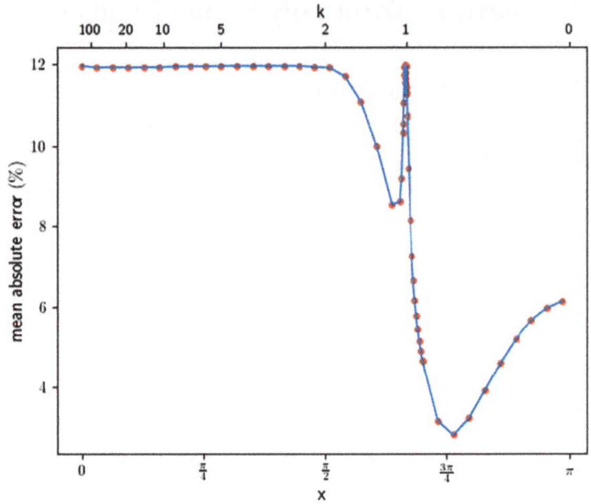

Fig. 3 Regression to predict volume based on evaluation of the Jones polynomial at different phases. Using $q = e^{\frac{2\pi i}{k+2}} = e^{ix}$, the phase translates to a fractional Chern–Simons level. This figure appears in [23]

gives an error of only 2.86% on 1, 701, 903 hyperbolic knots up to sixteen crossings. Equating the phase to $q = e^{\frac{2\pi i}{k+2}}$, this corresponds to a fractional Chern–Simons level $k = \frac{2}{3}$. In Fig. 3, we plot the best fit of the form in (3.1) against the argument of the phase.

This is interpretable in terms of the analytically continued Chern–Simons theory [33]. The approximation formula works well for the levels k for which the geometric conjugate connection $\mathcal{A}_+$ makes a contribution to the Chern–Simons path integral, and the accuracy of the prediction increases with the fraction of the knots in the data set that receive such a contribution. The plateau at $k > 2$ corresponds to latent correlations in the data set, viz., equating the volume of every knot to the mean volume of knots in the data set gives a 12% error. The minimum occurs near $k = \frac{2}{3}$. At integer values of k, the path integral receives contributions only from $SU(2)$ valued critical points, meaning that we lose information from the geometric conjugate connection. This explains the spike at $k = 1$.

Working with a data set comprised of 535, 239 knots for which we know the Khovanov and Jones polynomials, the s-invariant, and for which we know the slice genus in 438, 295 cases, we find that the Khovanov polynomial correctly predicts both the Rasmussen s-invariant and the slice genus more than 98% of the time. The Jones polynomial, respectively, predicts the Rasmussen s-invariant and the slice genus correctly 95% and 97% of the time. There is no known way of calculating the slice genus from the Khovanov polynomial. The success of the Khovanov polynomial in predicting the s-invariant is somewhat less surprising. The knight move conjecture [28], which is false [43], is satisfied by all the knots in our data set. It states

$$\mathrm{Kh}(K; q, -q^{-4}) = q^s(q + q^{-1}) . \tag{3.2}$$

An evaluation of the Khovanov polynomial at $t = -q^{-4}$ is successful at predicting the s-invariant 99.77% of the time. Surprisingly, an evaluation at $t = -q^{-2}$ is successful 99.88% of the time. In fact, the polynomials often have particular forms:

$$\mathrm{Kh}(K; q, -q^{-2}) = \left(\frac{s}{2} + 1\right)q^{s-1} - \left(\frac{s}{2} - 1\right)q^{s+1}, \tag{3.3}$$

$$\mathrm{Kh}(K; q, -q^{-2}) = -q^{s-1} + \left(\frac{s}{2} + 5\right)q^{s+1} - \left(\frac{s}{2} + 2\right)q^{s+3}. \tag{3.4}$$

The exponents being dependent on s is a mathematical renormalization of the Khovanov polynomial. Why the s-invariant appears in the coefficients is theoretically unmotivated. However, the experiments suggest that the s-invariant is encoded in the Khovanov polynomial in more than one way. The ability of the Jones polynomial to predict the Rasmussen s-invariant and the slice genus is a mystery. The lower bound in the inequality (2.9) is saturated by $414,615$ knots, so it is also not clear from numerical experiments whether the network is learning s or g.

4 Prospectus

Ever since [44], knot invariants have been a subject of machine learning. Besides the papers we have reviewed here, other machine learning investigations on related data sets appear in [20, 45–47]. Our efforts point to deep relations in mathematics and physics. Perhaps there are algebraic formulas for the Rasmussen s-invariant and the slice genus. The knot invariants we examine are realized in quantum field theories in different dimensions. Understanding how these quantum field theories are related is work in progress. As well, [48] proposed a strategy for finding counterexamples to the smooth four dimensional Poincaré conjecture by looking for slice genus zero knots with certain properties. A neural network that accurately discriminates $g = 0$ knots may prove useful in this endeavor. Man and machine continue thinking about these problems.

Acknowledgements We thank Onkar Parrikar for collaboration on [22]. The work of JC and VJ is based on research supported in part by the South African Research Chairs Initiative of the National Research Foundation, grant number 78554. VJ is additionally supported by a Simons Foundation Mathematics and Physical Sciences Targeted Grant, 509116. AK is supported by the Simons Foundation through the It from Qubit Collaboration. These proceedings are based on a talk by VJ at the Nankai Symposium on Mathematical Dialogues held at the Chern Institute of Mathematics in August 2021. We are grateful to the organizers of this meeting.

References

1. G. Cybenko, *Approximation by superpositions of a sigmoidal function*, http://dx.doi.org/10.1007/BF02551274*Mathematics of Control, Signals and Systems***2** (Dec, 1989) 303–314.
2. K. Hornik, *Approximation capabilities of multilayer feedforward networks*, *Neural networks***4** (1991) 251–257.
3. K. Bull, Y.-H. He, V. Jejjala and C. Mishra, *Machine Learning CICY Threefolds*, http://dx.doi.org/10.1016/j.physletb.2018.08.008*Phys. Lett. B***785** (2018) 65–72, [arxiv.org/abs/1806.03121].
4. K. Bull, Y.-H. He, V. Jejjala and C. Mishra, *Getting CICY High*, http://dx.doi.org/10.1016/j.physletb.2019.06.067*Phys. Lett. B***795** (2019) 700–706, [arxiv.org/abs/1903.03113].
5. C. R. Brodie, A. Constantin, R. Deen and A. Lukas, *Machine Learning Line Bundle Cohomology*, http://dx.doi.org/10.1002/prop.201900087*Fortsch. Phys***68** (2020) 1900087, [arxiv.org/abs/1906.08730].
6. C. R. Brodie, A. Constantin, R. Deen and A. Lukas, *Index Formulae for Line Bundle Cohomology on Complex Surfaces*, http://dx.doi.org/10.1002/prop.201900086*Fortsch. Phys***68** (2020) 1900086, [arxiv.org/abs/1906.08769].
7. L. J. Dixon, *Some world sheet properties of superstring compactifications, on orbifolds and otherwise*, in *Summer Workshop in High-energy Physics and Cosmology*, 10, 1987.
8. W. Lerche, C. Vafa and N. P. Warner, *Chiral Rings in N=2 Superconformal Theories*, http://dx.doi.org/10.1016/0550-3213(89)90474-4*Nucl. Phys. B***324** (1989) 427–474.
9. P. Candelas, M. Lynker and R. Schimmrigk, *Calabi-Yau Manifolds in Weighted P(4)*, http://dx.doi.org/10.1016/0550-3213(90)90185-G*Nucl. Phys. B***341** (1990) 383–402.
10. B. R. Greene and M. R. Plesser, *Duality in Calabi-Yau Moduli Space*, http://dx.doi.org/10.1016/0550-3213(90)90622-K*Nucl. Phys. B***338** (1990) 15–37.
11. A. B. Givental, *Equivariant gromov - witten invariants*, arxiv.org/abs/alg-geom/9603021alg-geom/9603021.
12. B. H. Lian, K. F. Liu and S.-T. Yau, *Mirror principle. 1.*, *Asian J. Math.***1** (1997) 729–763, [arxiv.org/abs/alg-geom/9712011alg-geom/9712011].
13. B. H. Lian, K. F. Liu and S.-T. Yau, *Mirror principle. 2.*, *Asian J. Math.***3** (1999) 109–146, [arxiv.org/abs/math/9905006math/9905006].
14. M. Kreuzer and H. Skarke, *Complete classification of reflexive polyhedra in four-dimensions*, http://dx.doi.org/10.4310/ATMP.2000.v4.n6.a2*Adv. Theor. Math. Phys***4** (2000) 1209–1230, [arxiv.org/abs/hep-th/0002240hep-th/0002240].
15. W. Commons, *Trefoil and figure-eight knots*, 2008.
16. "The knot atlas: The take home database." http://katlas.org/wiki/The_Take_Home_Database.
17. C. Livingston and A. H. Moore, "Knotinfo: Table of knot invariants." URL: http://knotinfo.math.indiana.edu, November, 2021.
18. "The Mathematica package KnotTheory." http://katlas.org/wiki/The_Mathematica_Package_KnotTheory.
19. "KnotJob." https://www.maths.dur.ac.uk/users/dirk.schuetz/knotjob.html.
20. A. Davies, P. Velickovic, L. Buesing, S. Blackwell, D. Zheng, N. Tomasev et al., *Advancing mathematics by guiding human intuition with ai*, *Nature***600** (2021) 70–74.
21. A. Davies, A. Juhász, M. Lackenby and N. Tomasev, *The signature and cusp geometry of hyperbolic knots*, arxiv.org/abs/2111.15323.
22. V. Jejjala, A. Kar and O. Parrikar, *Deep Learning the Hyperbolic Volume of a Knot*, http://dx.doi.org/10.1016/j.physletb.2019.135033*Phys. Lett. B***799** (2019) 135033, [arxiv.org/abs/1902.05547].
23. J. Craven, V. Jejjala and A. Kar, *Disentangling a Deep Learned Volume Formula*, http://dx.doi.org/10.1007/JHEP06(2021)040*JHEP***06** (2021) 040, [arxiv.org/abs/2012.03955].
24. J. Craven, M. Hughes, V. Jejjala and A. Kar, *Learning knot invariants across dimensions*, arxiv.org/abs/2112.00016.
25. V. F. R. Jones, *A polynomial invariant for knots via von Neumann algebras*, http://dx.doi.org/10.1090/S0273-0979-1985-15304-2*Bull. Amer. Math. Soc. (N.S.)***12** (1985) 103–111.

26. E. Witten, *Quantum Field Theory and the Jones Polynomial*, http://dx.doi.org/10.1007/BF01217730*Commun. Math. Phys***121** (1989) 351–399.
27. M. Khovanov, *A categorification of the jones polynomial*, http://dx.doi.org/10.1215/S0012-7094-00-10131-7*Duke Math. J***101** (02, 2000) 359–426, [arxiv.org/abs/math/9908171math/9908171].
28. D. Bar-Natan, *On khovanov's categorification of the jones polynomial*, http://dx.doi.org/10.2140/agt.2002.2.337*Algebraic & Geometric Topology***2** (May, 2002) 337–370, [arxiv.org/abs/math/0201043math/0201043].
29. D. Gaiotto and E. Witten, *Knot Invariants from Four-Dimensional Gauge Theory*, http://dx.doi.org/10.4310/ATMP.2012.v16.n3.a5*Adv. Theor. Math. Phys***16** (2012) 935–1086, [arxiv.org/abs/1106.47891].
30. E. Witten, *Fivebranes and Knots*, arxiv.org/abs/1101.32161.
31. G. D. Mostow, *Quasi-conformal mappings in n-space and the rigidity of hyperbolic space forms*, *Publications Mathématiques de l'IHÉS***34** (1968) 53–104.
32. W. Thurston, *The Geometry and Topology of 3-Manifolds*, *Lecture Notes, Princeton University* (1978) .
33. E. Witten, *Analytic Continuation Of Chern-Simons Theory*, *AMS/IP Stud. Adv. Math.***50** (2011) 347–446, [arxiv.org/abs/1001.29331].
34. R. M. Kashaev, *The hyperbolic volume of knots from quantum dilogarithm*, http://dx.doi.org/10.1023/a:1007364912784*Letters in Mathematical Physics***39** (1997) 269–275, [arxiv.org/abs/arXiv:q-alg/9601025arXiv:q-alg/9601025].
35. H. Murakami and J. Murakami, *The colored Jones polynomials and the simplicial volume of a knot*, http://dx.doi.org/10.1007/BF02392716*Acta Math***186** (2001) 85–104, [arxiv.org/abs/arXiv:math/9905075arXiv:math/9905075].
36. H. Murakami, J. Murakami, M. Okamoto, T. Takata and Y. Yokota, *Kashaev's conjecture and the chern-simons invariants of knots and links*, *Experimental Mathematics***11** (2002) 427–435, [arxiv.org/abs/math/0203119math/0203119].
37. S. Gukov, *Three-dimensional quantum gravity, Chern-Simons theory, and the A polynomial*, http://dx.doi.org/10.1007/s00220-005-1312-y*Commun. Math. Phys***255** (2005) 577–627, [arxiv.org/abs/hep-th/0306165hep-th/0306165].
38. E. S. Lee, *An endomorphism of the Khovanov invariant*, http://dx.doi.org/10.1016/j.aim.2004.10.015*Adv. Math***197** (2005) 554–586, [arxiv.org/abs/math/0210213math/0210213].
39. J. Rasmussen, *Khovanov homology and the slice genus*, *Inventiones mathematicae***182** (2010) 419–447, [arxiv.org/abs/math/0402131math/0402131].
40. P. B. Kronheimer and T. S. Mrowka, *Gauge theory and rasmussen's invariant*, http://dx.doi.org/10.1112/jtopol/jtt008*Journal of Topology***6** (Apr, 2013) 659–674, [arxiv.org/abs/1110.12971].
41. S. Gukov, S. Nawata, I. Saberi, M. Stošić and P. Sułkowski, *Sequencing BPS Spectra*, http://dx.doi.org/10.1007/JHEP03(2016)004*JHEP***03** (2016) 004, [arxiv.org/abs/1512.07883].
42. G. Montavon, A. Binder, S. Lapuschkin, W. Samek and K.-R. Müller, *Layer-wise relevance propagation: an overview*, in *Explainable AI: interpreting, explaining and visualizing deep learning*, pp. 193–209. Springer, 2019.
43. C. Manolescu and M. Marengon, *The knight move conjecture is false*, *Proceedings of the American Mathematical Society***148** (2020) 435–439, [arxiv.org/abs/1809.09769].
44. M. C. Hughes, *A neural network approach to predicting and computing knot invariants*, *Journal of Knot Theory and Its Ramifications***29** (2020) 2050005, [arxiv.org/abs/1610.05744].
45. J. S. F. Levitt, M. Hajij and R. Sazdanovic, *Big data approaches to knot theory: Understanding the structure of the jones polynomial*, arxiv.org/abs/1912.10086.
46. S. Gukov, J. Halverson, F. Ruehle and P. Sułkowski, *Learning to Unknot*, arxiv.org/abs/2010.16263.
47. D. Paweł, D. Gurnari and R. Sazdanovic, *Knot invariants and their relations: a topological perspective*, arxiv.org/abs/2109.00831.
48. M. H. Freedman, R. E. Gompf, S. Morrison and K. Walker, *Man and machine thinking about the smooth 4-dimensional poincaré conjecture*, *Quantum Topology***1** (2010) 171–208, [arxiv.org/abs/0906.51770].

Holographic State Complexity from Group Cohomology

Bartłomiej Czech

Abstract Topological phases of matter are often described using auxiliary systems in one extra dimension. I review the one-dimensional cluster state—the simplest quantum state with Symmetry-Protected Topological (SPT) order—as a toy model of holographic duality, and inspect it for clues about defining holographic complexity. Group cohomology, which classifies SPT orders, is a viable candidate for a robust definition of complexity in gauge/gravity duality.

The concept of state complexity has taken research in the AdS/CFT correspondence by storm. Qualitatively, state complexity is understood as a CFT quantity, which is dual to the 'size' of the bulk AdS spacetime. Yet despite much effort, or perhaps because of it, there is no consensus definition of complexity on either side of the duality. In the bulk, 'sizes' of spacetimes can be quantified in multiple ways. The original and most popular options are the maximal volume of a spatial slice [1] and the action in the Wheeler-de Witt patch [2, 3], but recent work has revealed infinitely many other possibilities [4]. On the boundary, there is not even a unique formalism to define state complexity. The most common approach is circuit complexity [5–8], which counts (with some cost function) unitary gates necessary to convert a designated (usually unentangled) reference state into the desired state. But there are multiple alternatives, including path integral optimization [9–11], variants of information-geometric distances [12, 13], and others. Even if we accept circuit complexity as a preeminent paradigm, its inherent ambiguities pose a further question [14]: how to classify gate costs into universality classes according to their infrared (large complexity) behaviors?

The many unresolved differences and ambiguities involved in defining complexity call for a principled, rules-based approach. By what criteria should we judge a proposal for holographic complexity? I submit the following **desiderata**:

1. In the CFT, complexity should involve concepts, which play salient roles in bulk reconstruction. A nonexhaustive list of such concepts includes reduced states [15–18] and modular Hamiltonians [19, 20], local bulk operators [21–23],

B. Czech (✉)
Institute for Advanced Study, Tsinghua University, Beijing, China
e-mail: bartlomiej.czech@gmail.com

Y.-H. He et al. (eds.), *Nankai Symposium on Mathematical Dialogues*,
https://doi.org/10.1007/978-981-19-2328-9_14

modular Berry connections and curvatures [24–26], and others [27–30]. If we do not start with holographically meaningful ingredients, obtaining a holographically meaningful answer will be a miracle.
2. A definition of complexity should be faithful to the spirit of the original motivation for relating bulk 'sizes' to state complexity [1, 31]. Probably the best illustration of the original thinking is a tensor network: if counting tensors in a network enumerates steps necessary to prepare the state, and if a tensor network models a bulk spacetime, then state complexity should roughly go like bulk volume.
3. As in all physical modeling, a proposal is stronger if it involves fewer free parameters. In an ideal proposal, all parameters would be internally or dynamically determined.
4. Ideally, holographic complexity should mesh with some broader conceptual framework, which extends beyond AdS/CFT. For this desideratum, I see two main possibilities: One is bulk/boundary duality that is not gauge/gravity duality, often used for describing topological phases of matter [32–35]. Another one is complexity theory as studied by computer scientists. Other things being equal, a proposal for holographic complexity is stronger if it connects to one or both of these broader frameworks.

The purpose of this talk is to illustrate how these demands may be simultaneously met. Partly to situate holographic complexity in a broader framework, and partly for pure inspiration, I first consider an entirely different system: a one-dimensional array of qubits in an entangled state with symmetry-protected topological order (SPT). I argue that this system shows a strong analogy to the AdS/CFT correspondence. Under the analogy, the 'size' of the 'bulk' is quantified by a group cohomology relevant to the SPT order. I then outline how group cohomology may be relevant to defining complexity in gauge/gravity duality.

To avoid awkward generalities and an onslaught of definitions, I focus on a particular SPT-ordered state, called a cluster state [36]; see [37] for a pedagogical presentation. It is obtained (see Fig. 1) by acting with entangling gates

$$\text{CPhase}^{i,i+1} = \text{diag}\{1,1,1,-1\}^{i,i+1} = \mathbb{1}^{i,i+1} - 2\,|11\rangle^{i,i+1}\langle 11| \qquad (\text{in } \sigma_z^i \otimes \sigma_z^{i+1} \text{ basis}) \tag{1}$$

on each pair of neighboring spins, initialized in $+1$ eigenstates of σ_x, denoted $|+\rangle = \sigma_x|+\rangle$:

$$|\text{cluster}\rangle = \prod_i \text{CPhase}^{i,i+1} \bigotimes_i |+\rangle^i \tag{2}$$

It is most convenient to consider an even number $2K$ of spins arranged on a loop, so that the product in (2) contains $\text{CPhase}^{2K,2K+1} \equiv \text{CPhase}^{2K,1}$. Then $|\text{cluster}\rangle$ is the unique ground state of the Hamiltonian:

$$H = -\sum_i \sigma_z^i \sigma_x^{i+1} \sigma_z^{i+2} = -\left(\prod_i \text{CPhase}^{i,i+1}\right)\left(\sum_i \sigma_x^i\right)\left(\prod_i \text{CPhase}^{i,i+1}\right) \tag{3}$$

It follows that the state is stabilized by each operator $\sigma_x^i \sigma_z^{i+1} \sigma_x^{i+2}$ and by products of them. Using now that the number of spins is assumed even, we identify two special stabilizing operators:

$$\prod_{i \text{ even}} \sigma_z^i \otimes \sigma_x^{i+1} \otimes \sigma_z^{i+2} = \prod_{j \text{ odd}} \sigma_x^j \equiv S_{\text{odd}} \quad \text{and} \quad \prod_{i \text{ odd}} \sigma_z^i \otimes \sigma_x^{i+1} \otimes \sigma_z^{i+2} = \prod_{j \text{ even}} \sigma_x^j \equiv S_{\text{even}} \tag{4}$$

They generate an on-site $\mathbb{Z}_2 \otimes \mathbb{Z}_2$ symmetry of the state, which is not respected by the entangling gate in Eq. (2): $[\text{CPhase}^{i,i+1}, S_{\text{odd}}] \neq 0 \neq [\text{CPhase}^{i,i+1}, S_{\text{even}}]$. Thus, although it is possible to bring the cluster state to an unentangled (product) form with short range disentangling gates—as showcased by Eq. (2)—it is not possible to do so without breaking the on-site $\mathbb{Z}_2 \otimes \mathbb{Z}_2$ symmetry. This is 'Symmetry-Protected Topological order' (SPT).

I now argue that the cluster state is a toy model of holographic duality. The reasoning, which backs this assertion takes several steps. The first is to rewrite $|\text{cluster}\rangle$ as a Matrix Product State (MPS). A little play with Pauli matrices (see Fig. 1) reveals that its amplitudes in the $|\pm\rangle = \pm\sigma_x|\pm\rangle$ basis are:

$$\langle (s_1 s_2) \otimes (s_3 s_4) \otimes \ldots \otimes (s_{2K-1} s_{2K}) | \text{cluster} \rangle = 2^{1-2K} \, \text{Tr}\, A[s_1 s_2] A[s_3 s_4] \ldots A[s_{2K-1} s_{2K}] \tag{5}$$

$$\text{with} \quad A[++] = \mathbb{1} \quad \text{and} \quad A[+-] = -\sigma_z \quad \text{and} \quad A[-+] = -\sigma_x \quad \text{and} \quad A[--] = \sigma_x \sigma_z \tag{6}$$

The $(s_i s_{i+1})$ in (5) range over the set $\{++, +-, -+, --\}$ and denote $\sigma_x^i \otimes \sigma_x^{i+1}$ eigenstates. The objects $A[s_i s_{i+1}]$ are the 'matrices' in the matrix product description of $|\text{cluster}\rangle$.

The next step is to think about the $(s_i s_{i+1}) \in \{++, +-, -+, --\}$ as elements of $\mathbb{Z}_2 \times \mathbb{Z}_2$—the group that protects the topological order of $|\text{cluster}\rangle$. Observe that $A[s_i s_{i+1}]$ form a projective representation of the group:

$$A[g]A[h] = \omega(g, h)A[gh] \qquad \forall g, h \in \mathbb{Z}_2 \times \mathbb{Z}_2 \tag{7}$$

An example of a nontrivial ω, which makes Eq. (6) into a projective and not an ordinary (affine) representation, is $A[--]A[--] = -\mathbb{1} = -A[++]$, so $\omega(--, --) = -1$.

It is well known that both SPT phases of matter [34] and projective representations of groups are classified by the second group cohomology $H^2(G, U(1))$. For representations, this is easy to understand. First, associate $\omega(g, h)$ with an oriented two-simplex whose edges are marked (in the order fixed by the orientation) with $g \in G$, $h \in G$, and $gh \in G$; see Fig. 2. If the representation matrix $A[(gh)k] = A[g(hk)]$ is to be well-defined, we must have:

$$\omega(g, h)\omega(gh, k) = \omega(g, hk)\omega(h, k) \tag{8}$$

This means that $\log \omega(g, h)$ behaves like a closed two-form or—figuratively speaking—like the 'area' of a 'triangle' with sides g, h, gh. If we now choose to dress the MPS matrices $A[g]$ with g-dependent phases, $A[g] \to \mu(g)A[g]$, then $\omega(g, h) \to$

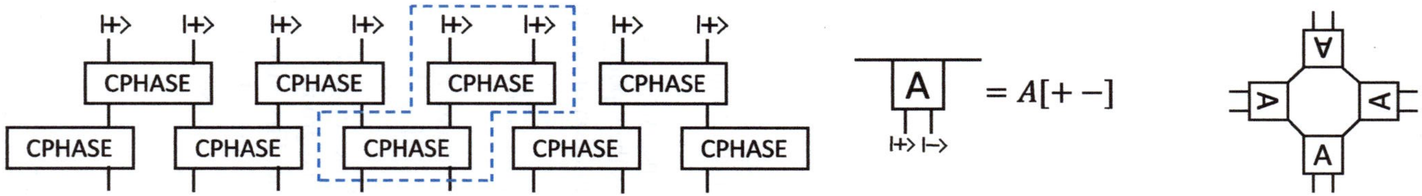

Fig. 1 The cluster state, here shown for eight qubits on a loop, as given by Eqs. (1) (left panel) and (5) (right panel). The highlighted part of the left panel is the 'matrix' of the MPS description (middle panel)

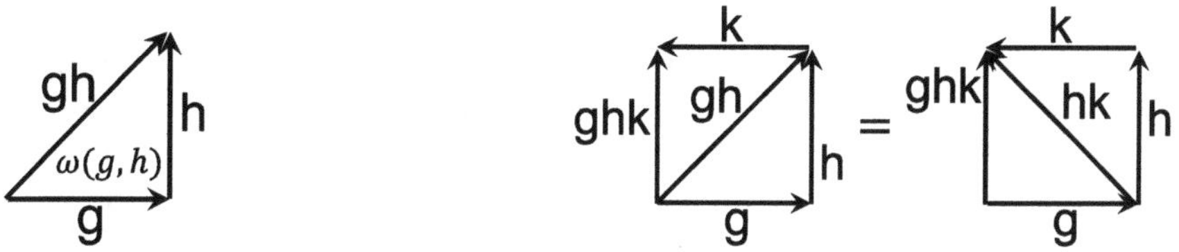

Fig. 2 The factor $\omega(g, h)$ in Eq. (7) can be visualized as a triangle with oriented edges, which run neither clockwise nor counterclockwise. Associativity of the group product mandates Eq. (8), visualized in the right panel. Therefore products of ω's are represented by triangulation-invariant two-complexes and $\log \omega$ behaves like a closed two-form

$\omega(g, h)\mu(g)\mu(h)/\mu(gh)$. Applied to Eq. (5), this operation dresses $|\text{cluster}\rangle$ with an irrelevant overall phase, so we would like to treat such ω's as equivalent. In group cohomology, this redefinition is understood as adding to the 'two-form' $\log \omega(g, h)$ the exterior derivative of a 'one-form' $\log \mu(g)$.

We are now ready to see the 'holographic' description of the cluster state. For this, substitute Eq. (7) in (5) and use the graphical representation of $\omega(g, h)$. For example, the circular cluster on $2K = 8$ qubits can be written as:

$$|\text{cluster}\rangle = \frac{1}{64} \sum_{f,g,h \in \mathbb{Z}_2 \times \mathbb{Z}_2} \left\{\omega(++, f)\, \omega(f, g)\, \omega(fg, h)\, \omega\big(fgh, (fgh)^{-1}\big)\right\}\, |f\rangle\, |g\rangle\, |h\rangle\, |(fgh)^{-1}\rangle, \quad (9)$$

where again we use the notation $|f\rangle$ with $f \in \mathbb{Z}_2 \times \mathbb{Z}_2$ for $\sigma_x^i \otimes \sigma_x^{i+1}$ eigenstates $|{+}{+}\rangle$, $|{+}{-}\rangle$, $|{-}{+}\rangle$, and $|{-}{-}\rangle$. This state is shown in Fig. 3.

Using Eq. (8) and Fig. 2, we see that the graphical representations of $|\text{cluster}\rangle$ in Fig. 3 are triangulation-invariant. In the spirit of holography, we interpret the complexes of triangles as path integrals of a bulk theory, which is tasked with preparing $|\text{cluster}\rangle$. The action of the bulk theory on a triangle with sides g, h and gh is given by $e^{-iS} = \omega(g, h)$ if the g, h-sides make a counterclockwise pattern, and $e^{+iS} = \omega(g, h)$ otherwise. On interior edges, the correctly normalized measure of the bulk path integral is $\mathcal{D}g = (1/4) \sum_{g \in \mathbb{Z}_2 \times \mathbb{Z}_2}$.

This bulk theory is a discrete topological field theory (TFT), where 'topological' connotes triangulation invariance. Every discrete SPT-ordered system can be prepared by a path integral of this type of TFT [34]. An MPS description of SPT phases [38], which we used in Eq. (5), was developed in generality based on the bulk TFT perspective. Continuum versions of these bulk theories, called equivariant topological field theories, are also well-studied [39].

I now highlight features of SPT holography, which are analogous to gauge/gravity duality:

- Gauge/gravity duality is often modeled with tensor networks such as MERA [40] or HaPPY [41]. The preparation of $|\text{cluster}\rangle$ in Eq. (2) is like a single layer of those networks; see the left panel of Fig. 1. There is only a single layer because SPT phases have only short range entanglement [34].

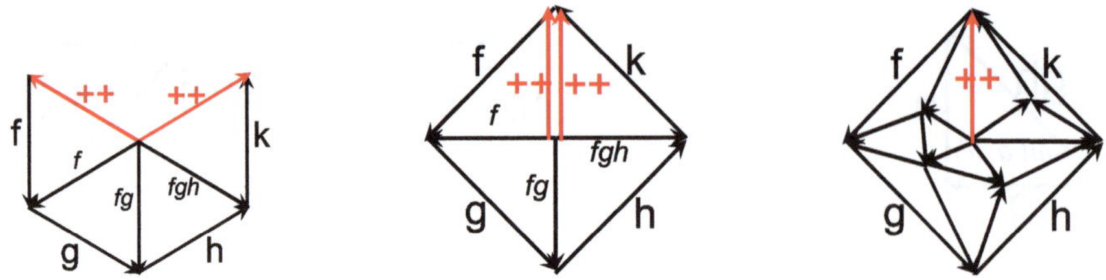

Fig. 3 The cluster state on eight qubits (also shown in Fig. 1) is prepared by the path integral of a discrete two-dimensional topological field theory with action $S = i \log \omega(g, h)$. I label the basis $|f\rangle\,|g\rangle\,|h\rangle\,|k\rangle$ used in Eq. (9). Interior edges, which are summed in the path integral, are marked in italics whenever they are uniquely determined. All three pictures are equivalent because of triangulation invariance (Fig. 2). The progression from the left to the right panel shows that the boundary condition $|{+}{+}\rangle\langle{+}{+}|$, which sets $k = (fgh)^{-1}$ and imposes Tr(...) in Eq. (5), is contractible in the sense that the path integral is effectively carried out over the half-sphere without punctures

- The MPS presentation of $|\text{cluster}\rangle$ involves a projective representation of the symmetry group. This is analogous to the gauge/gravity lore that global boundary symmetries are local bulk symmetries.
- The triangulation invariance of the TFT path integral is analogous to diffeomorphism invariance in gravity.
- Triangulation invariance also motivates a notion of local Renormalization Group analogous to holographic RG of gauge/gravity duality; see Fig. 4. The same triangle can be considered part of the preparation of the (ultraviolet) state, or excluded from the preparation of the (infrared) state and absorbed by a redefinition of the cutoff.
- As a slight generalization of $|\text{cluster}\rangle$, we can consider states [38] whose amplitudes are as in Eq. (5), but with the replacement Tr $(\ldots) \to$ Tr $(\Sigma \ldots)$ where $\Sigma = \sigma_x$, σ_z, or $\sigma_x\sigma_z$. Their path integral preparations, as shown in Fig. 3, would have one $++$ replaced with $+-$, $-+$, or $--$, so the rightmost panel of Fig. 3 would contain a defect-carrying bulk puncture. This is like a black hole in gauge/gravity duality. In SPT holographic duality, the product $fghk \in G$ equals the G-valued 'charge' of the defect; this is like Gauss's law in gravity.

In this analogy, there is an obvious quantity, which satisfies points 2–4 from our list of desiderata for holographic complexity. It is the action of the bulk TFT. Qualitatively speaking, it 'counts' the gates necessary to map the product state $\otimes_i |+\rangle^i$ to $|\text{cluster}\rangle$ in Eq. (2). But it does so in an unambiguous way because the complexity measure is constrained by the topological (triangulation) invariance of the bulk. Indeed, the action of the bulk is essentially the only sensible measure of complexity applicable to topological field theories [42]. At the same time, the TFT action is very much like the bulk spatial volume in gauge/gravity duality because—as I emphasize in Eq. (8) and in Fig. 2—the 'Lagrangian density' $i \log \omega$ behaves like a closed two-form. In this sense, the bulk action is the closest possible analogue for quantifying bulk sizes, which makes sense in a topological field theory.

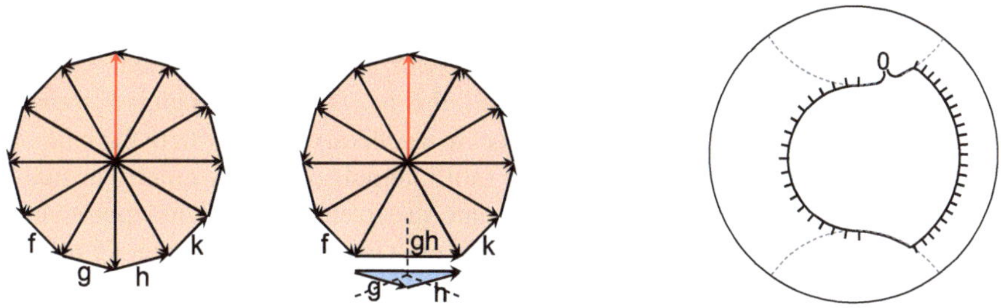

Fig. 4 Using triangulation invariance (Fig. 2), we can split the cluster state on 24 qubits ('ultraviolet,' left panel) into a cluster on 22 qubits ('infrared,' middle panel) and a gate, which fuses four 'ultraviolet' qubits in state $|g\rangle|h\rangle$ into two 'infrared' qubits in state $|gh\rangle$. This is a local application of holographic Renormalization Group in SPT holography. The right panel shows an independently derived [43] MPS form of the vacuum of a holographic CFT_2 at a variable UV cutoff, which is an amputated Wilson line network of $SL(2,\mathbb{R}) \times SL(2,\mathbb{R})$ Chern-Simons theory. Observe the similarity to the cluster state, with the $|0\rangle\langle 0|$ insertion functioning like $|{++}\rangle\langle{++}|$ in Fig. 3

The parallel between SPT phases and gauge/gravity duality validates the intuition behind the volume [1] and action [2] proposals. It does not, and cannot, lend them quantitative support because in SPT holography we have a $(d+1)$-dimensional action while the gauge/gravity proposals call for a $(d+2)$-dimensional action or a $(d+1)$-dimensional volume—which the TFT cannot see. A more compelling match is with path integral optimization when it ends up identifying constant mean-curvature slices of the bulk geometry [44, 45].

But a more daring lesson that might be drawn from SPT phases concerns complexity and group cohomology. For the cluster state, a cocycle of its on-site symmetry group supplies a measure of complexity, which captures the 'size' of the bulk description. Can the same be done in gauge/gravity duality? It seems that the answer is yes.

In using group cohomology to define a bulk density of complexity, we must have an on-site symmetry G acting at every location on the bulk cutoff surface. The boundary degrees of freedom must transform in an ordinary (affine) representation of G while the bulk degrees of freedom must transform projectively. We should also have a flat G-connection in the bulk; see Figs. 2 and 3. The group cohomology $H^{d+1}(G, M)$, which is relevant to describing states in d spatial boundary dimensions, may have $M = \mathbb{R}$ or $M = U(1)$. Then (logarithms of) cocycles drawn from $H^{d+1}(G, M)$ automatically give well-defined bulk densities because they are essentially closed $(d+1)$-forms.

These prerequisites are easily met by gauge/gravity duality. In fact, there is already an explicit example: the complexity of the ground state of a holographic CFT_2 at a variable UV cutoff, which my collaborators and I studied in [43]. There, we presented the ground state in an MPS-like form (see Fig. 4) as an amputated network of Wilson lines of the $SL(2,\mathbb{R}) \times SL(2,\mathbb{R})$ Chern-Simons theory, whose flat solutions describe locally AdS_3 spacetimes [46]. We then argued that the density of complexity in this setup should go like bulk area. But $e^{-A_\triangle}$, where $A_\triangle$ is the area of a triangle in hyperbolic space, is a cocycle of $SL(2,\mathbb{R})$—a fact we instantly recognize by looking back at Fig. 2.

Can group cohomology define a measure of complexity more broadly, beyond locally AdS_3 spaces? The key is to identify a group, which acts as an on-site symmetry everywhere on the cutoff surface. In [43], this is $SL(2, \mathbb{R})$—the global isometry group of two-dimensional hyperbolic space. In a general context, where the bulk has no global isometries, the only geometric (not internal) on-site symmetry available is the *local* isometry group, i.e. the Poincaré group. In the bulk, its cohomology gives rise to measures of complexity, which effectively produce integrals of bulk curvature invariants—and so agree well with the heuristics of holographic complexity. In the boundary language, generators of the bulk Poincaré symmetry are partly understood in terms of modular scrambling modes [26], modular zero modes including the modular Hamiltonian [20], and a subtle combination thereof in the case of bulk time translations [30]. Assuming this understanding can be advanced, defining holographic complexity from the cohomology $H^{d+1}(\text{Poincaré}, \mathbb{R})$ will have met desideratum 1.

Before closing, I briefly revisit desiderata 3 and 4. First, group cohomology is a tight algebraic structure. While it cannot fix a definition of holographic complexity uniquely, it imposes on it a set of stringent rules and eliminates spurious ambiguities. The second point concerns embedding holographic complexity in a broader framework. SPT phases in general, and the cluster state in particular, are resource states for Measurement-Based Quantum Computation [47]—a framework for doing quantum computing by measurements alone. There is evidence that group cohomology also quantifies the computer scientist's complexity of certain algorithms carried out in the MBQC or MBQC-like framework [43, 48]. If correct, this would be a remarkable reunification of the term 'complexity' with its original, non-holographic meaning.

Acknowledgements I thank Yang-Hui He for his kind invitation to present this talk at the 2021 Nankai Symposium on Mathematical Dialogues. The talk is based in part on two ongoing collaborations: one with Jan de Boer, Bowen Chen, Lampros Lamprou and Zi-zhi Wang, and one with Robert Raussendorf and Gabriel Wong. I have also benefitted greatly from conversations with Ling-Yan (Janet) Hung, as well as Jingyuan Chen and Runze Chi.

References

1. D. Stanford and L. Susskind, "Complexity and shock wave geometries," Phys. Rev. D **90**, no.12, 126007 (2014) [arXiv:1406.2678 [hep-th]].
2. A. R. Brown, D. A. Roberts, L. Susskind, B. Swingle and Y. Zhao, "Holographic complexity equals bulk action?," Phys. Rev. Lett. **116**, no.19, 191301 (2016) [arXiv:1509.07876 [hep-th]].
3. A. R. Brown, D. A. Roberts, L. Susskind, B. Swingle and Y. Zhao, "Complexity, action, and black holes," Phys. Rev. D **93**, no.8, 086006 (2016) [arXiv:1512.04993 [hep-th]].
4. A. Belin, R. C. Myers, S. M. Ruan, G. Sárosi and A. J. Speranza, "Complexity equals anything?," [arXiv:2111.02429 [hep-th]].
5. S. Aaronson, "The complexity of quantum states and transformations: From quantum money to black holes," [arXiv:1607.05256 [quant-ph]].
6. R. Jefferson and R. C. Myers, "Circuit complexity in quantum field theory," JHEP **10**, 107 (2017) [arXiv:1707.08570 [hep-th]].

7. S. Chapman, J. Eisert, L. Hackl, M. P. Heller, R. Jefferson, H. Marrochio and R. C. Myers, "Complexity and entanglement for thermofield double states," SciPost Phys. **6**, no.3, 034 (2019) [arXiv:1810.05151 [hep-th]].
8. A. R. Brown and L. Susskind, "Complexity geometry of a single qubit," Phys. Rev. D **100**, no.4, 046020 (2019) [arXiv:1903.12621 [hep-th]].
9. P. Caputa, N. Kundu, M. Miyaji, T. Takayanagi and K. Watanabe, "Anti-de Sitter space from optimization of path integrals in conformal field theories," Phys. Rev. Lett. **119**, no.7, 071602 (2017) [arXiv:1703.00456 [hep-th]].
10. B. Czech, "Einstein equations from varying complexity," Phys. Rev. Lett. **120** (2018) no.3, 031601 [arXiv:1706.00965 [hep-th]].
11. P. Caputa, N. Kundu, M. Miyaji, T. Takayanagi and K. Watanabe, "Liouville action as path-integral complexity: From continuous tensor networks to AdS/CFT," JHEP **11**, 097 (2017) [arXiv:1706.07056 [hep-th]].
12. S. Chapman, M. P. Heller, H. Marrochio and F. Pastawski, "Toward a definition of complexity for quantum field theory states," Phys. Rev. Lett. **120** (2018) no.12, 121602 [arXiv:1707.08582 [hep-th]].
13. A. Belin, A. Lewkowycz and G. Sárosi, "Complexity and the bulk volume, a new York time story," JHEP **03** (2019), 044 [arXiv:1811.03097 [hep-th]].
14. A. R. Brown, M. H. Freedman, H. W. Lin and L. Susskind, "Effective geometry, complexity, and universality," [arXiv:2111.12700 [hep-th]].
15. S. Ryu and T. Takayanagi, "Holographic derivation of entanglement entropy from AdS/CFT," Phys. Rev. Lett. **96**, 181602 (2006) [arXiv:hep-th/0603001 [hep-th]].
16. S. Ryu and T. Takayanagi, "Aspects of holographic entanglement entropy," JHEP **08**, 045 (2006) [arXiv:hep-th/0605073 [hep-th]].
17. V. E. Hubeny, M. Rangamani and T. Takayanagi, "A covariant holographic entanglement entropy proposal," JHEP **07**, 062 (2007) [arXiv:0705.0016 [hep-th]].
18. B. Czech, J. L. Karczmarek, F. Nogueira and M. Van Raamsdonk, "The gravity dual of a density matrix," Class. Quant. Grav. **29** (2012), 155009 [arXiv:1204.1330 [hep-th]].
19. T. Faulkner, A. Lewkowycz and J. Maldacena, "Quantum corrections to holographic entanglement entropy," JHEP **11** (2013), 074 [arXiv:1307.2892 [hep-th]].
20. D. L. Jafferis, A. Lewkowycz, J. Maldacena and S. J. Suh, "Relative entropy equals bulk relative entropy," JHEP **06** (2016), 004 [arXiv:1512.06431 [hep-th]].
21. A. Hamilton, D. N. Kabat, G. Lifschytz and D. A. Lowe, "Local bulk operators in AdS/CFT: A boundary view of horizons and locality," Phys. Rev. D **73**, 086003 (2006) [arXiv:hep-th/0506118 [hep-th]].
22. X. Dong, D. Harlow and A. C. Wall, "Reconstruction of bulk operators within the entanglement wedge in gauge-gravity duality," Phys. Rev. Lett. **117** (2016) no.2, 021601 [arXiv:1601.05416 [hep-th]].
23. J. Cotler, P. Hayden, G. Penington, G. Salton, B. Swingle and M. Walter, "Entanglement wedge reconstruction via universal recovery channels," Phys. Rev. X **9** (2019) no.3, 031011 [arXiv:1704.05839 [hep-th]].
24. B. Czech, L. Lamprou, S. Mccandlish and J. Sully, "Modular Berry connection for entangled subregions in AdS/CFT," Phys. Rev. Lett. **120** (2018) no.9, 091601 [arXiv:1712.07123 [hep-th]].
25. B. Czech, J. De Boer, D. Ge and L. Lamprou, "A modular sewing kit for entanglement wedges," JHEP **11** (2019), 094 [arXiv:1903.04493 [hep-th]].
26. J. De Boer and L. Lamprou, "Holographic order from modular chaos," JHEP **06** (2020), 024 [arXiv:1912.02810 [hep-th]].
27. T. Takayanagi and K. Umemoto, "Entanglement of purification through holographic duality," Nature Phys. **14**, no.6, 573–577 (2018) [arXiv:1708.09393 [hep-th]].
28. Y. Chen, X. Dong, A. Lewkowycz and X. L. Qi, "Modular flow as a disentangler," JHEP **12**, 083 (2018) [arXiv:1806.09622 [hep-th]].
29. J. Kirklin, "The holographic dual of the entanglement wedge symplectic form," JHEP **01** (2020), 071 [arXiv:1910.00457 [hep-th]].

30. D. L. Jafferis and L. Lamprou, "Inside the hologram: Reconstructing the bulk observer's experience," [arXiv:2009.04476 [hep-th]].
31. T. Hartman and J. Maldacena, "Time evolution of entanglement entropy from black hole interiors," JHEP **05**, 014 (2013) [arXiv:1303.1080 [hep-th]].
32. I. Runkel, J. Fjelstad, J. Fuchs and C. Schweigert, "Topological and conformal field theory as Frobenius algebras," Contemp. Math. **431** (2007), 225–248 [arXiv:math/0512076 [math.CT]].
33. M. Levin and X.-G. Wen, "String-net condensation: A physical mechanism for topological phases," Phys. Rev. B **71** (2005), 045110.
34. X. Chen, Z. C. Gu, Z. X. Liu and X. G. Wen, "Symmetry protected topological orders and the group cohomology of their symmetry group," Phys. Rev. B **87** (2013) no.15, 155114 [arXiv:1106.4772 [cond-mat.str-el]].
35. L. Kong, T. Lan, X. G. Wen, Z. H. Zhang and H. Zheng, "Classification of topological phases with finite internal symmetries in all dimensions," JHEP **09** (2020), 093 [arXiv:2003.08898 [math-ph]].
36. H. J. Briegel and R. Raussendorf, "Persistent entanglement in arrays of interacting particles," Phys. Rev. Lett. **86**, 910 (2001).
37. B. Yoshida, "Topological phases with generalized global symmetries," Phys. Rev. B **93** (2016) no.15, 155131 [arXiv:1508.03468 [cond-mat.str-el]].
38. K. Shiozaki and S. Ryu, "Matrix product states and equivariant topological field theories for bosonic symmetry-protected topological phases in (1+1) dimensions," JHEP **04** (2017), 100 [arXiv:1607.06504 [cond-mat.str-el]].
39. R. Dijkgraaf and E. Witten, "Topological gauge theories and group cohomology," Commun. Math. Phys. **129** (1990), 393
40. G. Vidal, "Entanglement renormalization," Phys. Rev. Lett. **99**, no.22, 220405 (2007) [arXiv:cond-mat/0512165 [cond-mat]].
41. F. Pastawski, B. Yoshida, D. Harlow and J. Preskill, "Holographic quantum error-correcting codes: Toy models for the bulk/boundary correspondence," JHEP **06** (2015), 149 [arXiv:1503.06237 [hep-th]].
42. J. Couch, Y. Fan and S. Shashi, "Circuit complexity in topological quantum field theory," [arXiv:2108.13427 [hep-th]].
43. B. Chen, B. Czech and Z. z. Wang, "Query complexity and cutoff dependence of the CFT_2 ground state," Phys. Rev. D **103** (2021) no.2, 026015 [arXiv:2004.11377 [hep-th]].
44. T. Takayanagi, "Holographic spacetimes as quantum circuits of path-integrations," JHEP **12**, 048 (2018) [arXiv:1808.09072 [hep-th]].
45. J. Boruch, P. Caputa, D. Ge and T. Takayanagi, "Holographic path-integral optimization," JHEP **07** (2021), 016 [arXiv:2104.00010 [hep-th]].
46. E. Witten, "(2+1)-Dimensional gravity as an exactly soluble system," Nucl. Phys. B **311**, 46 (1988)
47. R. Raussendorf and H.-J. Briegel, "A one-way quantum computer," Phys. Rev. Lett. **86**, 5188 (2001)
48. B. Czech, G. Wong, and R. Raussendorf, *in preparation*.

Duality in Hyperkähler Geometry

Andrew Dancer

2000 Mathematics Subject Classification 53C26 · 53D20 · 14L24

1 Introduction

In this article we discuss Kähler and hyperkähler structures, and report on joint work with A. Bourget, J. Grimminger, A. Hanany, F. Kirwan and Z. Zhong on symplectic duality for certain hyperkähler spaces (the implosion spaces).

2 Kähler and Hyperkähler Structures

The basic geometric object on a smooth manifold is a metric, which will be Riemannian (positive definite) for our purposes. This abstracts the idea of an inner product from Euclidean geometry–at each point of the manifold we have an inner product on the tangent space, varying smoothly as we move around the manifold.

Kähler and hyperkähler geometry arise when we try to introduce complex and quaternionic structures that are compatible in a suitable sense with the metric g.

A complex structure is an endomorphism I of the tangent bundle satisfying $I^2 = -1$, together with a certain integrability condition (vanishing of the Nijenhuis tensor). We require that the complex structure be length-preserving–this is the Hermitian condition

$$g(IX, IY) = g(X, Y)$$

A. Dancer (✉)
Jesus College, Oxford, UK
e-mail: andrew.dancer@maths.ox.ac.uk

Y.-H. He et al. (eds.), *Nankai Symposium on Mathematical Dialogues*,
https://doi.org/10.1007/978-981-19-2328-9_15

and implies, via

$$g(IX, Y) = g(I(IX), IY) = -g(X, IY)$$

that

$$\omega(X, Y) := g(IX, Y)$$

defines an alternating 2-form. We obtain a deeper compatibility between metric and complex structure by further imposing the Kähler condition

$$d\omega = 0.$$

The 2-form ω thus defines a symplectic structure on M. The Kähler condition actually implies that I is covariant constant with respect to the Levi-Civita connexion of the metric, hence parallel transport is complex-linear and the holonomy reduces to the unitary group. Examples include smooth projective varieties, that is, complex submanifolds of complex projective space. Noncompact examples include smooth affine varieties, or more generally complex submanifolds of $\mathbb{C}^n$.

We now consider a quaternionic analogue of this picture. A *hyperkähler* structure is a quadruple (g, I, J, K) where g is a Riemannian metric and I, J, K are complex structures, each Kähler with respect to g, and satisfying the quaternionic relations

$$IJ = K = -JI \tag{2.1}$$

Parallel transport is now quaternion-linear, implementing a reduction of holonomy to the symplectic group.

Fixing a complex structure, say I, we obtain a holomorphic symplectic structure with holomorphic 2-form

$$\Omega = \omega_J + i\omega_K$$

This induces a parallel trivialisation of the canonical bundle, so hyperkähler manifolds are in fact Calabi-Yau, that is, Ricci-flat Kähler. We can also view this as saying that a holonomy reduction to the symplectic group a fortiori implies a reduction to a subgroup of the special unitary group.

Not every holomorphic symplectic space admits a hyperkähler structure, but many examples do–complex coadjoint orbits, complex representation varieties for fundamental groups of Riemann surfaces, and (at least on a neighbourhood of the zero section) cotangent bundles of analytic Kähler manifolds. This has in fact been a good heuristic for searching for hyperkähler structures, with the proviso that showing that a holomorphic symplectic structure actually comes from a hyperkähler structure is usually quite difficult. Typically the proof requires an application of twistor theory or, for the best global results, the hyperkähler quotient construction, of which more is said below.

In contrast to the Kähler situation, we cannot usually produce nontrivial examples by a submanifold construction. Instead, the best way to produce global examples is via the hyperkähler quotient construction [9]. If a group K acts on a hyperkähler space

M preserving the hyperkähler structure then, in good cases, we have a momentum map $\mu : M \to Lie(K)^* \otimes \mathbb{R}^3$, and the quotient $\mu^{-1}(0,0,0)/K$ is hyperkähler.

The power of this construction lies in the fact that even a linear action on flat quaternionic space $\mathbb{H}^n$ often yields highly nontrivial examples. For example the Eguchi-Hanson space arises as a hyperkähler quotient of $\mathbb{H}^2$ by a circle action.

3 Symplectic Duality

Symplectic duality is a correspondence between certain holomorphic symplectic varieties (in most cases these are actually hyperkähler, at least in a suitable stratified sense). We are typically dealing with affine varieties, in particular hyperkähler cones with an $Sp(1)$ action rotating complex structures, and deformations of these where the $Sp(1)$ action is broken to a circle action.

Physically, the duality is between Higgs and Coulomb branches of a 3-dimensional $N = 4$ supersymmetric theory. The Higgs branch often arises as a hyperkähler quotient by a linear action. The duality is supposed to exchange deformation parameters with parameters for the Cartan algebra of the symmetry group. See [2] for example for a mathematical approach to this subject.

One case where the picture is fairly well understood is *hypertoric varieties*, that is, hyperkähler quotients of flat quaternionic space by compact tori. We consider the hyperkähler quotient of $\mathbb{H}^d$ by a $(d-n)$-dimensional subtorus N of T^d. We therefore have $d - n$ deformation parameters and n Abelian symmetry parameters (coming from the global symmetry group $T^n = T^d/N$). As in the case of classical toric varieties, this data can be encoded in terms of combinatorial convex geometry. The hypertoric variety has quaternionic dimension $n = d - (d - n)$.

In this context, symplectic duality dualises the exact sequence of tori, so we replace T^n by the dual torus $\hat{N}$ of N, and N by $\hat{T}^n$, so that the roles of n and $d - n$ are swapped. This is a manifestation of a phenomenon in convex geometry and linear programming called *Gale duality*. Note that the dimension of the hypertoric is changed under the duality.

The case of hyperkähler quotients by a non-Abelian group K is much harder to analyse. The Coulomb branch should now be birational to the quotient $T^*(\hat{T}_{\mathbb{C}})/W$ by the Weyl group of the cotangent bundle of the dual complex maximal torus of K. This quotient represents the classical moduli space, while the birational transformations are to be interpreted as quantum corrections.

String theorists have developed powerful techniques to work out properties of symplectic duals, such as dimension, global symmetry and Hilbert series. For a unitary quiver, each gauge node gives a unitary group with 1-dimensional centre, and hence one deformation parameter. In the dual, this gives one parameter in the Cartan algebra of the symmetry group. A more refined analysis shows that the nodes which satisfy a certain combinatorial condition called ‘balancing’ actually give the Dynkin diagram of a subgroup of the symmetry group. In this sense we may see how to thicken up the Cartan algebra to a reductive group of symmetries.

The chiral ring (coordinate ring of the variety) is particularly important as since our spaces are affine varieties they are determined by the ring as its spectrum. The associated Hilbert series is computed via the *monopole formula* [4], which from a physical point of view counts monopole operators with Dirac singularities.

We have been studying symplectic duality for a class of holomorphic symplectic varieties called universal implosions [6], inspired by the implosion construction in real symplectic geometry due to Guillemin et al. [8]. Given a complex reductive group $G = K_{\mathbb{C}}$, its universal implosion is a space of dimension $\dim G + \text{rank } G$ with an action of $G \times T_{\mathbb{C}}$, such that the reductions by the complex torus $T_{\mathbb{C}}$ give the Kostant varieties of the Lie algebra of G [3]. In particular reduction at level zero gives the nilpotent cone. We can view the implosion as a master space for generating the various Kostant varieties.

The implosion may be viewed as a nonreductive Geometric Invariant Theory quotient $(G \times \mathfrak{n}^{\circ})//N$ where N is the unipotent radical of G (see [7] for further discussion of this quotient). This may be interpreted as a holomorphic-symplectic quotient of the cotangent bundle T^*G by N. If $G = SL(n, \mathbb{C})$ then we also have a quiver variety description as a hyperkähler quotient at level zero.

This suggests that the implosions should fit into the symplectic duality picture. In our papers [1, 5] we have produced candidates for the duals of the implosions for unitary and orthogonal groups. The key idea is to start with the quiver for the nilpotent cone and then to replace the top-dimensional flavour node by a bouquet of Abelian nodes. The bouquet is chosen so as to preserve balancing conditions for the original nodes, thus generating a global symmetry G in the Coulomb branch. Meanwhile the bouquet nodes (which are usually unbalanced) generate the Abelian symmetry $T_{\mathbb{C}}$. Using the monopole formula one may check that the Coulomb branch has the correct dimension and global symmetry group, and that the reduction by the torus action yields the nilpotent cone, as desired.

Conjecturally, there should be a similar picture for implosions associated to general parabolic subgroups of G, with the unipotent N replaced by the commutator of the parabolic. We are currently investigating possible symplectic duals in this more general situation.

References

1. A. Bourget, A. Dancer, J. Grimminger, A. Hanany, F. Kirwan and Z. Zhong *Orthosymplectic implosions* JHEP August 2021.
2. A. Braverman, M. Finkelberg and H.Nakajima Towards a mathematical definition of Coulomb branches of 3-dimensional N=4 gauge theories, II. Adv. Theor. Math. Phys. **22** 1071–1147 (2018).
3. N. Chriss and V. Ginzburg, *Representation theory and complex geometry*, Birkhauser, Boston (1997).
4. S. Cremonesi, A. Hanany and A. Zeffironi, *Monopole operators and Hilbert series of Coulomb branches of 3d N=4 gauge theories* JHEP **1** (2014).
5. A. Dancer, A.Hanany and F.Kirwan, *Symplectic duality and implosions*: arXiv:2004.09620; in Adv. Theor. Math. Phys. **25** 1367–1387 (2021).

6. A. Dancer, F. Kirwan and A. Swann, *Implosion for hyperkähler manifolds*, Compositio Mathematica **149** 592–630 (2013).
7. V. Ginzburg and S. Riche, *Differential operators on* G/U *and the affine Grassmannian*, J. Math. Jussieu **14** 493–575 (2015).
8. V. Guillemin, L. Jeffrey and R. Sjamaar, *Symplectic implosion*, Transformation Groups **7** 155–184 (2002).
9. N.J. Hitchin, A. Karlhede, U. Lindström and M.Roček. Hyperkähler metrics and supersymmetry. Comm. Math. Phys. **108** 535-589 (1987).

A Sheep in Wolf's Clothing: From Group Theory to a Novel Antiviral Strategy

Pierre-Philippe Dechant

Abstract The recent pandemic has led to a particular interest in mathematical and computational virology. I have previously given an introduction to virus structure and assembly, in particular the group theory aspect of structure and the modelling aspects of packaging signal-mediated assembly. Here I will focus on three more recent developments in these two areas. Firstly, based on the mechanistic understanding of the viral assembly instruction manual, we have recently proposed a new type of antiviral agent, a Therapeutic Interfering Particle, that parasitises the virus itself and misdirects its assembly. This could open up a whole new approach to antiviral therapy and immunisation. The second example is the demonstration of the usefulness of data science techniques in this field, complementary to the modelling simulations, that allow the machine learning of structure in the data from simulations and experiments. Thirdly, the recent discovery of giant viruses leads to renewed interest in geometric models of virus structure, with potentially predictive power via a scaling relation and quantised geometric blueprints.

Keywords Coxeter groups · Root systems · Viruses · Virus structure · Therapeutic interfering particles · Molecular dynamics · Mathematical biology · Packaging signals · Antivirals · Machine learning · Data science · Geometry · Symmetry · Evolution · Fitness landscape · Virus assembly · Stochastic simulations

Mathematics Subject Classification (2010) 52B10 · 52B15 · 20F55 · 17B22 · 97M10 · 97M60 · 37N25 · 92B20 · 92C42 · 37M05

Proceedings of the 2021 Nankai Mathematical Dialogues.

P.-P. Dechant (✉)
School of Science, Health and Technology, York St John University, Lord Mayors Walk, UK
e-mail: ppd22@cantab.net

School of Mathematics, University of Leeds, Leeds, UK

Department of Mathematics, York Cross-disciplinary Centre for Systems Analysis, University of York, Heslington, UK

Y.-H. He et al. (eds.), *Nankai Symposium on Mathematical Dialogues*,
https://doi.org/10.1007/978-981-19-2328-9_16

1 Introduction

The last 4 years since the previous Nankai Symposium have of course seen a viral pandemic and a wealth of new research into viruses: a lot specifically pandemic-related, as well as some more foundational research. In the talk for the last Nankai Symposium I presented an introduction to the language of symmetry and its applications to virus structure, dynamics and assembly (see the companion paper [11]). Since some of the introductory material is already contained in this other paper, in this article I will focus on the more recent developments, after giving a brief summary of the main points.

1.1 Virus Structure

Icosahedral tilings are of crucial importance to structures in biology (e.g. viruses), chemistry (e.g. fullerenes) and biomedicine and nanotechnology (e.g. artificial virus-like particles (VLPs)), which all use repeated identical subunits. Watson and Crick pointed out the importance of polyhedral symmetry for the structure of small 'spherical' viruses, necessitated by the fundamentally discrete building blocks in biology [8]. Shortly after, Caspar and Klug [5] presented a systematic construction of icosahedral tilings with the same type of building block, which are not symmetry-equivalent (different orbits of the icosahedral group) but are 'essentially the same' ('quasi-equivalent') as tiles, or building blocks. This construction was based on triangular building blocks subtriangulating an icosahedron, by essentially superimposing an inflated icosahedral net on a triangular lattice (or its dual hexagonal lattice). But other tile shapes are of course possible, since there are well-known icosahedral tilings in terms of pentagons (the dodecahedron), golden rhombuses (the rhombic triacontahedron) or kites (the deltoidal hexecontahedron). Essentially, these can also just be constructed in terms of superimposing an (inflated) icosahedral net on different types of lattices. In particular, the work of Twarock [35] has clarified possible generalisations: that of different, single tile shapes (still quasiequivalent), or a mix of different tiles (not quasiequivalent). In recent work, [37] suggested a generalisation and systematisation of the idea of superimposing inflated icosahedral nets on different types of hexagonal lattices, e.g. those that give rise to the icosahedral Archimedean solids and their duals. These have mixtures of different tile shapes, such as the hexagon and pentagon from the truncated icosahedron, the pentagon and triangle from the icosidodecahedron, triangle, square and pentagon from the rhombicosidodecahedron and so on, reflecting more general biological constraints and data. More comprehensive reviews of virus structure are e.g. in [13] as well as the companion article [11]. For the purposes of these proceedings, we focus on one recent development regarding the structure of giant viruses, which look essentially like chiral polyhedra consisting of pentagons and triangles (Sect. 4).

1.2 Virus Assembly

A polyhedral structure consisting of identical tile shapes raises the question of how such a structure can assemble [15, 42]. The order of the addition of building blocks onto the growing polyhedral shell essentially defines a Hamiltonian path on the dual polyhedron (see Fig. 1). In the case of viruses, the situation is more complicated because of the biophysical interactions between the building blocks and the thermodynamics of the assembly process, which exponentially favours certain assembly pathways over others. Secondly, in a large class of viruses, the genomic material (single-stranded RNA) is actually involved in the assembly process. This confers the advantage that the virus selectively packages the correct genetic information inside. The way this happens mechanistically is that on top of encoding the genetic information for genes/proteins, the RNA contains a different code, which is effectively an 'assembly instruction manual' (Fig. 2) [6, 26, 27, 31, 32]. The RNA has multiple sites with a similar motif that can form hairpin secondary structures, called packaging signals (PS). These can bind to the coat protein (CP) which is the geometric building block for the polyhedral shell (the virus capsid). Sometimes it is even only upon this binding that the protein gets turned into the correct shape [30, 33, 34]. Using the RNA as an assembly substrate thus has the advantage that the RNA can bind coat protein building blocks, and once in proximity, these can then bind to each other and complete the polyhedral shell (Fig. 3). This biases the assembly process in a favourable direction, speeds it up, and ensures virus-specific encapsidation [16, 17]. These RNA-CP binding sites are in symmetry-equivalent positions at the inside of the capsid shell (as dictated by the symmetry of the capsid), such that a degree of

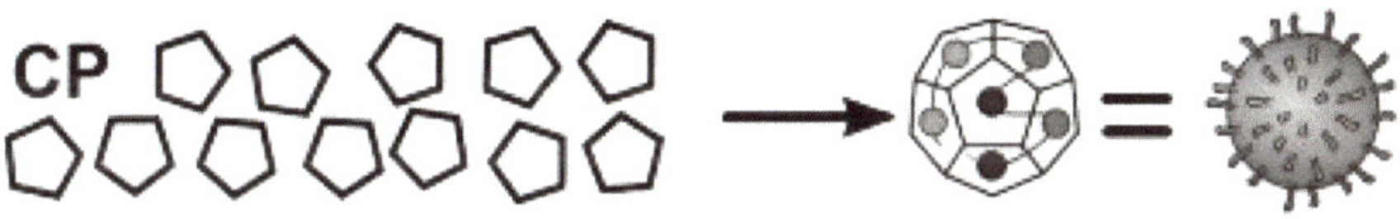

Fig. 1 The order of addition of the 12 pentagonal building blocks defines a Hamiltonian path. The completely assembled polyhedral shell schematically denotes a virus capsid, which contains and protects the viral RNA genome

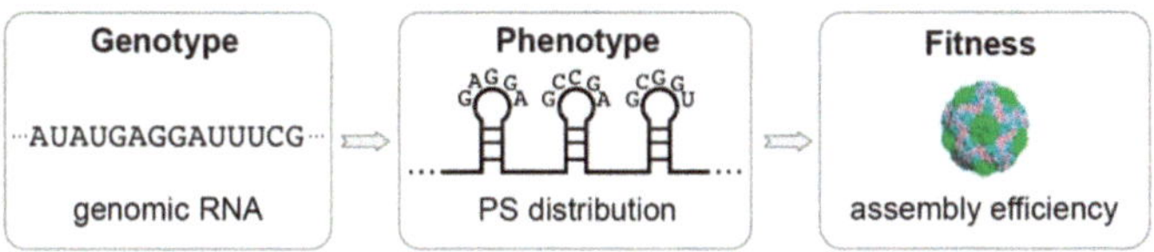

Fig. 2 The viral RNA genome (genotype) contains of course information to make proteins. But it also contains sequences that form stem loops (phenotype) that act as packaging signals (PS) by binding coat protein which helps recruit these building blocks onto the growing capsid. How well these PSs collectively and cooperatively support the assembly of the virus capsid determines the assembly efficiency (fitness)

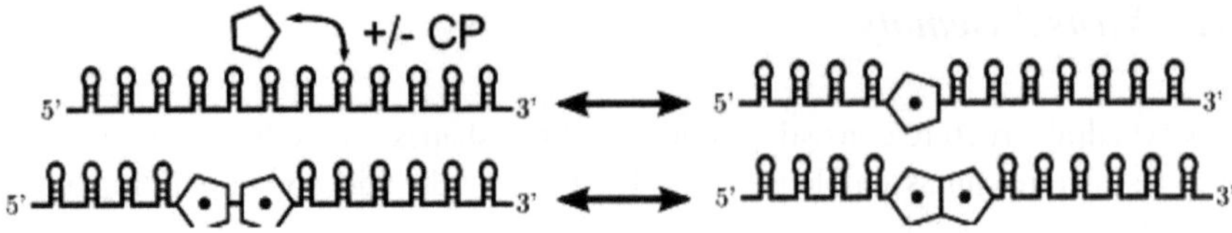

Fig. 3 The PSs in the RNA genome can reversibly bind coat protein. Once in closer proximity on this RNA substrate, these coat proteins are more likely to (reversibly) bind to each other and drive the growing capsid to completion

symmetry propagates into the particle interior, which was previously assumed to be just disordered.

This paradigm shift to assembly via multiple dispersed packaging signals constitutes a step change in the understanding of assembly for a large class of viruses, as well as for potential applications in nanotechnology and biomedicine. Once the assembly mechanism is understood, the instruction manual can be repurposed and optimised such as for the construction of VLPs or other nanoparticles, or interfered with via drugs such as in antiviral therapy. In this paper, we will look at two recent developments concerning packaging signal-mediated assembly. The first is our recent proposal of Therapeutic Interfering Particles (Sect. 2), which use this mechanistic understanding of the assembly process to undermine viral assembly by using a new type of antiviral agent. Secondly, a toy model of such an assembly process for a dodecahedral capsid has recently been studied in [4, 36] via extensive stochastic simulations. We have shown [12] that patterns in the data from these simulations can be very effectively machine learned (Sect. 3).

2 Going Viral–Therapeutic Interfering Particles

The mechanistic understanding of PS-mediated viral assembly opens up the possibility of interfering with this process in an antiviral approach. However, the number of PSs, CPs and viral particles generally is daunting to try to inhibit these RNA-CP interactions en masse. This dilemma begged for a different approach.

I have been dazzled by the simple genius of the polymerase chain reaction (PCR) since I did a school placement at the Institute of Immunology and my national service at the Institute for Virology at the University Hospital in my home town. PCRs have been ubiquitous for testing in the pandemic: they allow the detection of trace amounts of viral genomic material (RNA or DNA) via a chain reaction that leads to exponential amplification. This is achieved through various cycles of doubling: double-stranded DNA has its strands separated, and polymerase then synthesises the missing complementary strands, leading to two intact copies of dsDNA, thus doubling the genomic material in that cycle. Doing this through dozens of cycles turns microscopic amounts of viral material into measurable, macroscopic amounts.

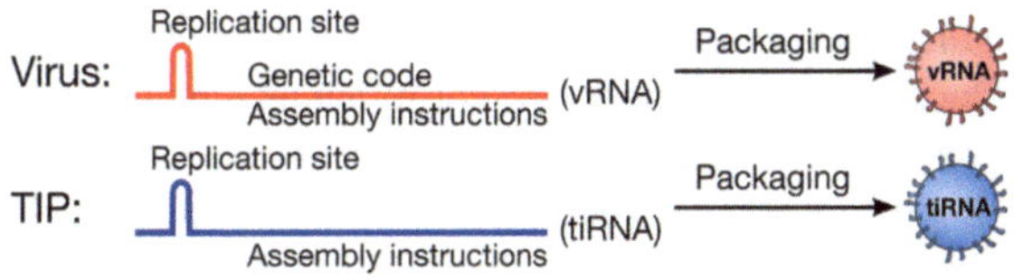

Fig. 4 Therapeutic Interfering Particles can be constructed that have an optimised assembly code, but do not contain the harmful viral genetic code of the real virus. The TIPs freeload by being replicated by the virus and siphon off a fraction of the coat protein that it produces, turning a significant amount of viral resources into innocuous versions of the viral particles

Of course, viruses themselves often exhibit exponential growth. I have been fascinated by the idea of harnessing exponential growth in order to *combat* viruses. What if viruses themselves amplified the agents that combat them? This led me to the idea of Therapeutic Interfering Particles (TIPs). These particles are antiviral agents that piggy-back onto the viruses, and harness their polymerase to copy the TIPs along with the actual virus. What the TIPs do is to misdirect viral assembly and to sequester a large fraction of CP produced by the virus into innocuous particles; these look like virus (so that the immune system can produce antibodies and T-cells) but they are actually harmless, like a sheep in wolf's clothing. In addition, these TIPs could also be spread along viral transmission routes, as a 'spreadable vaccine or treatment'.

More concretely, the understanding of the above PS-mediated assembly mechanism allows the engineering of artificial RNA sequences that have the following properties. Firstly, they lack protein-coding information, so are harmless by themselves. Secondly, they have an optimised PS assembly manual, so they assemble more efficiently and quickly than the real virus. Thirdly, they can get copied by viral polymerase, so in the presence of virus these antiviral agents get copied (see Fig. 4). We have proposed and modelled such TIPs recently [2, 19]. The outcome of this intracellular competition between wild-type (WT) virus and TIPs is that the TIPs tend to outcompete WT virus for several reasons. Firstly, they have been designed to assemble more efficiently. Secondly, they are shorter than WT sequences and can be copied faster by viral polymerase. Thirdly, viral RNA gets translated into protein in the ribosomes, whilst therapeutic interfering RNA does not code for proteins, so tiRNA is more available for being copied as well as for assembling particles.

Defective viral genomes routinely arise naturally through copying errors, for instance when viral polymerase falls off the genome that is being copied early [14, 21, 25, 28, 40]. Some of these can also interfere with wild-type virus in a similar competition scenario to the above, turning them into *defective interfering genomes and particles*. They are associated with an attenuation of the viral infection, but also a possible role in viral persistence as well as stimulating unspecific antiviral responses. Fragments that are extremely short have a replicative advantage, but may not be able to package on their own, but rather are co-packaged with wild-type. Viruses assembling via multiple dispersed packaging signals may not easily give rise to defective interfering genomes that are actually assembly-competent, leading to their immediate removal. Our proposal of TIPs may therefore give rise to particles that interfere more with WT since they can assemble on their own, and outcompete virus even more strongly. Once WT is no longer present to act as a helper virus by providing polymerase and capsid protein, the TIPs are likely to be removed by the immune system, stimulating the antiviral response even further (specific and non-specific).

Of course, more research needs to be done into engineering their design, understanding their dynamics and ensuring their safety (e.g. that TIPs can't co-package WT, or have the improved assembly instruction manual captured by WT). But this newly proposed antiviral strategy therefore opens up exciting new possibilities for our antiviral arsenal by harnessing exponential growth for antiviral agents, potentially beating viruses at their own game.

3 Machine Learning Virus Assembly Models

Our second example also concerns the understanding of the PS-mediated assembly (PSMA) mechanism. The simplest mathematical model for an icosahedral virus capsid is a dodecahedron, consisting of 12 pentagons. In the PSMA paradigm, this would be complemented by 12 packaging signals in the viral genome that each bind to coat protein with a certain binding strength. The way these binding strengths vary within the genome determines the assembly efficiency of that viral genome as it assembles via the Hamiltonian path (see Fig. 5).

The full fitness landscape of all such phenotypes for packaging signals with 3 different binding strengths (strong, medium, weak) was computed in [9, 36]. This work implemented extensive stochastic simulations using a Gillespie algorithm to model the biophysics of the assembly process: as determined by the binding strength each packing signal/stem loop can reversibly bind coat protein (and fall apart again), and similarly, coat proteins could bind each other and fall apart again (see Fig. 3). Each simulation was given enough resources to make 2000 virus capsids, and the fitness of each 'genetic sequence' was measured by how many capsids were successfully assembled out of this maximum of 2000. These substantial computations were performed on a supercomputer over several weeks. The resulting dataset was simply a map from a set of 12-dimensional input vectors describing the binding strengths of the 12 PSs to the scalar giving the overall assembly efficiency (output vectors).

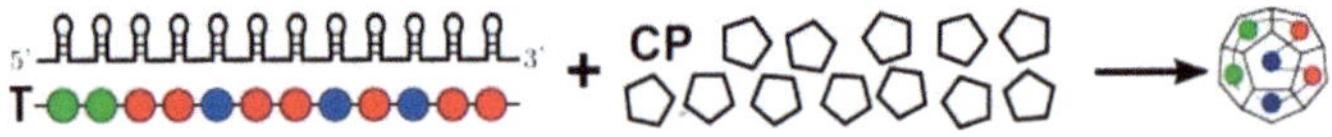

Fig. 5 The level of binding strengths with which individual PSs in a genome can bind CP determines different assembly efficiencies (and thus fitness) for that genome. The whole fitness landscape for 12 PSs with 3 different binding strengths has been explored via stochastic simulations in [36]. This data set mapping input binding strength to assembly efficiency (output) contains a lot of structure that can be easily machine learnt [12]

This is a typical problem in machine learning: there is a set of input vectors and output vectors that are related by an expensive map. Neural networks are trying to approximate this map in a much quicker and more manageable way by learning from part of the data. In [12] we therefore took this data set for the toy model and machine learnt it. The stochastic simulations had taken several weeks on a supercomputer, whilst we could machine learn patterns in this data in minutes.

This opens up exciting opportunities in future: one could try to model more realistic viral assembly processes by combining simulations with machine learning. The many more component parts of more detailed geometric models would make exhaustive simulations prohibitive at present–but the fitness landscape could be partially explored, with machine learning 'filling in the gaps'.

4 Data Mining the World of Giant Viruses

Our final example concerns novel developments in virus structure. Recent years have seen the discovery of unexpectedly large viruses of bacteria (jumbophages) as well as eukaryotes (giant viruses) [7, 20]. In particular, the latter raise some interesting questions regarding structure and assembly. They have an unprecedented range of triangulation numbers that are much larger than anything seen before.

Whilst Caspar-Klug theory in general allows triangulation numbers of $T = h^2 + hk + k^2$ for non-negative integers h and k, in practice their range has historically been very limited. However, recent discoveries have found triangulation numbers as large as 499, as well as a series of viruses that all have $h = 7$ and different k [1, 3, 18, 22, 24, 29, 38, 39, 41]. The reason for this could be the following. Instead of having 20 large exactly triangular building blocks, these viruses appear to have little triangular hooks in the corners of the triangles, which effectively form the 5-fold axes (Fig. 6 left). Similar hooks can be seen in icosahedral lamp shades, holding the building blocks in the right position. The outcome is that the tile shape is either a large triangle with three small triangles stuck to it, or a large triangle (trisymmetron) with the little hooks together forming pentagons at the 5-fold axes (pentasymmetrons) (Fig. 6 middle). This ties in with the above-mentioned developments in allowing mixtures of different tile shapes. In particular, if these viruses assemble from prefabricated pentasymmetrons and trisymmetrons (they certainly break apart into them) then we are back to the idea of the tile order in assembly giving a Hamiltonian path on the polyhedron, either just with the trisymmetrons, or a Hamiltonian path across the 'Archimedean' polytope, i.e. visiting two different tile shapes.

There is a second interesting feature in these giant viruses. In Caspar-Klug theory, one starts with a hexagonal lattice, and can construct different triangulation capsids by scaling the icosahedral net one superimposes onto this hexagonal lattice. In this approach, the fundamental hexagonal building block is the same size. This is typically not the case for different viruses: they tend to use different building blocks and different triangulations with no relationship among them.

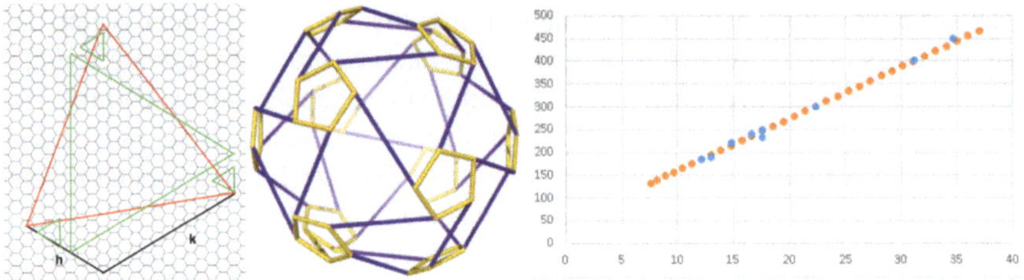

Fig. 6 Left: A Caspar-Klug triangle (red) with $h = 7$ and, here, $k = 12$ (black) can also be viewed as a large triangle with base length $k + 3$ with three smaller triangles that form little hooks at the corners of the triangle, which form the pentasymmetrons around the 5-fold axes (green). Middle: this gives rise to a chiral polytope made of triangles (trisymmetrons) and pentagons (pentasymmetrons). Right: Assuming that giant viruses with $h = 7$ may be evolutionarily related and the fundamental tiling unit highly conserved predicts a scaling relation between capsids with different triangulation numbers: a quantised family of possible geometries is predicted (orange), indexed by different k but with h fixed to 7, with particle diameter growing as $\sqrt{T}$. This appears to fit well with observed giant viruses (blue)

However, if we suppose that we could build different triangulation capsids based on the same building block: we would expect the capsid size to scale with the square root of the triangulation number:

$$d \sim \sqrt{T}$$

This is because T is effectively a measure of area and thus quadratic in the edge length, whilst particle size scales linearly with the edge length.

Coming up with a gene that is a suitable hexagonal geometric building block is tricky. Such a geometric tiling unit may be much more strongly evolutionarily conserved than other aspects of the viral life cycle. Giant viruses are interesting in that their major capsid proteins may be very similar to each other, hinting that they are evolutionarily related and that such a building block is strongly conserved. This would then imply a scaling relation between different giant viruses with different triangulation numbers but approximately the same fundamental building block (Fig. 6 right).

In fact, this seems to be the case for the giant viruses of the family $k^2 + 7k + 49$ above. This family has a range of k, but also some gaps that don't exist or have simply not been observed yet. Moreover, the determination of structure and thus triangulation number is notoriously difficult. There are giant viruses whose sizes are known but not their triangulation numbers. Assuming our scaling relation holds for these, we can *predict* which viral geometry they should have, with a specific triangulation number. Having a putative triangulation number may also make structure determination easier. Furthermore, having a quantised spectrum of 'allowed' particle sizes may help direct more targeted searches for giant viruses and better identification in future.

5 Conclusions

Early work in Mathematical Virology has focused on geometry, symmetry and virus structure. It is interesting that recent developments demonstrate that these ideas are still interesting and relevant, such as Archimedean tilings and giant viruses. Furthermore, deeper insights into virus structure have allowed investigation of virus assembly. The symmetry from the capsid surface is to some extent imprinted onto parts of the RNA via the RNA-CP contact [10, 23]. This has led to new mechanistic insights into virus assembly via multiple dispersed packaging signals. These can be repurposed for nanotechnology as well as antiviral approaches via the TIPs we have recently proposed.

This 'real-world' topic of viruses, heightened by the COVID pandemic, illustrates the unity of Mathematics (across Pure and Applied) and its relevance to applications in our world and society: from group theory, to stochastic and ODE modelling, to machine learning, with connections to exceptional geometries, Clifford algebras and ADE correspondences [11]. Having *Mathematical Dialogues* connecting seemingly different areas in our increasingly fragmented world of mathematical specialisation will be crucial to the relevance and voice of the Mathematical Sciences in society: the future of Mathematics will have a strong interdisciplinary component.

Dedication Dedicated to the memory of Jiří Patera and Uwe Grimm.

Acknowledgements I would like to thank Yang-Hui He, Reidun Twarock, Richard Bingham, Susan Stepney and Thomas Klose for interesting discussions. Many thanks also to the organisers of the ICERM Semester Programme organisers on 'Nonlinear Algebra in Applications' at Brown University, as well as the organisers of the 'Nankai Symposium on Mathematical Dialogues' for an excellent online meeting. I look forward to returning to the Chern Institute again in the future.

References

1. Julien Andreani, Jacques Yaacoub Bou Khalil, Madhumati Sevvana, Samia Benamar, Fabrizio Di Pinto, Idir Bitam, Philippe Colson, Thomas Klose, Michael G Rossmann, Didier Raoult, et al. Pacmanvirus, a new giant icosahedral virus at the crossroads between asfarviridae and faustoviruses. *Journal of virology*, 91(14):e00212–17, 2017.
2. Thomas R Aunins, Katherine A Marsh, Gitanjali Subramanya, Susan L Uprichard, Alan S Perelson, and Anushree Chatterjee. Intracellular hepatitis C virus modeling predicts infection dynamics and viral protein mechanisms. *J. Virol.*, 92(11), 2018.
3. Leena H Bajrai, Samia Benamar, Esam I Azhar, Catherine Robert, Anthony Levasseur, Didier Raoult, and Bernard La Scola. Kaumoebavirus, a new virus that clusters with faustoviruses and asfarviridae. *Viruses*, 8(11):278, 2016.
4. Richard J Bingham, Eric C Dykeman, and Reidun Twarock. RNA virus evolution via a quasispecies-based model reveals a drug target with a high barrier to resistance. *Viruses*, 9(11):347, 2017.
5. D. L. D. Caspar and A. Klug. Physical principles in the construction of regular viruses. *Cold Spring Harbor Symp. Quant. Biol.*, 27:1–24, 1962.

6. Rebecca Chandler-Bostock, Carlos P Mata, Richard J Bingham, Eric C Dykeman, Bo Meng, Tobias J Tuthill, David J Rowlands, Neil A Ranson, Reidun Twarock, and Peter G Stockley. Assembly of infectious enteroviruses depends on multiple, conserved genomic rna-coat protein contacts. *PLoS Pathogens*, 16(12):e1009146, 2020.
7. Jean-Michel Claverie, Hiroyuki Ogata, Stéphane Audic, Chantal Abergel, Karsten Suhre, and Pierre-Edouard Fournier. Mimivirus and the emerging concept of "giant" virus. *Virus research*, 117(1):133–144, 2006.
8. F. H. C. Crick and J. D. Watson. The structure of small viruses. *Nature*, 177:473–475, 1956.
9. Harel Dahari, Jennifer E Layden-Almer, Eric Kallwitz, Ruy M Ribeiro, Scott J Cotler, Thomas J Layden, and Alan S Perelson. A mathematical model of hepatitis C virus dynamics in patients with high baseline viral loads or advanced liver disease. *Gastroenterology*, 136(4):1402–1409, 2009.
10. Xinghong Dai, Zhihai Li, Mason Lai, Sara Shu, Yushen Du, Z Hong Zhou, and Ren Sun. In situ structures of the genome and genome-delivery apparatus in a single-stranded RNA virus. *Nature*, 541(7635):112–116, 2017.
11. Pierre-Philippe Dechant. Root systems & Clifford algebras: from symmetries of viruses to E_8 & ADE correspondences. *inSTEMM Journal*, 2022.
12. Pierre-Philippe Dechant and Yang-Hui He. Machine-learning a virus assembly fitness landscape. *PloS one*, 16(5):e0250227, 2021.
13. Pierre-Philippe Dechant and Reidun Twarock. Models of viral capsid symmetry as a driver of discovery in virology and nanotechnology. *The Biochemist*, 43(1):20–24, 2021.
14. Nigel J Dimmock and Andrew J Easton. Defective interfering influenza virus RNAs: time to reevaluate their clinical potential as broad-spectrum antivirals? *Journal of virology*, 88(10):5217–5227, 2014.
15. Eric C Dykeman, Peter G Stockley, and Reidun Twarock. Building a viral capsid in the presence of genomic RNA. *Physical Review E*, 87(2):022717, 2013.
16. Eric C Dykeman, Peter G Stockley, and Reidun Twarock. Solving a Levinthal's paradox for virus assembly identifies a unique antiviral strategy. *Proceedings of the National Academy of Sciences*, 111(14):5361–5366, 2014.
17. Eric Charles Dykeman. A model for viral assembly around an explicit RNA sequence generates an implicit fitness landscape. *Biophysical Journal*, 113(3):506–516, 2017.
18. Qianglin Fang, Dongjie Zhu, Irina Agarkova, Jagat Adhikari, Thomas Klose, Yue Liu, Zhenguo Chen, Yingyuan Sun, Michael L Gross, James L Van Etten, et al. Near-atomic structure of a giant virus. *Nature communications*, 10(1):1–11, 2019.
19. Farzad Fatehi, Richard J Bingham, Pierre-Philippe Dechant, Peter G Stockley, and Reidun Twarock. Therapeutic interfering particles exploiting viral replication and assembly mechanisms show promising performance: a modelling study. *Scientific reports*, 11(1):1–10, 2021.
20. Jianfei Hua, Alexis Huet, Carlos A Lopez, Katerina Toropova, Welkin H Pope, Robert L Duda, Roger W Hendrix, and James F Conway. Capsids and genomes of jumbo-sized bacteriophages reveal the evolutionary reach of the HK97 fold. *mBio*, 8(5):e01579–17, 2017.
21. Alice S Huang and David Baltimore. Defective viral particles and viral disease processes. *Nature*, 226(5243):325, 1970.
22. Thomas Klose, Dorine G Reteno, Samia Benamar, Adam Hollerbach, Philippe Colson, Bernard La Scola, and Michael G Rossmann. Structure of faustovirus, a large dsdna virus. *Proceedings of the National Academy of Sciences*, 113(22):6206–6211, 2016.
23. Charles M Knobler and William M Gelbart. How and why RNA genomes are (partially) ordered in viral capsids. *Current Opinion in Virology*, 52:203–210, 2022.
24. Yu G Kuznetsov, James R Gurnon, James L Van Etten, and Alexander McPherson. Atomic force microscopy investigation of a chlorella virus, pbcv-1. *Journal of structural biology*, 149(3):256–263, 2005.
25. Tomaz B Manzoni and Carolina B López. Defective (interfering) viral genomes re-explored: impact on antiviral immunity and virus persistence. *Future Virol.*, 13(7):493–503, 2018.
26. Nikesh Patel, Simon J White, Rebecca F Thompson, Richard Bingham, Eva U Weiß, Daniel P Maskell, Adam Zlotnick, Eric C Dykeman, Roman Tuma, Reidun Twarock, et al. HBV RNA

pre-genome encodes specific motifs that mediate interactions with the viral core protein that promote nucleocapsid assembly. *Nature microbiology*, 2:17098, 2017.
27. Nikesh Patel, Emma Wroblewski, German Leonov, Simon EV Phillips, Roman Tuma, Reidun Twarock, and Peter G Stockley. Rewriting nature's assembly manual for a ssRNA virus. *Proceedings of the National Academy of Sciences*, page 201706951, 2017.
28. Ulfert Rand, Sascha Young Kupke, Hanna Shkarlet, Marc Dominique Hein, Tatjana Hirsch, Pavel Marichal-Gallardo, Luka Cicin-Sain, Udo Reichl, and Dunja Bruder. Antiviral activity of influenza A virus defective interfering particles against SARS-CoV-2 replication in vitro through stimulation of innate immunity. *Cells*, 10(7):1756, 2021.
29. Dorine Gaëlle Reteno, Samia Benamar, Jacques Bou Khalil, Julien Andreani, Nicholas Armstrong, Thomas Klose, Michael Rossmann, Philippe Colson, Didier Raoult, and Bernard La Scola. Faustovirus, an asfarvirus-related new lineage of giant viruses infecting amoebae. *Journal of Virology*, 89(13):6585–6594, 2015.
30. O. Rolfsson, K. Toropova, N. Ranson, and P. Stockley. Mutually-induced conformational switching of RNA and coat protein underpins efficient assembly of a viral capsid. *J Mol Biol*, 401:309–322, 2010.
31. Shabih Shakeel, Eric C Dykeman, Simon J White, Ari Ora, Joseph JB Cockburn, Sarah J Butcher, Peter G Stockley, and Reidun Twarock. Genomic RNA folding mediates assembly of human parechovirus. *Nature Communications*, 8(1):5, 2017.
32. H Stewart, RJ Bingham, SJ White, EC Dykeman, C Zothner, AK Tuplin, Peter G Stockley, R Twarock, and M Harris. Identification of novel RNA secondary structures within the hepatitis C virus genome reveals a cooperative involvement in genome packaging. *Scientific reports*, 6:22952, 2016.
33. Peter G Stockley, Neil A Ranson, and Reidun Twarock. A new paradigm for the roles of the genome in ssRNA viruses. *Future Virology*, 8(6):531–543, 2013.
34. Peter G Stockley, Simon J White, Eric Dykeman, Iain Manfield, Ottar Rolfsson, Nikesh Patel, Richard Bingham, Amy Barker, Emma Wroblewski, Rebecca Chandler-Bostock, et al. Bacteriophage MS2 genomic RNA encodes an assembly instruction manual for its capsid. *Bacteriophage*, 6(1):e1157666, 2016.
35. R. Twarock. The architecture of viral capsids based on tiling theory. *J. Theor. Medicine*, 6:87–90, 2005.
36. Reidun Twarock, Richard J Bingham, Eric C Dykeman, and Peter G Stockley. A modelling paradigm for RNA virus assembly. *Current opinion in virology*, 2018.
37. Reidun Twarock and Antoni Luque. Structural puzzles in virology solved with an overarching icosahedral design principle. *Nature communications*, 10(1):1–9, 2019.
38. Chuan Xiao, Matthias G Fischer, Duer M Bolotaulo, Nancy Ulloa-Rondeau, Gustavo A Avila, and Curtis A Suttle. Cryo-em reconstruction of the cafeteria roenbergensis virus capsid suggests novel assembly pathway for giant viruses. *Scientific reports*, 7(1):5484, 2017.
39. Xiaodong Yan, Paul R Chipman, Tonje Castberg, Gunnar Bratbak, and Timothy S Baker. The marine algal virus PpV01 has an icosahedral capsid with $T = 219$ quasisymmetry. *Journal of Virology*, 79(14):9236–9243, 2005.
40. Shun Yao, Anoop Narayanan, Sydney A Majowicz, Joyce Jose, and Marco Archetti. A synthetic defective interfering SARS-CoV-2. *PeerJ*, 9:e11686, 2021.
41. Genki Yoshikawa, Romain Blanc-Mathieu, Chihong Song, Yoko Kayama, Tomohiro Mochizuki, Kazuyoshi Murata, Hiroyuki Ogata, and Masaharu Takemura. Medusavirus, a novel large DNA virus discovered from hot spring water. *Journal of Virology*, 93(8), 2019.
42. Adam Zlotnick. To build a virus capsid: an equilibrium model of the self assembly of polyhedral protein complexes. *Journal of molecular biology*, 241(1):59–67, 1994.

[illegible] genome encodes specific motifs that mediate interactions with the viral core protein that [illegible] nucleocapsid assembly. Nature microbiology, 2:17088, 2017.

27. [illegible] Patel, Emma Wroblewski, German Leonov, Simon EV Phillips, Roman Tuma, Reidun Twarock, and Peter G Stockley. [illegible]

[illegible]

2d Supersymmetric Gauge Theories, D-Branes and Trialities

Sebastián Franco

Abstract Engineering quantum field theories in String Theory in terms of branes is a powerful approach for understanding their dynamics. We review recent progress in the realization of $2d$ $\mathcal{N} = (0, 2)$ gauge theories in terms of branes. We discuss Brane Brick Models, a new class of Type IIA brane configurations which are T-dual to D1-branes over singular toric Calabi-Yau 4-folds. They fully encode the infinite class of $2d$ $\mathcal{N} = (0, 2)$ quiver gauge theories on the worldvolume of the D1-branes and significantly streamline their connection to the probed geometries. As an application, we explain how these constructions provide a brane realization of triality. We also comment on the realization of $2d$ $\mathcal{N} = (0, 1)$ theories via Spin(7) orientifolds. This note is based on the author's talk at the Nankai Symposium on Mathematical Dialogues celebrating the 110th anniversary of the birth of Prof. S.-S. Chern.

1 Introduction

Understanding the dynamics of quantum field theories in various dimensions is an active area of research. For this purpose, engineering quantum field theories in terms of branes in String and M-Theory is often a powerful tool. In this note, we will review the realization of $2d$ $\mathcal{N} = (0, 2)$ gauge theories on the world volume of D1-branes probing singular toric Calabi-Yau (CY) 4-folds [1, 2], as schematically shown in Fig. 1.

S. Franco (✉)
Physics Department, The City College of the CUNY, 160 Convent Avenue, New York, NY 10031, USA
e-mail: sfranco@ccny.cuny.edu

Physics Program and Initiative for the Theoretical Sciences, The Graduate School and University Center, The City University of New York, 365 Fifth Avenue, New York, NY 10016, USA

Y.-H. He et al. (eds.), *Nankai Symposium on Mathematical Dialogues*,
https://doi.org/10.1007/978-981-19-2328-9_17

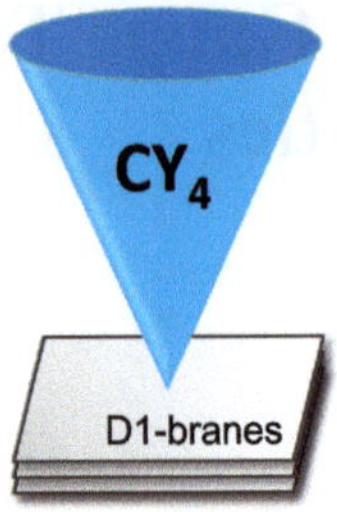

Fig. 1 D1-branes probing a CY_4

2 Brane Brick Models

Brane brick models are obtained from D1-branes at CY_4 singularities by T-duality. We refer the reader to [2–5] for detailed presentations. A brane brick model is a Type IIA brane configuration consisting of D4-branes wrapping a 3-torus $\mathbb{T}^3$ and suspended from an NS5-brane that wraps a holomorphic surface Σ intersecting with $\mathbb{T}^3$. The holomorphic surface Σ is the zero locus of the Newton polynomial defined by the toric diagram of the CY_4. The basic ingredients of the brane setup are summarized in Table 1. The (246) directions are compactified on a $\mathbb{T}^3$. The $2d$ gauge theory lives on the two directions (01) common to all the branes.

Brane brick models, or equivalently their dual periodic quivers, fully encode the $2d$ $\mathcal{N} = (0, 2)$ quiver gauge theories on the worldvolume of D1-branes probing toric CY 4-folds. Namely, they summarize not only the quivers but also the J- and E-terms. The dictionary between brane brick models and gauge theories is summarized in Table 2.

Table 1 Brane brick model configuration

	0	1	2	3	4	5	6	7	8	9
D4	×	×	×	·	×	·	×	·	·	·
NS5	×	×	——— Σ ———						·	·

Table 2 Dictionary between brane brick models and $2d$ $\mathcal{N} = (0, 2)$ gauge theories

Brane brick model	Gauge theory	Periodic quiver
Brick	Gauge group	Node
Oriented face between bricks i and j	Bifundamental chiral field from node i to node j	Oriented (black) arrow from node i to node j
Unoriented square face between bricks i and j	Bifundamental Fermi field between nodes i and j	Unoriented (red) line between nodes i and j
Edge	Interaction by J- or E-term	Plaquette encoding a J- or an E-term

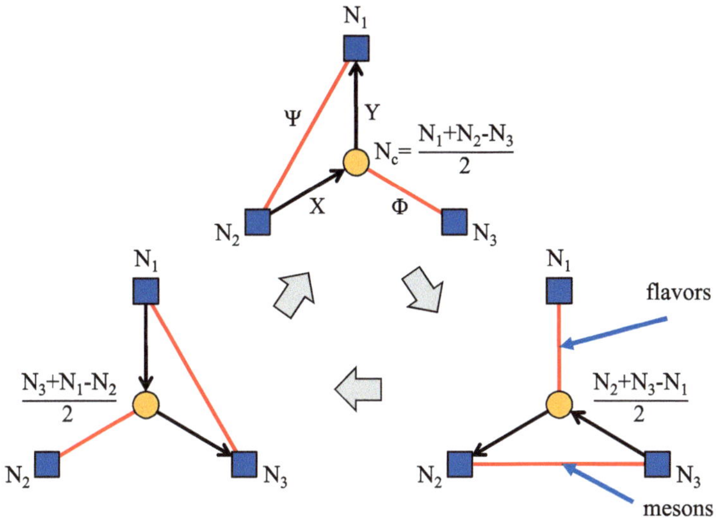

Fig. 2 Triality of $2d$ $\mathcal{N} = (0, 2)$ SQCD. Black arrows and red lines represent chiral and Fermi fields, respectively

Brane brick models reduce the computation of the underlying CY_4 geometry starting from the gauge theory to a combinatorial problem, which is based on a generalization of perfect matchings [3]. Conversely, several efficient algorithms for determining the brane brick models associated to a given geometry have been developed [2, 5–8].

3 Triality

A new IR equivalence for $2d$ $\mathcal{N} = (0, 2)$ gauge theory was discovered in [9]. In its simplest incarnation, which is shown in Fig. 2, it relates three SQCD-like theories and was therefore dubbed *triality*.[1] In more general theories, triality can be locally applied to an individual gauge group, with the rest of the theory acting as a spectator. Applying triality to the same gauge group three consecutive times results in the original theory.

[1] In the figure, yellow nodes indicate $SU(N_c)$ gauge groups. The theories on D1-branes have instead $U(N_c)$ gauge groups. A $U(N_c)$ version of triality was also introduced in [9]. It differs from the $SU(N_c)$ triality in Fig. 2 by the presence of additional Fermi fields in the determinant representation of the gauge group for the cancellation of the Abelian anomaly. It is expected that Abelian anomalies of gauge theories on D1-branes are cancelled via a generalized Green-Schwarz mechanism. For this reason, the determinant Fermi fields are not present in those theories and triality reduces to the one considered here.

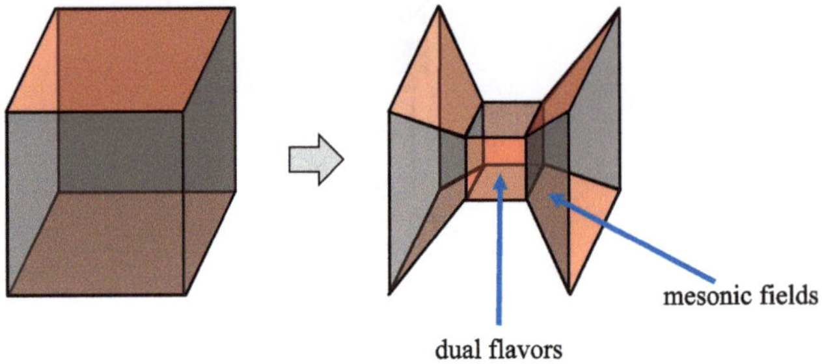

Fig. 3 The cube move is a local transformation of a brane brick models realizing $\mathcal{N} = (0, 2)$ triality

For brane brick models, triality is elegantly realized by the local transformation shown in Fig. 3, which we denote *cube move*. More precisely, the cube move is the simplest representative of a family of local transformations implementing triality [4]. All these moves leave the underlying toric CY_4 invariant.

The correspondence between D1-brane probing toric CY 4-folds and brane brick models can be understood systematically using mirror symmetry [10, 11]. The mirror configuration consists of D5-branes wrapping 4-spheres and the gauge theory is determined by how they intersect. In this context, triality is realized in terms of geometric transitions in the mirror geometry [11]. More generally, mirror symmetry leads to a beautiful geometric unification of dualities in different dimensions, where the order of the duality is $n - 1$ for a CY n-fold [11, 12].

4 Spin(7) Orientifolds

As usual, it is desirable to investigate theories with less supersymmetry. The next step corresponds to $2d$ $\mathcal{N} = (0, 1)$, namely minimally supersymmetric, theories. These theories are extremely interesting, since they are barely supersymmetric and live at the borderline between non-SUSY theories and others with higher amounts of SUSY, for which powerful tools such as holomorphy become applicable. Due to the reduced SUSY, they enjoy a broad range of interesting dynamics. While there has been recent progress in their understanding, they remain relatively unexplored.

The geometric engineering of $2d$ $\mathcal{N} = (0, 1)$ gauge theories on D1-branes probing singularities was initiated in [13], where a new class of backgrounds, denoted *Spin(7) orientifolds* was introduced. These orientifolds are quotients of CY 4-folds by a combination of an anti-holomorphic involution leading to a Spin(7) cone and worldsheet parity. They naturally connect the view of $\mathcal{N} = (0, 1)$ theories as real slices of $\mathcal{N} = (0, 2)$ theories and Joyce's geometric construction of Spin(7) manifolds starting from CY 4-folds [14]. This geometric construction provides a new approach for studying $2d\ \mathcal{N} = (0, 1)$ theories. It also gives a new perspective on the recently proposed triality of $2d\ \mathcal{N} = (0, 1)$ gauge theories [15], in which triality translates into the fact that multiple gauge theories correspond to the same underlying orientifold [16].

Acknowledgements I would like to thank the organizers of the Nankai Symposium on Mathematical Dialogues for putting together such an exciting meeting and for the opportunity to present my work. I am indebted to my wonderful colleagues, Cyril Closset, Dongwook Ghim, Jirui Guo, Azeem Hasan, Sangmin Lee, Alessandro Mininno, Gregg Musiker, Rak-Kyeong Song, Angel Uranga, Cumrun Vafa, Daisuke Yokoyama and Xingyang Yu, for enjoyable collaborations on the topics presented in this note. This research was supported by the U.S. National Science Foundation grants PHY-1820721, PHY-2112729 and DMS-1854179.

References

1. H. Garcia-Compean and A. M. Uranga, *Brane box realization of chiral gauge theories in two-dimensions*, http://dx.doi.org/10.1016/S0550-3213(98)00725-1*Nucl. Phys. B***539** (1999) 329–366, [http://arxiv.org/abs/hep-th/9806177].
2. S. Franco, D. Ghim, S. Lee, R.-K. Seong and D. Yokoyama, *2d (0,2) Quiver Gauge Theories and D-Branes*, http://dx.doi.org/10.1007/JHEP09(2015)072*JHEP***09** (2015) 072, [http://arxiv.org/abs/1506.03818].
3. S. Franco, S. Lee and R.-K. Seong, *Brane Brick Models, Toric Calabi-Yau 4-Folds and 2d (0,2) Quivers*, http://dx.doi.org/10.1007/JHEP02(2016)047*JHEP***02** (2016) 047, [http://arxiv.org/abs/1510.01744].
4. S. Franco, S. Lee and R.-K. Seong, *Brane brick models and 2d (0, 2) triality*, http://dx.doi.org/10.1007/JHEP05(2016)020*JHEP***05** (2016) 020, [http://arxiv.org/abs/1602.01834].
5. S. Franco, S. Lee, R.-K. Seong and C. Vafa, *Brane Brick Models in the Mirror*, http://dx.doi.org/10.1007/JHEP02(2017)106*JHEP***02** (2017) 106, [http://arxiv.org/abs/1609.01723].
6. S. Franco, S. Lee and R.-K. Seong, *Orbifold Reduction and 2d (0,2) Gauge Theories*, http://dx.doi.org/10.1007/JHEP03(2017)016*JHEP***03** (2017) 016, [http://arxiv.org/abs/1609.07144].
7. S. Franco and A. Hasan, *3d printing of 2d $\mathcal{N} = (0, 2)$ gauge theories*, http://dx.doi.org/10.1007/JHEP05(2018)082*JHEP***05** (2018) 082, [http://arxiv.org/abs/1801.00799].
8. S. Franco and A. Hasan, *Calabi-Yau products: graded quivers for general toric Calabi-Yaus*, http://dx.doi.org/10.1007/JHEP02(2021)174*JHEP***02** (2021) 174, [http://arxiv.org/abs/2004.13765].
9. A. Gadde, S. Gukov and P. Putrov, *(0, 2) trialities*, http://dx.doi.org/10.1007/JHEP03(2014)076*JHEP***03** (2014) 076, [http://arxiv.org/abs/1310.0818].
10. M. Futaki and K. Ueda, *Tropical Coamoeba and Torus-Equivariant Homological Mirror Symmetry for the Projective Space*, http://dx.doi.org/10.1007/s00220-014-2155-1*Commun. Math. Phys***332** (2014) 53–87.
11. S. Franco, S. Lee, R.-K. Seong and C. Vafa, *Quadrality for Supersymmetric Matrix Models*, http://dx.doi.org/10.1007/JHEP07(2017)053*JHEP***07** (2017) 053, [http://arxiv.org/abs/1612.06859].
12. S. Franco and G. Musiker, *Higher Cluster Categories and QFT Dualities*, http://dx.doi.org/10.1103/PhysRevD.98.046021*Phys. Rev. D***98** (2018) 046021, [http://arxiv.org/abs/1711.01270].
13. S. Franco, A. Mininno, A. M. Uranga and X. Yu, *2d $\mathcal{N} = (0, 1)$ gauge theories and Spin(7) orientifolds*, http://dx.doi.org/10.1007/JHEP03(2022)150*JHEP***03** (2022) 150, [http://arxiv.org/abs/2110.03696].
14. D. Joyce, *A new construction of compact 8-manifolds with holonomy* Spin(7), *J. Diff. Geom.***53** (1999) 89–130, [http://arxiv.org/abs/math/9910002].

15. S. Gukov, D. Pei and P. Putrov, *Trialities of minimally supersymmetric 2d gauge theories*, http://dx.doi.org/10.1007/JHEP04(2020)079*JHEP***04** (2020) 079, [http://arxiv.org/abs/1910.13455].
16. S. Franco, A. Mininno, A. M. Uranga and X. Yu, *Spin(7) orientifolds and 2d $\mathcal{N}$ = (0, 1) triality*, http://dx.doi.org/10.1007/JHEP01(2022)058*JHEP***01** (2022) 058, [http://arxiv.org/abs/2112.03929].

On a Matrix Element Representation of Special Functions Associated with Toric Varieties

A. A. Gerasimov, D. R. Lebedev, and S. V. Oblezin

Abstract We develop representation theory approach to the study of special functions associated with toric varieties. In particular we show that the corresponding special functions are given by matrix elements of certain non-reductive Lie algebras.

With toric manifolds and their duals one can naturally associate generalized hypergeometric functions introduced by Gelfand-Kapranov-Zelevinski [3]. These functions arise in description of quantum cohomology of toric manifolds and its mirror dual (see e.g. [2]). In this note we consider a special instance of these functions associated with toric varieties obtained by Hamiltonian reduction of $\mathbb{C}^N$ (see e.g. [1]). Precisely let us fix the standard symplectic structure on $\mathbb{C}^N$

$$\Omega = \frac{\imath}{2}\sum_{i=1}^{N} dz_i \wedge d\bar{z}_i = \frac{1}{2}\sum_{i=1}^{N} d|z_i|^2 \wedge d\mathrm{Arg}(z_i), \tag{1}$$

and consider the Hamiltonian action of the compact torus $T^n = (S^1)^n$ on $\mathbb{C}^N$. The corresponding toric manifold $Z = \mathbb{C}^N//T^n$ is the Hamiltonian reduction under action of T^n. The torus $T^N = (S^1)^N$ and hence its subtorus $T^n \subset T^N$ act on Z with the following momentum maps

$$\mathbb{C}^N \longrightarrow \mathrm{Lie}(T^N)^*, \qquad \mathbb{C}^N \longrightarrow \mathrm{Lie}(T^n)^*. \tag{2}$$

Contribution to Proceedings of the Nankai Symposium on Mathematical Dialogue 2021.

A. A. Gerasimov (✉) · D. R. Lebedev
Laboratory for Quantum Field Theory and Information, Institute for Information Transmission Problems, RAS, Moscow, Russia
e-mail: anton.a.gerasimov@gmail.com

A. A. Gerasimov
Interdisciplinary Scientific Center J.-V. Poncelet (CNRS UMI 2615), Moscow, Russia

D. R. Lebedev
Moscow Center for Continuous Mathematical Education, Moscow, Russia

S. V. Oblezin
School of Mathematical Sciences, University of Nottingham, Nottingham, UK

Y.-H. He et al. (eds.), *Nankai Symposium on Mathematical Dialogues*,
https://doi.org/10.1007/978-981-19-2328-9_18

Define a natural projection

$$m\,:\quad \mathrm{Lie}(T^N)^* \longrightarrow \mathrm{Lie}(T^n)^*\,, \tag{3}$$

which is dual to the inclusion $T^n \subset T^N$. Let $\{E_i,\ i\in I\}$ indexed by $I=\{1,\dots,N\}$ be a basis in $\mathrm{Lie}(T^N)^*$ and let $\{e_\alpha,\ \alpha\in A\}$ indexed by $A=\{1,\dots,n\}$ be a basis in $\mathrm{Lie}(T^n)^*$. Then we have explicitly

$$m(E_j)=\sum_{\alpha=1}^{n} e_\alpha m_j^\alpha, \qquad j\in I\,. \tag{4}$$

Let $\{v_k,\ k\in \overline{A}\}$ indexed by $\overline{A}=\{1,\dots,N-n\}$ be a basis in the kernel $\mathrm{Ker}(m)$ of (3):

$$v_k=\sum_{i=1}^{N} v_k^i E_i \in \mathrm{Lie}(T^N)^*, \qquad \sum_{i=1}^{N} m_i^\alpha v_k^i\,=0, \qquad \alpha\in A,\quad k\in\overline{A}\,. \tag{5}$$

Let $Z_1,\dots,Z_N$ be cohomology classes dual to the coordinate hyperplane divisors in Z given by intersects of the coordinate hyperplanes in $\mathbb{C}^N$ with Z. Then in the quantum cohomology $QH^*(Z,q)$ the following relations hold (see [2]):

$$\prod_{j=1}^{N} Z_j^{m_j^\alpha}=q_\alpha, \qquad \alpha\in A, \tag{6}$$

where $q_\alpha=e^{x^\alpha}$ are quantum deformation parameters. These relations are deformations of the classical relations for $q_\alpha=0$ corresponding to the sets of hyperplanes with zero intersections.

The T^N-equivariant mirror dual to the toric manifold Z is defined by the data $(Y_q,\mathcal{F},\omega_q)$ where Y_q is an affine space given by the set of solutions of (6), $\mathcal{F}$ is given by

$$\mathcal{F}(t,\lambda)=\sum_{i=1}^{N}(-t_i+\imath\lambda_i\ln t_i), \qquad \omega_q=\prod_{\alpha=1}^{n}\delta(f_\alpha)\,\frac{dt_1\wedge\cdots\wedge dt_N}{t_1\cdots t_N}, \tag{7}$$

where

$$f_\alpha=e^{-x^\alpha}\prod_{j=1}^{N} t_j^{m_j^\alpha}-1, \qquad \alpha\in A. \tag{8}$$

The special function associated with these data is then given by the following period

$$\Psi_{\lambda_1,\ldots,\lambda_N}(q_1,\cdots,q_n)=\int\limits_{\mathbb{R}_+^N}\omega_q\ \prod_{i=1}^N t_i^{\imath\lambda_i}e^{-t_i}\,. \tag{9}$$

The integral representation (9) allows interpretation as a matrix element of a suitable Lie algebra. Namely, consider the semi-direct sum $\mathcal{L}_N$ of abelian Lie algebra $(\mathfrak{gl}_1)^N$ and the Heisenberg algebra of dimension $2N+1$ defined by the set generators $\{\mathcal{E}_i,\mathcal{F}_i,\mathcal{H}_i,\ i\in I\}$, central element $\mathcal{C}$ and relations

$$[\mathcal{E}_i,\mathcal{H}_j]=\delta_{ij}\mathcal{E}_i,\qquad [\mathcal{F}_i,\mathcal{H}_j]=-\delta_{ij}\mathcal{F}_i,\qquad [\mathcal{E}_i,\mathcal{F}_j]=\mathcal{C}\delta_{ij},\qquad i,j\in I. \tag{10}$$

We fix a generic irreducible representation $\pi_{\lambda,c}$ in a suitable space of functions in $t=(t_1,\cdots,t_N)\in\mathbb{R}_+^N$ with images of the generators in this representation given by

$$E_i=c\partial_{t_i},\qquad F_i=t_i,\qquad H_i=t_i\partial_{t_i}+\imath\lambda_i,\qquad C=c, \tag{11}$$

where $\lambda_i\in\mathbb{R},\ c\in\mathbb{R}_+$. In the following for convenience we take $c=1$. Define left and right vectors by the following conditions

$$\begin{gathered}E_i|\psi_R\rangle=-|\psi_R\rangle,\qquad i\in I\,;\\ \textstyle\sum_{i=1}^N v_k^i H_i|\psi_L\rangle=0,\ k\in\overline{A},\qquad \prod_{j=1}^N F_j^{m_j^\alpha}|\psi_L\rangle=|\psi_L\rangle,\ \alpha\in A,\end{gathered} \tag{12}$$

where $\{v_k,\,k\in\overline{A}\}$ are given by (5). Now introduce the following matrix element

$$\Psi_\lambda(e^y):=\Psi_{\lambda_1,\ldots,\lambda_N}(e^{y^1},\ldots,e^{y^N})=\langle\psi_L|e^{\sum\limits_{j=1}^N y^jH_j}|\psi_R\rangle, \tag{13}$$

with the non-degenerate pairing $\langle\,,\,\rangle$ given by

$$\langle a\,,\,b\rangle:=\int\limits_{\mathbb{R}_+^N}\prod_{i=1}^N d^\times t_i\ \overline{a(t)}\,b(t),\qquad d^\times t_i=\frac{dt_i}{t_i}\,. \tag{14}$$

Theorem 1 *The matrix element* (13) *allows the following integral representation*

$$\Psi_\lambda(e^y)=\int\limits_{\mathbb{R}_+^N}\prod_{i=1}^N d^\times t_i\ \prod_{\alpha=1}^n\delta\Big(e^{-\sum\limits_{j=1}^N y^jm_j^\alpha}\prod_{j=1}^N t_j^{m_j^\alpha}-1\Big)\prod_{j=1}^N t_j^{\imath\lambda_j}\,e^{-t_j}\,, \tag{15}$$

and satisfies the equations:

$$\Big\{\prod_{j=1}^N\prod_{k=0}^{m_j^\alpha-1}(\imath\lambda_j+k-\partial_{y^j})-e^{\sum\limits_{j=1}^N y^jm_j^\alpha}\Big\}\Psi_\lambda(e^y)=0,\qquad \alpha\in A, \tag{16}$$

$$\sum_{j=1}^{N} v_k^j \partial_{y^j} \cdot \Psi_\lambda(e^y) = 0, \qquad k \in \overline{A}, \tag{17}$$

with v_k, $k \in \overline{A}$ *defined in* (5).

Proof Explicitly for the vectors $\psi_{L,R}$ we have

$$|\psi_R\rangle = e^{-\sum\limits_{i=1}^{N} t_i}, \qquad |\psi_L\rangle = \prod_{i=1}^{N} t_i^{-\imath\lambda_i} \prod_{\alpha=1}^{n} \delta\Big(\prod_{j=1}^{N} t_j^{m_j^\alpha} - 1\Big). \tag{18}$$

Substitution of (18) into (13) after a simple change of variables leads to (15). For the second assertion, consider the following polynomials of generators acting trivially in representation $\pi_{\lambda,c}$ defined by (11)

$$\mathcal{C}_\alpha = \prod_{j=1}^{N} (F_j E_j)^{m_j^\alpha} - \prod_{j=1}^{N} C^{m_j^\alpha} (H_j - \imath\lambda_j)^{m_j^\alpha}, \qquad \alpha \in A. \tag{19}$$

Insertion of these expressions into the matrix element (13) then leads to the first set of equations (16). In addition we have the following set of equations (17) obtained by inserting the linear combinations $\sum_{j=1}^{N} v_k^j H_j$ into the matrix element (13) and taking into account the defining relation (12). ■

Corollary 1 *The matrix element* (13) *effectively depends on a smaller number of variables* $x^\alpha = \sum_{i=1}^{N} y^i m_i^\alpha$, $\alpha \in A$ *and allows the following representation*

$$\Psi_{\lambda_1,\dots,\lambda_N}(e^{x_1}, \dots, e^{x_n}) = \int\limits_{\mathbb{R}_+^N} \prod_{i=1}^{N} d^\times t_i \prod_{\alpha=1}^{n} \delta\Big(e^{-x^\alpha} \prod_{j=1}^{N} t_j^{m_j^\alpha} - 1\Big) \prod_{j=1}^{N} t_j^{\imath\lambda_j} e^{-t_j}. \tag{20}$$

This function coincides with the mirror period (9) *and solves the set of equations for* $\alpha \in A$

$$\Big\{\prod_{j=1}^{N} \prod_{k=0}^{m_j^\alpha - 1} (\imath\lambda_j + k - \sum_{\beta=1}^{n} m_j^\beta \partial_{x^\beta}) - e^{x^\alpha}\Big\} \Psi_{\lambda_1,\dots,\lambda_N}(e^{x_1}, \dots, e^{x_n}) = 0. \tag{21}$$

A simple example of the proposed construction is provided by mirror correspondence for complex projective space $Z = \mathbb{P}^\ell$. Projective vector space allows the standard Hamiltonian reduction representation $\mathbb{P}^\ell = \mathbb{C}^{\ell+1}//S^1$, where the action is defined as

$$S^1 : \ \mathbb{C}^{\ell+1} \longrightarrow \mathbb{C}^{\ell+1}, \qquad e^{\imath\theta} : (z_1, \dots, z_{\ell+1}) \longmapsto (z_1 e^{\imath\theta}, \dots, z_{\ell+1} e^{\imath\theta}). \tag{22}$$

Therefore we have $N = \ell + 1$, $n = 1$ and the relation matrix is given by $||m_j^\alpha|| = (1, \dots, 1)$. The resulting representation for the mirror period (9) reads

$$\Psi_{\lambda_1,\ldots,\lambda_{\ell+1}}(e^x) = \int\limits_{\mathbb{R}_+^N} \prod_{i=1}^{\ell+1} d^\times t_i \; \delta\Big(e^{-x}\prod_{j=1}^{\ell+1} t_j - 1\Big) \prod_{j=1}^{\ell+1} t_j^{\imath\lambda_j}\, e^{-t_j}. \tag{23}$$

By Theorem 1, (23) coincides with the following matrix element of the Lie algebra $\mathcal{L}_{\ell+1} = (\mathfrak{gl}_1)^{\ell+1} \uplus \mathcal{H}^c_{2(\ell+1)}$ in the generic representation (11):

$$\Psi_{\lambda_1,\ldots,\lambda_{\ell+1}}(e^x) = \langle\psi_L| e^{x\sum\limits_{i=1}^{\ell+1} a_i H_i} |\psi_R\rangle, \qquad \sum_i a_i = 1, \tag{24}$$

where the left and right vectors are defined by

$$\begin{gathered} E_i|\psi_R\rangle = -|\psi_R\rangle, \qquad i \in \{1,\ldots,\ell+1\}; \\ (H_k - H_{\ell+1})|\psi_L\rangle = 0, \quad k \in \{1,\ldots,\ell\}, \qquad F_1\cdots F_{\ell+1}|\psi_L\rangle = |\psi_L\rangle. \end{gathered} \tag{25}$$

Particular choice of $\{a_i,\ i \in I\}$ is irrelevant due to (25). The defining equations for (23) are obtained by using the following operator acting by zero in the representation (11):

$$\mathcal{C}_{\ell+1} = \prod_{i=1}^{\ell+1} F_i E_i - C^{\ell+1} \prod_{i=1}^{\ell+1} (H_i - \imath\lambda_i). \tag{26}$$

The resulting equation

$$\Big\{\prod_{i=1}^{\ell+1}(\imath\lambda_i - \partial_x) - e^x\Big\}\,\Psi_\lambda(e^x) = 0, \tag{27}$$

is in agreement with the proper equivariant counting of holomorphic maps of $\mathbb{P}^1$ to $\mathbb{P}^\ell$ (see e.g. [4]).

Acknowledgements The research of the second author was supported by RSF grant 16-11-10075. The work of the third author was partially supported by the EPSRC grant EP/L000865/1.

References

1. M. Audin, Torus Actions on Symplectic Manifolds, 2nd Revised Ed., Progr. in Math. 93 , Birkhäuser-Verlag-Basel-Berlin, 2004.
2. V. Batyrev, *Quantum cohomology rings of toric manifolds*, in Journées de géometrie algébrique d'Orsay, Juillet 1992, Astérisque 218 (1993).
3. I.M. Gelfand, M.M. Kapranov, A.V. Zelevinsky, *Hypergeometric functions and toric varieties*, Funct. Anal. Appl. 23:2 (1989) 12–26.
4. A. Gerasimov, D. Lebedev, S. Oblezin, *Parabolic Whittaker functions and topological field theories I*, Commun. Number Theory Phys. 5:2 (2011) 135–201.

Univalent Foundations

Daniel R. Grayson

Homotopy type theory, together with the partition of types into levels and the univalence axiom developed by Vladimir Voevodsky, provides both a new logical foundation for mathematics, called Univalent Foundations, and a formal language usable with computers for checking the proofs mathematicians make. As a foundation, it replaces set theory with a framework where propositions and sets are defined in terms of a more primitive notion called *type*—in this framework the notion of symmetry arises at the most basic level—from the logic. As a formal language, it encodes the axioms of mathematics and the rules of logic simultaneously, and it promises to make the extraction of algorithms and values from constructive proofs easy. As a mathematical topic, it offers an intriguing range of open problems at all levels of accessibility.

Voevodsky started thinking about computer formalization of mathematics in 2002. In an email to me that year he included this query:

> I am thinking again about the applications of computers to pure math. Do you know of anyone working in this area? I mean mostly some kind of a computer language to describe mathematical structures, their properties and proofs in such a way that ultimately one may have mathematical knowledge archived and logically verified in a fixed format.

In his February, 2010, talk at CMU Voevodsky delivered to us a world where proofs work like the following one.

Theorem 1 *Suppose G and H are isomorphic groups, and G is free. Then H is free.*

Proof Pick an isomorphism $\theta : G \xrightarrow{\cong} H$. It provides an equality $\lambda : G \xrightarrow{=} H$. In particular, G and H are equal, so they share the same properties. Thus, since G is free, so is H.

D. R. Grayson (✉)
Tianjin, China
e-mail: danielrichardgrayson@gmail.com

Y.-H. He et al. (eds.), *Nankai Symposium on Mathematical Dialogues*,
https://doi.org/10.1007/978-981-19-2328-9_19

That proof illustrates the novel feature of this world, that there is a link between higher structures and logic: isomorphisms between objects give rise to equalities, and, as a matter of logic, equal objects share the same properties. In particular, every statement of the language is invariant under isomorphism, and the dream of category theorists from its inception is thereby fulfilled!

In that talk, Voevodsky presented the *Univalence Axiom* in the context of type theory, and sketched a proof that the formal reasoning system supported by it is at least as consistent as traditional reasoning based on the Zermelo-Fraenkel axioms for set theory. This involves constructing a model of the formal language in topology, with the *types* of type theory modeled by topological spaces, and with the equalities modeled by *fibrations* over the unit interval. Intuitively, such a fibration can be visualized by thinking of the unit interval as parametrizing an interval of time, and thinking of the fibration as the graph of a *movie*, where the frames of the movie are the fibers of the projection map, and the changes to the fibers are *gradual*, as in the following illustration, which shows a disk being gradually shrunk to a point.

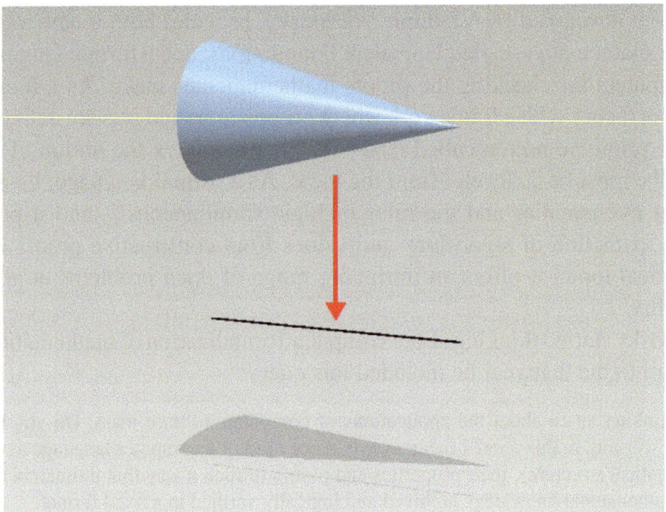

The crucial idea for the proof that the model validates the Univalence Axiom was provided by Pete Bousfield, in 2006, in the following email message – one must show that a homotopy equivalence between two spaces gives rise to a movie (i.e., a fibration) that changes one of them gradually into the other.

```
Date:    Mon, 01 May 2006 10:10:30 CDT
To:      Peter May <may@math.uchicago.edu>
From:    "A. Bousfield" <bous@uic.edu>
Subject: Re: Simplicial question

Dear Peter,

I think that the answer to Voevodsky's basic question is "yes," and I'll
try to sketch a proof.
```

```
Since the Kan complexes X and Y are homotopy  equivalent, they share the
same minimal complex M, and we have trivial fibrations X -> M  and Y -> M
by Quillen's main lemma in "The geometric realization of a Kan fibration."
Thus  X + Y -> M + M is also a trivial fibration where "+" gives the
disjoint union.  We claim that the composition of  X + Y -> M + M with the
inclusion  M + M >--> M x Delta^1 may be factored as the composition of an
inclusion X + Y >--> E with a trivial fibration  E -> M x Delta^1 such that
the counterimage of  M + M is  X + Y.  We may then obtain the desired
fibration

E  ->  M x Delta^1  ->  Delta^1

whose fiber over 0 is X and whose fiber over 1 is Y.
...
```

Voevodsky named his new reasoning system *Univalent Foundations*, and much recent research is devoted to elaborating it.

In my talk at Nankai, which, because I came down with a cold, was actually just a video of me giving part of an earlier talk at the Institute for Advanced Study, I try to clarify these developments.

Voevodsky's style of working was to work on one thing to the exclusion of all else. His project at his death in September, 2017, was to publish all the details of his proof that there is a model of Univalent Foundations in topology. He dedicated himself to this to hasten the time when all research articles in mathematics are accompanied by machine-verifiable formal proofs, and thus mistakes in published works no longer occur.

I close with a picture of him on a fossil hunt with me in Svalbard in 2014—a fond memory, indeed.

Applications of Numerically Solving Polynomial Systems

Jonathan D. Hauenstein

Abstract The problem of solving systems of polynomial equations is ubiquitous throughout science and engineering. The mathematical subject of numerical algebraic geometry consists of a collection of approaches for numerically solving polynomial systems with one foundational technique being homotopy continuation. This short manuscript summarizes using homotopy continuation on two different problems. In the first problem, homotopy continuation is used to approximate a critical parameter value where two solutions of a parameterized differential equation merge together. In the second problem, homotopy continuation is used to compute critical points of a sum of squares best fit function for given data.

Keywords Numerical algebraic geometry · Applied algebraic geometry · Homotopy continuation · Sum of squares best fit

1 Introduction

Computing and analyzing the solution set of a system of nonlinear polynomial equations is a classical problem forming the foundation of the mathematical subject of algebraic geometry. Since systems of polynomial equations are ubiquitous throughout science and engineering, there are many applications of solving polynomial systems such as biology [13, 32], chemistry [1, 10, 12, 20, 30], dynamical systems [11, 17, 22, 27], physics [16, 18, 23], kinematics [7, 14, 15, 26, 28, 33, 34], and control [9, 24, 29] to list a few. For solving univariate polynomials equations of degree at most 4, there exists formulas for expressing the solutions in radicals in terms of the coefficients, such as the quadratic formula

Research supported in part by NSF CCF 1812746 and NSF CMMI 2041789.

J. D. Hauenstein (✉)
Department of Applied and Computational Mathematics and Statistics, University of Notre Dame, Notre Dame, IN, USA
e-mail: hauenstein@nd.edu
URL: http://hauenstein.phd

Y.-H. He et al. (eds.), *Nankai Symposium on Mathematical Dialogues*,
https://doi.org/10.1007/978-981-19-2328-9_20

$$ax^2 + bx + c = 0 \quad \Longrightarrow \quad x = \frac{-b \pm \sqrt{b^2 - 4ac}}{2a}. \tag{1}$$

The impossibility of having similar explicit formulas for arbitrary univarite polynomials of degree at least 5, let alone for multivariate systems, has necessitated the development of numerical approaches for approximating solutions. For example, Newton's method is widely used to numerically approximate a solution since it is locally quadratically convergent near nonsingular solutions. That is, if one has a reasonably good guess of a nonsingular solution, then one can obtain highly accurate approximations in relatively few Newton iterations. In essence, the problem of solving is transformed into finding reasonably good guesses.

The mathematical subject of numerical algebraic geometry, e.g., see [6, 31], consists of a collection of approaches for numerically computing and analyzing solution sets to systems of polynomial equations. One of the foundational techniques in numerical algebraic geometry is homotopy continuation which turns the problem of solving into path tracking. Path tracking traditionally consists of applying a predictor-corrector scheme. The predictor is often based on numerically approximating a solution to an initial value problem which yields a reasonably good guess to utilize a corrector, e.g., Newton's method, to remove local error and obtain accurate approximations.

The rest of this paper applies homotopy continuation to polynomial systems arising in various applications. In Sect. 2, homotopy continuation is used to identify a parameter value where two distinct solutions merge together for a parameterized polynomial system arsing from discretizing an ordinary differential equation. In Sect. 3, critical points of an objective function obtained via a sum of squares best fit of given data are computed using homotopy continuation.

2 Merging Solutions

Homotopies are typically constructed to "end" at the system of interest so that efficient algorithms called endgames, e.g., see [6, Chap. 3] for an overview, can be used to accurately approximate the endpoints. An an illustration of using homotopy continuation, consider the following parameterized two-point boundary value problem for $\lambda > 0$ from [2, Sect. 3.2]:

$$\begin{aligned} y''(t) &= -\lambda(1 + y(t)^2) \quad \text{for } 0 < t < 1, \\ y(0) &= 0, \\ y(1) &= 0. \end{aligned} \tag{2}$$

It is known [2, 19] that there exists $\lambda^* > 0$ such that (2) has two solutions for $\lambda \in (0, \lambda^*)$, unique solution for $\lambda = \lambda^*$, and no solutions for $\lambda \in (\lambda^*, \infty)$. Figure 1 plots the two solutions for $\lambda = 3$ and $\lambda = 4$.

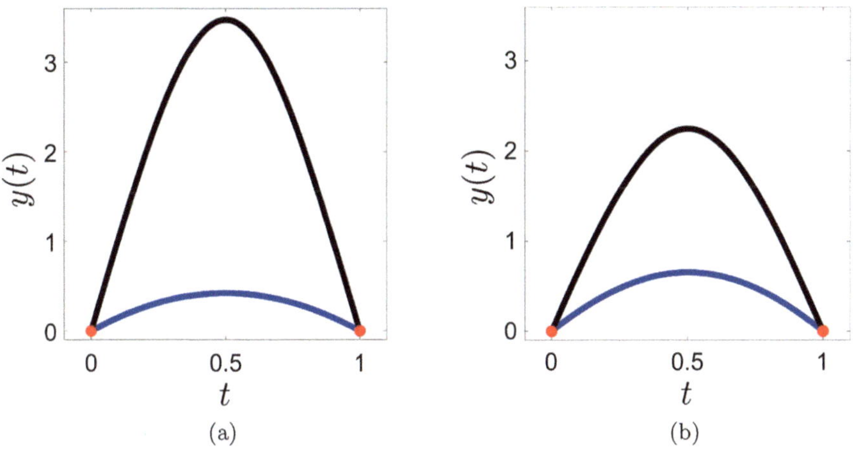

Fig. 1 Plot of the two solutions to (2) for **a** $\lambda = 3$ and **b** $\lambda = 4$

One approach for computing λ^* is to construct a homotopy that forces the two solutions, say $y_{\lambda,1}(t)$ and $y_{\lambda,2}(t)$, corresponding to the same value of λ to merge together. Hence, by treating λ as a variable, the two solutions will merge together precisely when λ is equal to λ^*. In particular, for $\lambda_0 = 4$, let $y_{\lambda_0,1}$ and $y_{\lambda_0,2}$ be the two solutions depicted in Fig. 1b so that, at $s = 1$, $(y_{\lambda_0,1}, y_{\lambda_0,2}, \lambda)$ is the start point of the following homotopy:

$$H(y_{\lambda,1}, y_{\lambda,2}, \lambda; s) = \left[\begin{array}{c} y''_{\lambda,1}(t) + \lambda(1 + y_{\lambda,1}(t)^2) \quad \text{for } 0 < t < 1 \\ y_{\lambda,1}(0) \\ y_{\lambda,1}(1) \\ \hline y''_{\lambda,2}(t) + \lambda(1 + y_{\lambda,2}(t)^2) \quad \text{for } 0 < t < 1 \\ y_{\lambda,2}(0) \\ y_{\lambda,2}(1) \\ \hline \|y_{\lambda,1} - y_{\lambda,2}\|_2^2 - s \cdot \|y_{\lambda_0,1} - y_{\lambda_0,2}\|_2^2 \end{array}\right] = 0.$$

Therefore, when $s = 0$, this homotopy "ends" with $\lambda = \lambda^*$ and $y_{\lambda,1} = y_{\lambda,2}$. Since this homotopy is formulated in terms of differential equations, we can approximate via polynomial equations, for example, by discretizing using equally spaced grid points and applying a second order central difference scheme. Hence, one can treat $y_{\lambda,1}$ and $y_{\lambda,2}$ as vectors so that the last equation in the homotopy simply corresponds with the square of the standard Euclidean norm of vectors. Utilizing `Bertini` [5], Table 1, which matches [8, Table 1], compares the grid spacing Δt with the computed value of λ^* showing that $\lambda^* \approx 4.755$.

Table 1 Comparison of grid spacing Δt and corresponding value of λ^*

Δt	λ^*
1/10	4.734384294
1/20	4.749878424
1/40	4.753696808
1/80	4.754647901
1/160	4.754885455
1/320	4.754944829
1/640	4.754959672
1/1280	4.754963383
1/2560	4.754964310

3 Critical Points From sum of Squares Best Fit

A classical problem is to compute the best fit line of the form $y = mx + b$ to a collection of data points (x_i, y_i) for $i = 1, \ldots, N$ where N is sufficiently large. Thus, one is aiming to compute m and b which minimizes

$$F(m, b) = \sum_{i=1}^{N} (mx_i + b - y_i)^2.$$

Since the gradient vector ∇F with respect to m and b is a full rank linear system, there is a unique line of best fit for generic data corresponding with the unique solution of $\nabla F = 0$. The following considers more general problems of best fit.

Suppose that $f(x; p)$ is a polynomial in $x \in \mathbb{R}^n$ and $p \in \mathbb{R}^k$. Then, given a collection of data $(x_i, y_i) \in \mathbb{R}^n \times \mathbb{R}$ and weights $w_i \in \mathbb{R}_{>0}$ for $i = 1, \ldots, N$ where N is sufficiently large, consider the weighted sum of squares best fit function

$$F(p) = \sum_{i=1}^{N} w_i \cdot (f(x_i; p) - y_i)^2.$$

The set of critical points of $F(p)$ satisfy $\nabla F(p) = 0$ which is a system of k polynomials in k variables. Hence, there is a generic number of isolated nonsingular solutions to $\nabla F(p) = 0$ which we denote as SOSdegree(f).

Example 1 For $f(x; p) = p_1 x + p_2$ associated with constructing the best fit line, SOSdegree$(f) = 1$. In fact, by viewing this polynomial as a linear span of the monomials x and 1, this can be generalized to any linear span of distinct multivariate monomials. For example, suppose that $x^{\alpha_1}, \ldots, x^{\alpha_k}$ is a list of multivariate monomials with $\alpha_i \neq \alpha_j$ for $i \neq j$ and

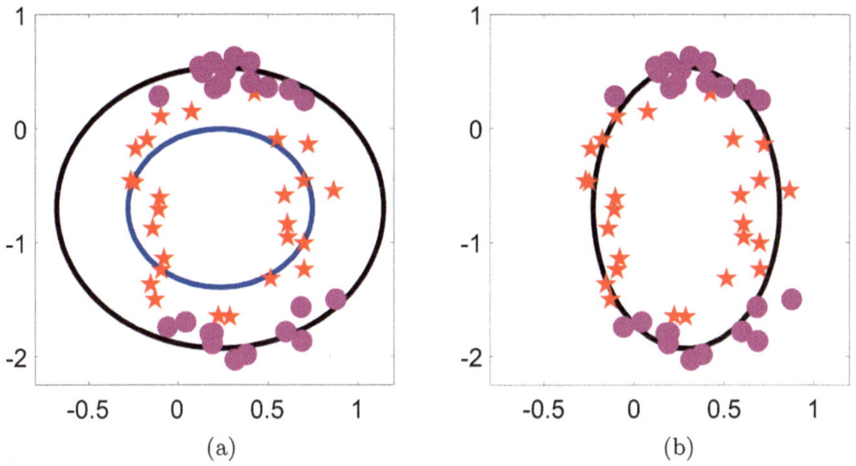

Fig. 2 Computing sum of squares best fit for the same 50 data points using **a** two ellipses and **b** one ellipse

$$f(x; p) = \sum_{j=1}^{k} p_j x^{\alpha_j},$$

then SOSdegree$(f) = 1$.

One can compute SOSdegree(f), for example, by counting the number of solutions to the polynomial system $\nabla F = 0$ obtained using homotopy continuation for generic data. Then, one can utilize a parameter homotopy [25] to deform from the generic data to the given data tracking SOSdegree(f) number of paths. This approach has already been used, for example, in approximate kinematics synthesis of mechanisms [3, 4], and machine learning [21]. We conclude with an illustration of this approach on some data points in the plane.

Example 2 Consider the 50 data points shown in Fig. 2 and

$$f(x; p) = p_1(x_1 - p_2)^2 + p_3(x_2 - p_4)^2.$$

First, consider computing the critical parameters p of the sum of squares best fit with equal weights, i.e., $w_i = 1$, such that $y_i = 0.65^2 = 0.4225$ on the stars and $y_i = 1.15^2 = 1.3225$ on the dots from Fig. 2a. To accomplish this, we first compute SOSdegree$(f) = 33$ using homotopy continuation in `Bertini` [5] and then perform a parameter homotopy which deforms from the generically selected data to this given data yielding 33 critical points of the sum of squares best fit. Of these, 11 are real and, by analyzing the Hessian matrix, there are 3 that are local minima. The one which is the global minimum is shown in Fig. 2a.

For the second problem, on the same set of 50 data points, consider computing the critical parameters p of the sum of squares best fit with equal weights, i.e., $w_i = 1$, and equal output, i.e., $y_i = 1$. Performing a parameter homotopy to this special case results in 25 critical points which is less than the generic count. Of these, 9 are real and, by analyzing the Hessian matrix, there is a unique local minimum which is the global minimum that is shown in Fig. 2b.

References

1. Al-Khateeb, A.N., Powers, J.M., Paolucci, S., Sommese, A.J., Diller, J.A., Hauenstein, J.D., Mengers, J.D.: One-dimensional slow invariant manifolds for spatially homogenous reactive systems. The Journal of Chemical Physics **131**(2), 024118 (2009)
2. Allgower, E.L., Bates, D.J., Sommese, A.J., Wampler, C.W.: Solution of polynomial systems derived from differential equations. Computing **76**(1-2), 1–10 (2006)
3. Baskar, A., Liu, C., Plecnik, M., Hauenstein, J.D.: Designing rotary linkages for polar motions. In: 2021 IEEE/RSJ International Conference on Intelligent Robots and Systems (IROS). pp. 1384–1391 (2021)
4. Baskar, A., Plecnik, M., Hauenstein, J.D.: Computing saddle graphs via homotopy continuation for the approximate synthesis of mechanisms. Mech. Mach. Theory 176, 104932 (2022). Preprint, available at www.nd.edu/~jhauenst/preprints/bphSaddleGraphs.pdf
5. Bates, D.J., Hauenstein, J.D., Sommese, A.J., Wampler, C.W.: Bertini: Software for numerical algebraic geometry. Available at bertini.nd.edu
6. Bates, D.J., Hauenstein, J.D., Sommese, A.J., Wampler, C.W.: Numerically solving polynomial systems with Bertini, Software, Environments, and Tools, vol. 25. Society for Industrial and Applied Mathematics (SIAM), Philadelphia, PA (2013)
7. Brake, D., Hauenstein, J., Murray, A., Myszka, D., Wampler, C.: The complete solution of mixed Burmester synthesis probles for four-bar linkages. Journal of Mechanisms and Robotics **8**, 041018 (2016)
8. Collins, J., Hauenstein, J.D.: A singular value homotopy for finding critical parameter values. Applied Numerical Mathematics **161**, 233–243 (2021)
9. Coss, O., Hauenstein, J., Hong, H., Molzahn, D.: Locating and counting equilibria of the kuramoto model with rank-one coupling. SIAM Journal on Applied Algebra and Geometry **2**(1), 45–71 (2018)
10. Dickenstein, A.: Biochemical reaction networks: an invitation for algebraic geometers. In: Mathematical Congress of the Americas, Contemp. Math., vol. 656, pp. 65–83. Amer. Math. Soc., Providence, RI (2016)
11. Dickenstein, A., Pérez Millán, M., Shiu, A., Tang, X.: Multistationarity in structured reaction networks. Bull. Math. Biol. **81**(5), 1527–1581 (2019)
12. Feliu, E., Helmer, M.: Multistationarity and bistability for Fewnomial chemical reaction networks. Bull. Math. Biol. **81**(4), 1089–1121 (2019)
13. Gross, E., Harrington, H.A., Rosen, Z., Sturmfels, B.: Algebraic systems biology: a case study for the Wnt pathway. Bull. Math. Biol. **78**(1), 21–51 (2016)
14. Hauenstein, J., Sherman, S., Wampler, C.: Exceptional stewart–gough platforms, segre embeddings, and the special euclidean group. SIAM Journal on Applied Algebra and Geometry **2**(1), 179–205 (2018)
15. Hauenstein, J., Wampler, C., Pfurner, M.: Synthesis of three-revolute spatial chains for body guidance. Mechanism and Machine Theory **110**, 61–72 (2017)
16. Hauenstein, J., He, Y.H., Mehta, D.: Numerical elimination and moduli space of vacua. J. High Energy Phys. (9), 083 (2013)
17. Ho, K.L., Harrington, H.A.: Bistability in apoptosis by receptor clustering. PLoS Comput. Biol. **6**(10), e1000956, 9 (2010)
18. Huang, R., Rao, J., Feng, B., He, Y.H.: An algebraic approach to the scattering equations. J. High Energy Phys. (12), 056 (2015)

19. Laetsch, T.: On the number of solutions of boundary value problems with convex nonlinearities. J. Math. Anal. Appl. **35**, 389–404 (1971)
20. Mehta, D., Chen, T., Hauenstein, J.D., Wales, D.J.: Communication: Newton homotopies for sampling stationary points of potential energy landscapes. The Journal of Chemical Physics **141**(12), 121104 (2014)
21. Mehta, D., Chen, T., Tang, T., Hauenstein, J.D.: The loss surface of deep linear networks viewed through the algebraic geometry lens. IEEE Transactions on Pattern Analysis and Machine Intelligence (2021), to appear
22. Mehta, D., Daleo, N.S., Dörfler, F., Hauenstein, J.D.: Algebraic geometrization of the Kuramoto model: equilibria and stability analysis. Chaos **25**(5), 053103 (2015)
23. Mehta, D., Kastner, M.: Stationary point analysis of the one-dimensional lattice Landau gauge fixing functional, aka random phase XY Hamiltonian. Ann. Physics **326**(6), 1425–1440 (2011)
24. Mehta, D., Molzahn, D.K., Turitsyn, K.: Recent Advances in Computational Methods for the Power Flow Equations. In: 2016 American Control Conference (ACC). pp. 1753–1765 (2016)
25. Morgan, A.P., Sommese, A.J.: Coefficient-parameter polynomial continuation. Appl. Math. Comput. **29**(2, part II), 123–160 (1989)
26. Myszka, D.H., Murray, A.P., Wampler, C.W.: Computing the Branches, Singularity Trace, and Critical Points of Single Degree-of-Freedom, Closed-Loop Linkages. Journal of Mechanisms and Robotics **6**(1) (12 2013)
27. Obatake, N., Shiu, A., Tang, X., Torres, A.: Oscillations and bistability in a model of ERK regulation. J. Math. Biol. **79**(4), 1515–1549 (2019)
28. Plecnik, M.M., Fearing, R.S.: Finding Only Finite Roots to Large Kinematic Synthesis Systems. Journal of Mechanisms and Robotics **9**(2) (2017)
29. Rahimian, S.K., Jalali, F., Seader, J.D., White, R.E.: A robust homotopy continuation method for seeking all real roots of unconstrained systems of nonlinear algebraic and transcendental equations. Industrial & Engineering Chemistry Research **50**(15), 8892–8900 (2011)
30. Sidky, H., Liddell, A.C., Mehta, D., Hauenstein, J.D., Whitmer, J.K.: Algebraic geometric method for calculating phase equilibria from fundamental equations of state. Industrial & Engineering Chemistry Research **55**(43), 11363–11370 (2016)
31. Sommese, A.J., Wampler, II, C.W.: The numerical solution of systems of polynomials arising in engineering and science. World Scientific Publishing Co. Pte. Ltd., Hackensack, NJ (2005)
32. Song, H., Smolen, P., Av-Ron, E., Baxter, D.A., Byrne, J.H.: Bifurcation and singularity analysis of a molecular network for the induction of long-term memory. Biophysical Journal **90**(7), 2309 – 2325 (2006)
33. Tsai, L.W., Morgan, A.P.: Solving the Kinematics of the Most General Six- and Five-Degree-of-Freedom Manipulators by Continuation Methods. Journal of Mechanisms, Transmissions, and Automation in Design **107**(2), 189–200 (1985)
34. Wampler, C.W., Morgan, A.P., Sommese, A.J.: Complete Solution of the Nine-Point Path Synthesis Problem for Four-Bar Linkages. Journal of Mechanical Design **114**(1), 153–159 (1992)

19. Lastoch, I.: On the number of solutions of boundary value problems with convex nonlinearities, J. Math. Anal. Appl. 35, 389–404 (1971)
20. Mehta, D., Chen, T., Hauenstein, J.D., Wales, D.J.: Communication: Newton homotopies for sampling stationary points of potential energy landscapes. [illegible]
21. [illegible]
[illegible]

Machine Learning for Hilbert Series

Edward Hirst

Abstract Hilbert series are a standard tool in algebraic geometry, and more recently are finding many uses in theoretical physics. This summary reviews work applying machine learning to databases of Hilbert series to determine geometric properties of the underlying algebraic variety; and was prepared for the proceedings of the Nankai Symposium on Mathematical Dialogues, 2021.

1 Introduction

Mathematically, Hilbert series (HS) provide a means of counting regular functions on varieties. Physically, they are used in a range of scenarios, including to count BPS operators [1–6], determine moduli spaces [7–10], find standard model invariants [11–13], deal with string compactifications [14], construct effective Lagrangians [15–20] and basis-invariant objects [21, 22], as well as for a range of other quiver-related computations [23–33].

In parallel, machine learning (ML) is fast becoming a definitive methodology for data analysis and pattern recognition. The most common tool being neural networks (NNs), which are many-parameter non-linear functions used to learn properties of a dataset. Beyond supervised methods, principal component analysis (PCA) is an established unsupervised method, used for examining clustering in datasets. PCA puts focus on a data representation's dominant principal components, which determine the directions the variance is largest and are found through diagonalisation of the data's covariance matrix. ML has also been successfully applied to a variety of physically motivated scenarios, including: Calabi-Yau manifolds [34–42], polytopes [43, 44], graph theory [45], knot theory [46, 47], amoebae [48], brane webs [49], conformal field theories [50], integrability [51], Seiberg duality among quivers [52], cluster algebras [53], and the related dessin d'enfant Galois orbits [54].

E. Hirst (✉)
Department of Mathematics, City, University of London, London, UK
e-mail: edward.hirst@city.ac.uk

London Institute for Mathematical Sciences, Royal Institution, London, UK

Y.-H. He et al. (eds.), *Nankai Symposium on Mathematical Dialogues*,
https://doi.org/10.1007/978-981-19-2328-9_21

In this summary, work from [55] using neural networks to learn geometric properties of a variety from coefficients in its respective HS is described. The work was carried out in `python` with the use of `Tensorflow` [56] for machine learning implementation.

2 Defining Hilbert Series

Within algebraic geometry HS encode information about a complex variety X, embedded in complex weighted projective space $\mathbb{P}_{\mathbb{C}}(p_0^{q_0}, \ldots, p_s^{q_s})$, such that weight p_i occurs q_i times. The variety's coordinate ring is denoted $R = \mathbb{C}[x_0, \ldots, x_k]/I$, for complex variables x_i, and homogeneous ideal I generated by the polynomials defining the variety. Any specific embedding of the variety then induces a grading on this coordinate ring [57], which can be written $R = \bigoplus_{i \geq 0} R_i$. The HS is the generating function for the dimensions of these graded pieces of R,

$$H(t; X) = \sum_{i=0}^{\infty} (\dim_{\mathbb{C}} R_i) t^i \ ; \tag{2.1}$$

such that $\dim_{\mathbb{C}} R_i$ defines the number of independent degree i polynomials on the variety, and t is a dummy variable used to keep track of the grading order (known as the fugacity in the context of quivers) [58].

This concept can be generalised to any n-multiple grading of the ring, such that $R = \bigoplus_{i_1, i_2, \ldots, i_n \geq 0} R_{i_1, i_2, \ldots, i_n}$, giving HS

$$H(t_1, t_2, \ldots, t_n; X) = \sum_{i_1, i_2, \ldots, i_n = 0}^{\infty} (\dim_{\mathbb{C}} R_{i_1, i_2, \ldots, i_n}) t_1^{i_1} t_2^{i_2} \ldots t_n^{i_n} \ ; \tag{2.2}$$

which in the physics literature is called the 'refined' HS and each fugacity is a gauge invariant operator. Note it is simple to unrefine by setting $t_k = t \ \forall \ k$.

In the context of supersymmetric gauge theory in string theory, when one knows the matter content and superpotential defining the theory they often wish to compute the geometric information about the moduli space of vacua. Whilst in the case of the moduli space being toric one can determine it from the combinatorics of the toric diagram [59], where the moduli space is not toric the process is more difficult. Hence in general, single-trace gauge invariant operators must be counted order by order to construct the HS to then learn about this moduli space and the underlying variety it is a cone over.

The process of computing a HS is nicely exemplified with the prototypical example of the clover quiver in $\mathcal{N} = 1$ supersymmetry, see Fig. 1, with $U(N)$ gauge group and superpotential given by tracing over a certain choice of cycle combinations: $W = \mathrm{Tr}(X_1 X_2 X_3 - X_1 X_3 X_2)$. The X_k are the chiral supermultiplets in a representation

Fig. 1 The quiver for a $\mathcal{N} = 1$ $U(N)$ gauge theory with three adjoint chiral multiplets

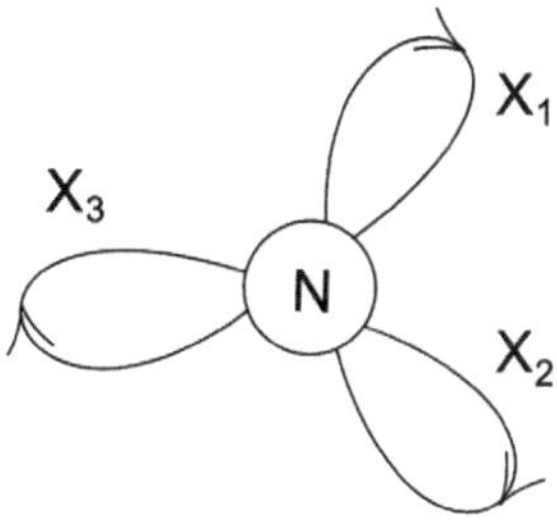

(here adjoint) of the gauge group, and can be related to the fugacities $t_k = \mathrm{Tr}(X_k)$ in this case as $\mathrm{Tr}(X_k)$ for $k \in \{1, 2, 3\}$ represent the 3 order 1 generators.

Higher order generators are found using the information of the superpotential choice, computing all the F-terms as $0 = \partial_{X_1} W = \partial_{X_2} W = \partial_{X_3} W$ shows all the 3 generators are freely commuting such that single trace operators can be constructed at each order from combining the operators in all ways.

Therefore in the refined case we have $\binom{n+2}{2} = \frac{(n+1)(n+2)}{2}$ single-trace operators at each order n, of the form $\mathrm{Tr}(X_1^a X_2^b X_3^c)$ such that (a, b, c) are all possible combinations that sum to n. Hence the HS starts

Order	#	Operators
0	1	1
1	3	$\mathrm{Tr}(X_1),\ \mathrm{Tr}(X_2),\ \mathrm{Tr}(X_3)$
2	6	$\mathrm{Tr}(X_1^2),\ \mathrm{Tr}(X_2^2),\ \mathrm{Tr}(X_3^2),\ \mathrm{Tr}(X_1X_2),\ \mathrm{Tr}(X_1X_3),\ \mathrm{Tr}(X_2X_3)$

for the first 3 orders. Consequently, if we consider the simplest case of $N = 1$ where the multiplet representations, X_k, are just complex numbers, the HS in the refined and unrefined cases are

$$H(t_1, t_2, t_3; X) = \frac{1}{(1 - t_1)(1 - t_2)(1 - t_3)}, \tag{2.3}$$

$$H(t; X) = \frac{1}{(1 - t)^3}, \tag{2.4}$$

respectively. Since there are 3 independent generators with no relations (seen via the trivial numerator), the moduli space is therefore simply $\mathbb{C}^3$. This can be shown by taking the plethystic logarithm of the HS where the first positive terms in the expansion are the generators, and the next are the relations (of which here there are none).

Note one may consider higher N where the moduli space is then the N-th symmetric product of $\mathbb{C}^3$, deriving from the permutations of the X_k matrices being invariant under the trace. Additionally, using the plethystic exponential creates multi-trace operators of the form Tr(...)Tr(...)...Tr(...), which are also valid gauge-invariant operators but are instead from analysis of these symmetric products of the moduli space [1].

In the work summarised here the focus is a single grading, and hence the unrefined HS, as in Eq. (2.1), which via Hilbert-Serre theorem [60] can be written in two equivalent forms:

$$H(t; X) = \frac{P(t)}{\prod_{i=0}^{s}(1 - t^{p_i})^{q_i}} , \tag{2.5}$$

$$H(t; X) = \frac{\tilde{P}(t)}{(1 - t^j)^{dim+1}} , \tag{2.6}$$

for polynomials P and $\tilde{P}$ in the fugacity t, and where the variety, X, has Gorenstein index j and dimension dim. From (2.6) one can tell if the variety is Gorenstein if and only if the numerator polynomial $\tilde{P}$ is palindromic [61]. Likewise one can also use the HS to distinguish whether a variety is a complete intersection, which happens when the expansion of the HS plethystic logarithm terminates at finite order [1, 2].

We can therefore see that our example HS in (2.4) is trivially Gorenstein, and also trivially a complete intersection as their are no relations (and hence also no syzygies) among the generators to occur in the plethystic logarithm expansion. One can also extract the dimension, however since the moduli space is formed from taking a specific cone over the variety its dimension is one larger than that given by equating (2.4) to (2.6). For our example it hence has a 3-dimensional $\mathbb{C}^3$ moduli space, which may be considered to be a cone over the 2-dimensional Fano $\mathbb{P}^2_{\mathbb{C}}$ variety.

3 Machine Learning

With data about the low order terms in the HS, one wishes to determine geometric information about the underlying algebraic variety. In the work summarised here, NN regressors were used to predict the defining weights (p_i) of the weighted projective space each variety is embedded in, and NN classifiers were used to learn the Gorenstein index j and dimension dim of the variety itself. Further to these, three binary classification investigations were performed with NN classifiers to determine whether input HS expansion coefficients came from varieties which were Gorenstein, were complete intersections, or were 'fake' such that they were generated from sampling as described below.

Data Information HS data used for this work was retrieved from the Graded Ring Database GRDB [62–64], consisting of HS associated to three-dimensional $\mathbb{Q}$-Fano varieties with Fano index one [65, 66]. Investigations also used generated 'fake' data, created through sampling the hyperparameters of the HS Eqs. (2.5) and (2.6) from distributions fitted to the 'real' GRDB data. The sampling discarded hyperparameter combinations which did not satisfy certain physical conditions, encouraging the data to be more representative; and also discarded any repeats of GRDB data.

For both HS styles of input data mentioned, which we call 'real' and 'fake' respectively, the HS closed forms were Taylor expanded and expansion coefficients saved. These lists of coefficients form vectors of integers, and make up the ML input for the investigations carried out. Two types of coefficient vector were used as inputs, one from the start of the series (coefficients 0–100) and one from deeper into it (coefficients 1000–1009). The reason to also learn with coefficients sampled from deeper in the series was to provide intuition on the importance of the variety's orbifold points, which have a greater relative significance than the 'initial part' at higher orders.

Output data depended on the investigation. For the regression of embedding weights, these were sorted 3-vectors of integers (sampled in the range $[1, 10]$). For the multi-classification the outputs were single integers in the range $[1, 5]$, which were the values of either the variety's Gorenstein index or its dimension respectively. The final binary classifications just outputted 0 or 1 dependent on whether the Gorenstein or complete intersection properties were respectively satisfied, or whether the data came from the GRDB or was generated by us.

NN Architectures For all investigations the NNs used were built of 4 dense layers of 1024 neurons, all with ReLU activation and 0.05 dropout factor. Training was in batches of 32 for 20 epochs using the Adam [67] optimiser to minimise either the regression log(cosh) loss function or the classification cross-entropy loss function. 5-fold cross-validation provided confidence on the metrics used to evaluate the learning via averaging and standard error.

In the regression investigation, mean absolute error (MAE) evaluated learning by computing the average absolute difference between each true embedding weight and the NN predicted embedding weight. This measure evaluates in the range $[0, \infty)$ with 0 indicating perfect learning, where the embedding weights are always correctly predicted. In the classification investigations, Matthew's correlation coefficient (MCC), which is an application of Pearson's correlation coefficient to binary variables, evaluated the learning. This measure evaluates in the range $[-1, 1]$ with 1 indicating perfect learning.

Results ML results are provided in Table 1 for each of these 5 investigations. Learning of the varieties' ambient space embedding weights predicted each weight within an average range of 1 from the correct values when using higher order coefficients. Additionally the geometric parameters of Eq. (2.6) were learnt exceptionally well, but now better from the lower order coefficients.

Each of the binary classifications also evaluated with MCC values exceeding 0.9, demonstrating strong learning. Both the Gorenstein and 'fake' (non-GRDB) properties were better identified with higher order coefficients, whilst the complete intersection property was only learnt from lower orders but performed equally well.

All these results demonstrate the success of ML in learning geometric properties of varieties from HS, and hopefully further encourages the use of ML techniques in the context of Hilbert series and beyond.

Table 1 Machine learning results for each of the investigations. Neural networks learnt geometric properties associated to an algebraic variety from coefficients in their Hilbert series expansions. Regression was assessed with mean absolute error (MAE) and classification with Matthew's correlation coefficient (MCC), training using 5-fold cross-validation to provide a standard error on the averaged measures

Investigation		Embedding weights	Gorenstein index	Dimension	Gorenstein property	Complete intersection	GRDB
Output ranges		$3 \times [1, 10]$	$[1, 5]$	$[1, 5]$	Binary	Binary	Binary
Measure		MAE	MCC	MCC	MCC	MCC	MCC
Orders	0–100	1.94 ± 0.11	0.916 ± 0.010	0.993 ± 0.006	0.717 ± 0.155	0.910 ± 0.022	0.717 ± 0.155
	1000–1009	1.04 ± 0.12	0.727 ± 0.022	0.822 ± 0.031	0.919 ± 0.073	–	0.919 ± 0.073

PCA Whilst all the learning so far uses NNs from supervised ML, PCA provides another avenue for analysis of this data from the conjugate field of unsupervised ML.[1]

In PCA, the covariance matrix is computed between all vector entries, which is then subsequently diagonalised and eigenvectors sorted in decreasing order of variance eigenvalue. Each data vector is then re-expressed in this eigenvector basis, which in this case of considering the order 0-100 coefficients gives vectors with 101 entries. Since this is too high-dimensional to visualise, the first 2 principal components of each datapoint are plotted to provide a means for examining the data's clustering, these plots can be seen for the 3 binary classification investigations in Fig. 2.

Here, 2d PCA was performed for each of the binary classification investigations, and the datapoints in each class plotted separately in different colours. Whilst there is no clear separation for the Gorenstein data, the complete intersection and 'real'/'fake' data do show some separation, indicating there is some linear structure that the NNs can take advantage of for classification. In particular the occurrence of this for the 'fake' data demonstrates the difficulty of generating representative HS.

[1] Supervised ML uses data were each input is paired with a known output and learns the functional relationship between them, unsupervised ML uses only input data with no known outputs to learn patterns in these inputs.

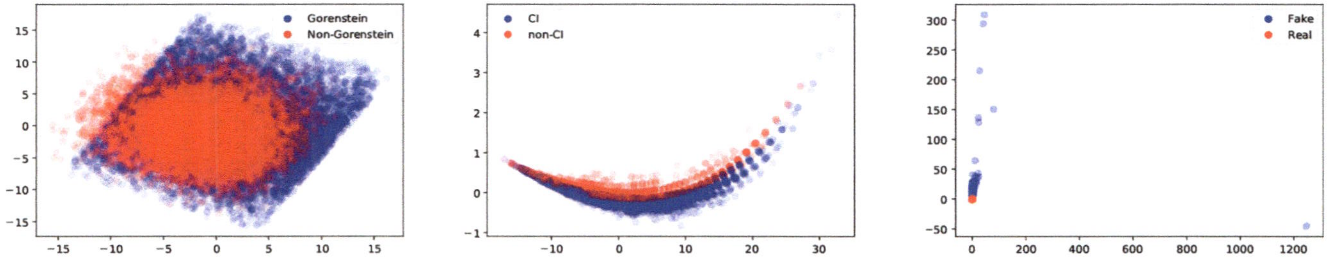

(a) Gorenstein (b) Complete Intersection (c) 'real' / 'fake'

Fig. 2 2d principal component analysis plots of the Hilbert series 0-100 coefficient data, labelled by each of the binary classification investigations respectively

Acknowledgements The author wishes to thank Jiakang Bao, Yang-Hui He, Johannes Hofscheier, Alexander Kasprzyk, and Suvajit Majumder for collaboration on this work; Jiakang Bao again for clarifying discussion; and also STFC for the Ph.D. studentship.

References

1. S. Benvenuti, B. Feng, A. Hanany, and Y.-H. He, "Counting BPS Operators in Gauge Theories: Quivers, Syzygies and Plethystics," http://dx.doi.org/10.1088/1126-6708/2007/11/050*JHEP***11** (2007) 050, arXiv:hep-th/0608050.
2. B. Feng, A. Hanany, and Y.-H. He, "Counting gauge invariants: The Plethystic program," http://dx.doi.org/10.1088/1126-6708/2007/03/090*JHEP***03** (2007) 090, arXiv:hep-th/0701063.
3. J. Gray, Y.-H. He, A. Hanany, N. Mekareeya, and V. Jejjala, "SQCD: a geometric aperçu," *Journal of High Energy Physics***2008** no. 05, (May, 2008) 099–099.
4. A. Hanany, N. Mekareeya, and G. Torri, "The Hilbert series of adjoint SQCD," *Nuclear Phys. B***825** no. 1-2, (2010) 52–97.
5. Y. Chen and N. Mekareeya, "The Hilbert series of U/SU SQCD and Toeplitz determinants," *Nuclear Phys. B***850** no. 3, (2011) 553–593.
6. N. Jokela, M. Järvinen, and E. Keski-Vakkuri, "New results for the SQCD Hilbert series," *J. High Energy Phys.* no. 3, (2012) 048, front matter+30.
7. S. Benvenuti, A. Hanany, and N. Mekareeya, "The Hilbert series of the one instanton moduli space," *J. High Energy Phys.* no. 6, (2010) 100, 40.
8. A. Hanany, N. Mekareeya, and S. S. Razamat, "Hilbert series for moduli spaces of two instantons," *J. High Energy Phys.* no. 1, (2013) 070, front matter + 48.
9. E. I. Buchbinder, A. Lukas, B. A. Ovrut, and F. Ruehle, "Instantons and Hilbert Functions," http://dx.doi.org/10.1103/PhysRevD.102.026019*Phys. Rev. D***102** no. 2, (2020) 026019, arXiv:1912.08358 [hep-th].
10. D. Forcella, A. Hanany, and A. Zaffaroni, "Master Space, Hilbert Series and Seiberg Duality," http://dx.doi.org/10.1088/1126-6708/2009/07/018*JHEP***07** (2009) 018, arXiv:0810.4519 [hep-th].
11. A. Hanany, E. E. Jenkins, A. V. Manohar, and G. Torri, "Hilbert series for flavor invariants of the Standard Model," *J. High Energy Phys.* no. 3, (2011) 096, 7.
12. L. Lehman and A. Martin, "Low-derivative operators of the Standard Model effective field theory via Hilbert series methods," *Journal of High Energy Physics***02** no. 81, (2016) .
13. Y. Xiao, Y.-H. He, and C. Matti, "Standard model plethystics," *Phys. Rev. D***100** (Oct, 2019) 076001.
14. V. Braun, "Counting points and Hilbert series in string theory," in *Strings, gauge fields, and the geometry behind*, pp. 225–235. World Sci. Publ., Hackensack, NJ, 2013.
15. L. Lehman and A. Martin, "Hilbert series for constructing lagrangians: Expanding the phenomenologist's toolbox," *Phys. Rev. D***91** (May, 2015) 105014.
16. B. Henning, X. Lu, T. Melia, and H. Murayama, "Hilbert series and operator bases with derivatives in effective field theories," *Comm. Math. Phys.***347** no. 2, (2016) 363–388.
17. A. Kobach and S. Pal, "Hilbert series and operator basis for NRQED and NRQCD/HQET," *Phys. Lett. B***772** (2017) 225–231.
18. Anisha, S. Das Bakshi, J. Chakrabortty, and S. Prakash, "Hilbert series and plethystics: paving the path towards 2HDM- and MLRSM-EFT," *J. High Energy Phys.* no. 9, (2019) 035, 113.
19. C. B. Marinissen, R. Rahn, and W. J. Waalewijn, "..., 83106786, 114382724, 1509048322, 2343463290, 27410087742, ... efficient Hilbert series for effective theories," *Phys. Lett. B***808** (2020) 135632, 7.
20. L. Graf, B. Henning, X. Lu, T. Melia, and H. Murayama, "2, 12, 117, 1959, 45171, 1170086, âŁ¦: a Hilbert series for the QCD chiral Lagrangian," http://dx.doi.org/10.1007/JHEP01(2021)142*JHEP***01** (2021) 142, arXiv:2009.01239 [hep-ph].

21. A. Trautner, "Systematic construction of basis invariants in the 2HDM," http://dx.doi.org/10.1007/JHEP05(2019)208*JHEP***05** (2019) 208, arXiv:1812.02614 [hep-ph].
22. A. Trautner, "On the systematic construction of basis invariants," http://dx.doi.org/10.1088/1742-6596/1586/1/012005*J. Phys. Conf. Ser***1586** no. 1, (2020) 012005, arXiv:2002.12244 [hep-ph].
23. S. Cremonesi, A. Hanany, and A. Zaffaroni, "Monopole operators and Hilbert series of Coulomb branches of 3d$\mathcal{N}=4$ gauge theories," *Journal of High Energy Physics***5** (2014) .
24. A. Bourget, J. F. Grimminger, A. Hanany, R. Kalveks, M. Sperling, and Z. Zhong, "Magnetic Lattices for Orthosymplectic Quivers," http://dx.doi.org/10.1007/JHEP12(2020)092*JHEP***12** (2020) 092, arXiv:2007.04667 [hep-th].
25. A. Bourget, J. F. Grimminger, A. Hanany, M. Sperling, and Z. Zhong, "Branes, Quivers, and the Affine Grassmannian," arXiv:2102.06190 [hep-th].
26. A. Bourget, S. Cabrera, J. F. Grimminger, A. Hanany, and Z. Zhong, "Brane Webs and Magnetic Quivers for SQCD," http://dx.doi.org/10.1007/JHEP03(2020)176*JHEP***03** (2020) 176, arXiv:1909.00667 [hep-th].
27. A. Bourget, J. F. Grimminger, A. Hanany, R. Kalveks, M. Sperling, and Z. Zhong, "Folding orthosymplectic quivers," http://dx.doi.org/10.1007/JHEP12(2021)070*JHEP***12** (2021) 070, arXiv:2107.00754 [hep-th].
28. C. Closset, S. Schafer-Nameki, and Y.-N. Wang, "Coulomb and Higgs Branches from Canonical Singularities: Part 0," http://dx.doi.org/10.1007/JHEP02(2021)003*JHEP***02** (2021) 003, arXiv:2007.15600 [hep-th].
29. S. Franco, A. Hanany, D. Martelli, J. Sparks, D. Vegh, and B. Wecht, "Gauge theories from toric geometry and brane tilings," http://dx.doi.org/10.1088/1126-6708/2006/01/128*JHEP***01** (2006) 128, arXiv:hep-th/0505211.
30. R. Argurio, M. Bertolini, S. Franco, E. García-Valdecasas, S. Meynet, A. Pasternak, and V. Tatitscheff, "Dimers, Orientifolds and Anomalies," http://dx.doi.org/10.1007/JHEP02(2021)153*JHEP***02** (2021) 153, arXiv:2009.11291 [hep-th].
31. S. Franco and D. Rodriguez-Gomez, "Quivers, Lattice Gauge Theories and Fractons," arXiv:2203.01335 [hep-th].
32. A. Dey, A. Hanany, N. Mekareeya, D. Rodríguez-Gómez, and R.-K. Seong, "Hilbert Series for Moduli Spaces of Instantons on C^2/Z_n," http://dx.doi.org/10.1007/JHEP01(2014)182*JHEP***01** (2014) 182, arXiv:1309.0812 [hep-th].
33. C. A. Florentino, "Plethystic exponential calculus and characteristic polynomials of permutations," http://dx.doi.org/10.47443/dml.2021.094*Discrete Mathematics Letters***8** (Oct, 2021) 22–29. http://dx.doi.org/10.47443/dml.2021.094.
34. Y.-H. He, "Deep-learning the landscape." arXiv:1706.02714 [hep-th], 2017.
35. Y.-H. He, "Machine-learning the string landscape," *Phys. Lett. B***774** (2017) 564–568.
36. J. Carifio, J. Halverson, D. Krioukov, and B. D. Nelson, "Machine learning in the string landscape," *J. High Energy Phys.* no. 9, (2017) 157, front matter+35.
37. D. Krefl and R.-K. Seong, "Machine learning of Calabi–Yau volumes," *Phys. Rev. D***96** no. 6, (2017) 066014, 8.
38. F. Ruehle, "Evolving neural networks with genetic algorithms to study the string landscape," *J. High Energy Phys.* no. 8, (2017) 038, front matter+19.
39. P. Candelas, A. Constantin, and C. Mishra, "Calabi-Yau Threefolds with Small Hodge Numbers," http://dx.doi.org/10.1002/prop.201800029*Fortsch. Phys***66** no. 6, (2018) 1800029, arXiv:1602.06303 [hep-th].
40. A. Ashmore, Y.-H. He, and B. A. Ovrut, "Machine learning Calabi-Yau metrics," *Fortschr. Phys.***68** no. 9, (2020) 2000068, 23.
41. D. S. Berman, Y.-H. He, and E. Hirst, "Machine learning Calabi-Yau hypersurfaces," http://dx.doi.org/10.1103/PhysRevD.105.066002*Phys. Rev. D***105** no. 6, (2022) 066002, arXiv:2112.06350 [hep-th].
42. J. Bao, Y.-H. He, E. Hirst, and S. Pietromonaco, "Lectures on the Calabi-Yau Landscape," arXiv:2001.01212 [hep-th].

43. J. Bao, Y.-H. He, E. Hirst, J. Hofscheier, A. Kasprzyk, and S. Majumder, "Polytopes and Machine Learning," arXiv:2109.09602 [math.CO].
44. P. Berglund, B. Campbell, and V. Jejjala, "Machine Learning Kreuzer-Skarke Calabi-Yau Threefolds," arXiv:2112.09117 [hep-th].
45. Y.-H. He and S.-T. Yau, "Graph Laplacians, Riemannian manifolds and their machine-learning." arXiv:2006.16619 [math.CO], 2020.
46. J. Craven, M. Hughes, V. Jejjala, and A. Kar, "Learning knot invariants across dimensions," arXiv:2112.00016 [hep-th].
47. S. Gukov, J. Halverson, F. Ruehle, and P. Sułkowski, "Learning to Unknot," http://dx.doi.org/10.1088/2632-2153/abe91f*Mach. Learn. Sci. Tech***2** no. 2, (2021) 025035, arXiv:2010.16263 [math.GT].
48. J. Bao, Y.-H. He, and E. Hirst, "Neurons on Amoebae," arXiv:2106.03695 [math.AG].
49. G. Arias-Tamargo, Y.-H. He, E. Heyes, E. Hirst, and D. Rodriguez-Gomez, "Brain Webs for Brane Webs," arXiv:2202.05845 [hep-th].
50. H.-Y. Chen, Y.-H. He, S. Lal, and M. Z. Zaz, "Machine Learning Etudes in Conformal Field Theories," arXiv:2006.16114 [hep-th].
51. S. Krippendorf, D. Lust, and M. Syvaeri, "Integrability Ex Machina," http://dx.doi.org/10.1002/prop.202100057*Fortsch. Phys***69** no. 7, (2021) 2100057, arXiv:2103.07475 [nlin.SI].
52. J. Bao, S. Franco, Y.-H. He, E. Hirst, G. Musiker, and Y. Xiao, "Quiver Mutations, Seiberg Duality and Machine Learning," http://dx.doi.org/10.1103/PhysRevD.102.086013*Phys. Rev. D***102** no. 8, (2020) 086013, arXiv:2006.10783 [hep-th].
53. P.-P. Dechant, Y.-H. He, E. Heyes, and E. Hirst, "Cluster Algebras: Network Science and Machine Learning," arXiv:2203.13847 [math.CO].
54. Y.-H. He, E. Hirst, and T. Peterken, "Machine-learning dessins d'enfants: explorations via modular and Seiberg–Witten curves," http://dx.doi.org/10.1088/1751-8121/abbc4f*J. Phys. A***54** no. 7, (2021) 075401, arXiv:2004.05218 [hep-th].
55. J. Bao, Y.-H. He, E. Hirst, J. Hofscheier, A. Kasprzyk, and S. Majumder, "Hilbert series, machine learning, and applications to physics," http://dx.doi.org/10.1016/j.physletb.2022.136966*Phys. Lett. B***827** (2022) 136966, arXiv:2103.13436 [hep-th].
56. M. Abadi *et al.*, "TensorFlow: Large-scale machine learning on heterogeneous systems," 2015. https://www.tensorflow.org/. Software available from tensorflow.org.
57. I. Dolgachev, "Weighted projective varieties," in *Group actions and vector fields (Vancouver, B.C., 1981)*, vol. 956 of *Lecture Notes in Math.*, pp. 34–71. Springer, Berlin, 1982.
58. M. Kreuzer and L. Robbiano, http://dx.doi.org/10.1007/3-540-28296-3*Computational Commutative Algebra 2*. 01, 2005.
59. A. Hanany and D. Vegh, "Quivers, tilings, branes and rhombi," http://dx.doi.org/10.1088/1126-6708/2007/10/029*JHEP***10** (2007) 029, arXiv:hep-th/0511063.
60. M. F. Atiyah and I. G. Macdonald, *Introduction to commutative algebra*. Addison-Wesley Publishing Co., Reading, Mass.-London-Don Mills, Ont., 1969.
61. R. P. Stanley, "Hilbert functions of graded algebras," *Advances in Mathematics***28** (1978) 57–83.
62. G. Brown and A. M. Kasprzyk, "The Graded Ring Database." Online. http://www.grdb.co.uk/.
63. G. Brown and A. M. Kasprzyk, "The Fano 3-fold database." *Zenodo*https://doi.org/10.5281/zenodo.5820338, 2022.
64. A. M. Kasprzyk, "Canonical toric fano threefolds," *Canadian Journal of Mathematics***62** no. 6, (2010) 1293–1309, arXiv:0806.2604 [math.AG].
65. S. Altınok, G. Brown, and M. Reid, "Fano 3-folds, $K3$ surfaces and graded rings," in *Topology and geometry: commemorating SISTAG*, vol. 314 of *Contemp. Math.*, pp. 25–53. Amer. Math. Soc., Providence, RI, 2002.
66. G. Brown and A. M. Kasprzyk, "Kawamata boundedness for Fano threefolds and the Graded Ring Database." arXiv:2201.07178 [math.AG], 2022.
67. D. P. Kingma and J. Ba, "Adam: A method for stochastic optimization." arXiv:1412.6980 [cs.LG], 2014.

Penrose Limit: A Stringy Regime in Holography

Min-xin Huang

Abstract This is a summary of my talk given online at the Nankai Symposium in Physics, Geometry and Number Theory, August 2021, based on two recent papers [1, 2].

The AdS/CFT correspondence [3–5] is a conceptual breakthrough in our understandings of quantum gravity, in particular provides a non-perturbative definition of string theory in AdS background in terms of $\mathcal{N}=4$ $SU(N)$ super-Yang-Mills theory. We consider the Penrose limit [6], which gives rise to another maximally supersymmetric background [7], known as the pp-wave or plane wave geometry

$$ds^2 = -4dx^+dx^- - \mu^2(\vec{r}^{\,2} + \vec{y}^{\,2})(dx^+)^2 + d\vec{r}^{\,2} + d\vec{y}^{\,2}, \tag{1}$$

where x^+, x^- are light cone coordinates, $\vec{r}, \vec{y}$ are 4-vectors, and the parameter μ measures the spacetime curvature as well as the Ramond-Ramond flux $F_{+1234} = F_{+5678} \sim \mu$. This appears to be a promising ground for quantitative explorations of the holographic duality in stringy regimes, as the dual theories on both sides can be either free or weakly coupled.

In the groundbreaking paper [8], Berenstein, Maldacena and Nastase (BMN) proposed a type of near-BPS operators, which correspond to the type IIB closed strings on the pp-wave background. The free string spectrum is correctly reproduced by gauge interactions as the planar conformal dimensions of BMN operators. The BMN scaling limit with large R-charge $J \sim \sqrt{N} \sim \infty$ appears to be the right Goldilocks limit in this situation, since a smaller R-charge would not provide finite string interactions in the strict $N \sim \infty$ limit, while a larger R-charge may blow up the strings into D-branes, known as giant gravitons, first studied in [9].

The Penrose limit provides a new twist to the holography story. In the celebrated AdS/CFT holographic dictionary in [5], the CFT lives at the boundary of a bulk AdS

M. Huang (✉)
Interdisciplinary Center for Theoretical Study, University of Science and Technology of China, Hefei, Anhui, China
e-mail: minxin@ustc.edu.cn

Peng Huanwu Center for Fundamental Theory, Hefei, Anhui, China

Y.-H. He et al. (eds.), *Nankai Symposium on Mathematical Dialogues*,
https://doi.org/10.1007/978-981-19-2328-9_22

space and its local operators couple to the boundary configurations of the AdS bulk fields. However, although the pp-wave background (1) comes from a Penrose limit of the AdS space, the geometry is rather different. As such, it is not clear how to directly apply the standard AdS holographic dictionary, particularly in the situations with finite string interactions. Our approach in some previous papers [10–14] is to consider another corner of the parameter space in the BMN limit, focusing on the free gauge theory. In this case, the string theory side becomes infinitely curved $\mu \sim \infty$, and strings are effectively infinitely long and tensionless, but can still have finite string interactions. Most interestingly, since the string spectrum is completely degenerate, the tensionless string can jump from one excited state to another without energy cost through a quantum unitary transition. It turns out that in this case the effective string coupling constant should be identified with a finite genus counting parameter $g := \frac{J^2}{N}$. Some higher genus BMN correlators were first computed in [15, 16].

A main goal of the theories of quantum gravity is to understand the physics in the regime beyond the reach of classical gravity, e.g. in the highly curved spacetime region near the black hole singularity. Certainly the effective action of classical gravity completely breaks down in the infinite curvature $\mu \sim \infty$ limit, but this is not necessarily a bad situation as we can instead probe the fundamental nature of spacetime. As the background remains maximally supersymmetric, the physics is much more amenable to studies than e.g. those near black hole or big bang singularities. Thanks to the helps from holographic duality, the physical string amplitudes can still be subjected to quantitative studies with some reasonable conjectures and indirect constrains such as unitarity, which is a fundamental principle that we expect to remain valid in the infinite curvature limit.

Let us introduce some notations. The BMN vacuum operator is simply proportional to $\mathrm{Tr}(Z^J)$ where Z is a complex scalar in the $\mathcal{N} = 4$ $SU(N)$ super-Yang-Mills theory. One can insert the four remaining real scalars into the trace with phases, corresponding to string modes in four of the eight transverse dimensions. The BMN operators are then denoted as $O^J_{(m_1,m_2,\cdots,m_k)}$, where the positive and negative integer modes represent the left and right moving stringy excited modes, while the zero modes are supergravity modes representing discretized momenta in the corresponding traverse direction. We will consider the case of string modes in different transverse directions, otherwise the BMN operators are no longer near-BPS and there may be some potential issues as discussed in [1]. Due to the closed string level matching condition $\sum_i m_i = 0$, the excited stringy states have at least two string modes. The BMN operators are properly normalized to be orthonormal at planar level, and the genus h two-point functions are proportional to g^{2h} as

$$\begin{aligned} \langle \bar{O}^J_{(m_1,m_2,\cdots,m_k)} O^J_{(n_1,n_2,\cdots,n_k)} \rangle_0 &= \delta_{m_1,n_1} \cdots \delta_{m_k,n_k}, \\ \langle \bar{O}^J_{(m_1,m_2,\cdots,m_k)} O^J_{(n_1,n_2,\cdots,n_k)} \rangle_h &\sim g^{2h}. \end{aligned} \tag{2}$$

As discussed in [1], the BMN two-point functions are real and symmetric, and there is a nice normalization relation summing over one set of mode numbers

$$\sum_{\sum_{i=1}^{k} n_k=0} \langle \bar{O}^J_{(m_1,m_2,\cdots,m_k)} O^J_{(n_1,n_2,\cdots,n_k)} \rangle_h = \frac{g^{2h}}{2^{2h}(2h+1)!}. \tag{3}$$

We may define a matrix element, summing up all genus contributions with a proper normalization by the all-genera formula of vacuum correlator

$$p_{(m_1,m_2,\cdots,m_k),(n_1,n_2,\cdots,n_k)} := \frac{g}{2\sinh(\frac{g}{2})} \sum_{h=0}^{\infty} \langle \bar{O}^J_{(m_1,m_2,\cdots,m_k)} O^J_{(n_1,n_2,\cdots,n_k)} \rangle_h, \tag{4}$$

so that it looks like a probability distribution

$$\sum_{\sum_{i=1}^{k} n_k=0} p_{(m_1,m_2,\cdots,m_k),(n_1,n_2,\cdots,n_k)} = 1. \tag{5}$$

To interpret the matrix elements as a probability distribution, they need to be non-negative. For the case of two string modes, the non-negativity at any genus can be easily proven since the two string modes are opposite numbers [13]. For the case of three string modes, we are not aware of a simple analytic proof of the non-negativity. Instead, we perform the detailed calculations and explicitly verify the non-negativity up to genus two [2].

The normalization relations and non-negativity with two and three string modes suggest a novel entry of the pp-wave holographic dictionary

$$p_{(m_1,\cdots,m_k),(n_1,\cdots,n_k)} = |\langle m_1,\cdots,m_k|\hat{U}(g)|n_1,\cdots,n_k\rangle|^2, \quad k=2,3, \tag{6}$$

where the operator $\hat{U}(g)$ describes the quantum unitary transition between the degenerate tensionless strings. The BMN single string states form a complete orthonormal basis of the Hilbert space under such finite string interactions. Of course, the probability interpretation only requires the matrix element (4) is non-negative. It would seem rather contrived to have negative two-point functions at some higher genus but the total contribution still manages to always remain non-negative. Expecting the same phenomenon as in the case of two string modes, we make the stronger conjecture that the BMN two-point functions with three string modes are also always non-negative separately at each genus.

This seemingly simple equation (6) has eluded research on the topic for many years, as the physical picture of string dynamics in the infinitely curved pp-wave background turns out to be drastically different from those familiar in flat spacetime or AdS space with large radius. In particular, there is no finite physical process of multiple particles or strings scattering to and from asymptotic region of spacetime, as the higher point functions always vanish in the strict $N \sim \infty$ limit. Instead, the tensionless string directly jumps from one excited state to another through a quantum unitary transition, much as in a S-matrix where the incoming and outgoing states have the same energy. Of course, the BMN higher point functions, particularly the cubic string vertices, are still very useful since infinitely many of them may combine to make a finite contribution in the string (loop) diagrams in [11, 12]. Since the only finite BMN two-point functions are always real and symmetric, unitarity of string interactions rules out the naive possibility that they are directly identified with quantum transition amplitudes on the string theory side [13]. So we arrive at the otherwise seemingly most natural conjecture (6).

There is no technical obstruction for our calculations on the free gauge theory side at any genus, however the available tools on the string theory side are very limited. Much progress for the calculations of higher genus critical superstring amplitudes focused on using the RNS (Ramond-Neveu-Schwarz) formalism in flat space, and is already quite difficult at genus two. Our conjecture (6) gives the norms of certain critical superstring amplitudes including all genus contributions. Of course, the string amplitudes are complex, and consist of the norms and phase angles. As discussed in [13], unitarity can in principle determine a large part of the phase angles, but not completely. Our studies thus provide a long term motivation for future research to develop techniques that can deal with string theory on highly curved background, with flux, and including highly excited stringy states, for the purposes of a direct verification of the conjecture (6) as well as the complete determination of string amplitudes including the phase angles. For the moment, our verifications of non-negativity on the gauge theory side provide indirect non-trivial evidence of the conjecture (6).

If our conjecture (6) is correct, to our knowledge, it would not only provide first examples of systematic calculations of (the norms of) the higher genus critical superstring amplitudes, but may also in principle gives exact complete results for any string coupling, due to the convergence of genus expansion, as mentioned in [14].

For the case of four string modes, it turns out that the genus one two-point functions can be negative [1]. There seems to be a rule forbidding the "crowdedness" of string modes, that we can not holographically use up all four remaining scalars to fully occupy the four transverse dimensions with $SO(4)$ rotational symmetry unbroken by the Ramond-Ramond flux in the pp-wave background. On the other hand, we confirm [1] that the factorization formulas [11, 12] are still valid for the case of four string modes, while the comparison with cubic string vertex [10, 17] is seen to be straightforwardly applied to any hypothetical number of string modes, not even restricted by the eight dimensions of transverse directions in the pp-wave background.

It is well known that string theory has a Hagedorn temperature inversely proportional to the string length, which may be interpreted as a maximal temperature or a temperature of phase transition. In our physical setting, the tensionless strings have effectively infinite length, so the Hagedorn temperature is zero. This is not necessarily a problem, since for example an extremal black hole emits no Hawking radiation, has zero temperature, but can still have finite event horizon and entropy. In [14], we consider the entropy of these tensionless BMN strings, which can be interpreted as the Von-Neumann entropy of a mixed BMN state due to decoherence. It would be interesting to see whether we can further study some interesting thermodynamics under such extreme conditions.

It is widely believed that strings may not be the right fundamental degrees of freedom to formulate the still mysterious M-theory, since there are non-perturbative objects like D-branes and M-branes. On the contrary, if our conjecture (6) is correct, it seems that in our very special setting with infinite spacetime curvature and Ramond-Ramond flux, the tensionless closed strings do provide the proper complete degrees of freedom for physics at any coupling constant. Although we focus on a highly unrealistic special situation, our studies may provide some insights for the non-perturbative formulation of string/M-theory on general backgrounds.

Acknowledgements I thank Bao-ning Du for collaborations on the papers, and Jun-Hao Li, Jian-xin Lu, Gao-fu Ren, Pei-xuan Zeng for helpful discussions. This work was supported in parts by the national Natural Science Foundation of China (Grants No.11947301 and No.12047502).

References

1. B.-n. Du and M.-x. Huang, "Free BMN Correlators With More Stringy Modes," https://doi.org/10.1007/JHEP03(2021)246JHEP 03 (2021) 246, arXiv:2101.07484 [hep-th].
2. B.-n. Du and M.-x. Huang, "Non-negativity of BMN two-point functions with three string modes," https://doi.org/10.1007/JHEP08(2021)006JHEP 08 (2021) 006, arXiv:2104.12502 [hep-th].
3. J. M. Maldacena, "The Large N limit of superconformal field theories and supergravity," https://doi.org/10.1023/A:1026654312961Int. J. Theor. Phys. 38 (1999) 1113–1133,Int. J. Theor. Phys. 38 (1999) 1113–1133, arXiv:hep-th/9711200.
4. S. Gubser, I. R. Klebanov, and A. M. Polyakov, "Gauge theory correlators from noncritical string theory," https://doi.org/10.1016/S0370-2693(98)00377-3Phys. Lett. B 428 (1998) 105–114, arXiv:hep-th/9802109.
5. E. Witten, "Anti-de Sitter space and holography," https://doi.org/10.4310/ATMP.1998.v2.n2.a2Adv. Theor. Math. Phys. 2 (1998) 253–291, arXiv:hep-th/9802150.
6. R. Penrose, *Any Space-Time has a Plane Wave as a Limit*, https://doi.org/10.1007/978-94-010-1508-0_23pp. 271–275. Springer Netherlands, Dordrecht, 1976. https://doi.org/10.1007/978-94-010-1508-0_23.
7. M. Blau, J. M. Figueroa-O'Farrill, C. Hull, and G. Papadopoulos, "A New maximally supersymmetric background of IIB superstring theory," http://dx.doi.org/10.1088/1126-6708/2002/01/047JHEP 01 (2002) 047, arXiv:hep-th/0110242.
8. D. E. Berenstein, J. M. Maldacena, and H. S. Nastase, "Strings in flat space and pp waves from N=4 superYang-Mills," https://doi.org/10.1088/1126-6708/2002/04/013JHEP 04 (2002) 013, arXiv:hep-th/0202021.
9. J. McGreevy, L. Susskind, and N. Toumbas, "Invasion of the giant gravitons from Anti-de Sitter space," https://doi.org/10.1088/1126-6708/2000/06/008JHEP 06 (2000) 008, arXiv:hep-th/0003075.
10. M.-x. Huang, "Three point functions of N=4 superYang-Mills from light cone string field theory in PP wave," https://doi.org/10.1016/S0370-2693(02)02345-6Phys. Lett. B 542 (2002) 255–260, arXiv:hep-th/0205311.
11. M.-x. Huang, "String interactions in PP wave from N=4 superYang-Mills," https://doi.org/10.1103/PhysRevD.66.105002Phys. Rev. D 66 (2002) 105002, arXiv:hep-th/0206248.
12. M.-x. Huang, "Higher Genus BMN Correlators: Factorization and Recursion Relations," https://doi.org/10.4310/ATMP.2012.v16.n2.a3Adv. Theor. Math. Phys. 16 no. 2, (2012) 421–503, arXiv:1009.5447 [hep-th].
13. M.-x. Huang, "Note on S-channel factorization in multitrace Berenstein-Maldacena-Nastase correlators," https://doi.org/10.1103/PhysRevD.101.026013Phys. Rev. D 101 no. 2, (2020) 026013, arXiv:1909.06995 [hep-th].

14. M.-x. Huang, "Entropy of Berenstein-Maldacena-Nastase Strings," https://doi.org/10.1103/PhysRevD.101.046027Phys. Rev. D 101 no. 4, (2020) 046027, arXiv:1910.02581 [hep-th].
15. C. Kristjansen, J. Plefka, G. Semenoff, and M. Staudacher, "A New double scaling limit of N=4 superYang-Mills theory and PP wave strings," https://doi.org/10.1016/S0550-3213(02)00749-6Nucl. Phys. B 643 (2002) 3–30, arXiv:hep-th/0205033.
16. N. R. Constable, D. Z. Freedman, M. Headrick, S. Minwalla, L. Motl, A. Postnikov, and W. Skiba, "PP wave string interactions from perturbative Yang-Mills theory," http://dx.doi.org/10.1088/1126-6708/2002/07/017JHEP 07 (2002) 017, arXiv:hep-th/0205089.
17. M. Spradlin and A. Volovich, "Superstring interactions in a p p wave background," https://doi.org/10.1103/PhysRevD.66.086004Phys. Rev. D 66 (2002) 086004, arXiv:hep-th/0204146.

Identifying Equivalent Calabi–Yau Topologies: A Discrete Challenge from Math and Physics for Machine Learning

Vishnu Jejjala, Washington Taylor, and Andrew P. Turner

Abstract We review briefly the characteristic topological data of Calabi–Yau threefolds and focus on the question of when two threefolds are equivalent through related topological data. This provides an interesting test case for machine learning methodology in discrete mathematics problems motivated by physics.

1 Calabi–Yau Threefolds

Calabi–Yau threefolds are a class of geometric spaces that have been studied intensively for several decades by mathematicians and physicists in the context of superstring compactification [1–3]. The key features of these real 6-dimensional manifolds are (1) that they admit a Ricci-flat Kähler metric, and thus solve the vacuum Einstein equations $R_{\mu\nu} = 0$, and (2) they are Kähler manifolds and thus have a complex structure compatible with supersymmetry. This makes them useful as geometries for compactifying superstring theory from ten dimensions to the physical four dimensions of space-time. Mathematically, the key feature of a Calabi–Yau variety is the vanishing of the canonical class $K = 0$ (up to torsion).

W. Taylor: This note is based on a talk by WT at the 2021 Nankai Symposium on Mathematical Dialogues in celebration of S. S. Chern's 110th anniversary

V. Jejjala (✉)
School of Physics, NITheCS, and CoE-MaSS, Mandelstam Institute for Theoretical Physics, University of the Witwatersrand, Johannesburg, South Africa
e-mail: vishnu@neo.phys.wits.ac.za

W. Taylor
Department of Physics, Center for Theoretical Physics, NSF AI Institute for Artificial Intelligence and Fundamental Interactions, Massachusetts Institute of Technology, Cambridge, MA, USA
e-mail: wati@mit.edu

A. P. Turner
Department of Physics and Astronomy, University of Pennsylvania, Philadelphia, PA, USA
e-mail: turnerap@sas.upenn.edu

Y.-H. He et al. (eds.), *Nankai Symposium on Mathematical Dialogues*,
https://doi.org/10.1007/978-981-19-2328-9_23

Despite extensive investigations on these geometries, a fundamental open question remains: *Is the number of topologically distinct Calabi–Yau threefolds finite or infinite?* [4].

Topologically, a theorem of Wall [5] states that a Calabi–Yau threefold X is uniquely determined by:

- $h^{1,1}, h^{2,1} \in \mathbb{Z}_+$: Hodge numbers;
- $C_{ijk} \in \mathbb{Z},\ i, j, k \in \{1, \ldots, h^{1,1}\}$: triple intersection numbers between divisors (complex codimension-one subvarieties) in the linear space $H^{1,1}(X, \mathbb{Z})$ (where $h^{1,1}(X) = \dim H^{1,1}(X, \mathbb{Z})$);
- $p_1(T_X) = 2c_2(T_X)$: The first Pontryagin class, a vector of $h^{1,1}$ integers.

Thus, the topological data of a Calabi–Yau threefold is characterized by a finite list of integer data. Fundamental questions to which the answers are not yet known are:

(1) Which sets of integer data are allowed?
(2) Which sets of integer data are equivalent under a basis change for $H^{1,1}$ (and hence describe the same geometry)?

We are focused here on the second of these questions.

Over the years, many large classes of Calabi–Yau threefolds have been constructed. The largest is the set of toric hypersurface Calabi–Yau threefolds realized through the Batyrev construction [6] as hypersurfaces in toric fourfolds; Kreuzer and Skarke have compiled a comprehensive list of 473,800,776 reflexive 4D polytopes that provide such constructions [7, 8]. Other constructions include complete intersection Calabi–Yau (CICY) threefolds [9] and generalized CICYs [10].

Regarding the finiteness question, recent work on elliptic Calabi–Yau threefolds, which are particularly relevant in the context of F-theory [11, 12], has given some fairly strong evidence that the number of distinct topological Calabi–Yau threefolds may be finite (up to birational equivalence under flops). An elliptic (or more generally genus one fibered) Calabi–Yau threefold admits a fibration $\pi : X \to B_2$ so that over an open subset of B_2 the fiber $\pi^{-1}(x)$ is an elliptic curve or genus one algebraic curve. It is known that the number of topological types of elliptic Calabi–Yau threefolds is finite [13, 14]. Furthermore, it has been found that almost all known Calabi–Yau threefold constructions admit an elliptic fibration in some flop phase. For example, 99.3% (7837/7890) of the CICY threefolds have an "obvious" elliptic or genus one fibration [15], and all have at least one such fibration when $h^{1,1} \geq 4$. For the larger class of toric hypersurface Calabi–Yau threefolds, a systematic analysis was carried out in [16], finding that 99.994% of the threefolds in the Kreuzer–Skarke database have a manifest elliptic fiber in some phase (for each polytope there are many toric varieties, and different triangulations give different phases related under flops). A graph of the Hodge numbers of the Calabi–Yau threefolds in the Kreuzer–Skarke database is shown in Fig. 1; the Hodge numbers of the 29,223 reflexive polytopes that give Calabi–Yau threefolds with no obvious elliptic fiber are shown in red.

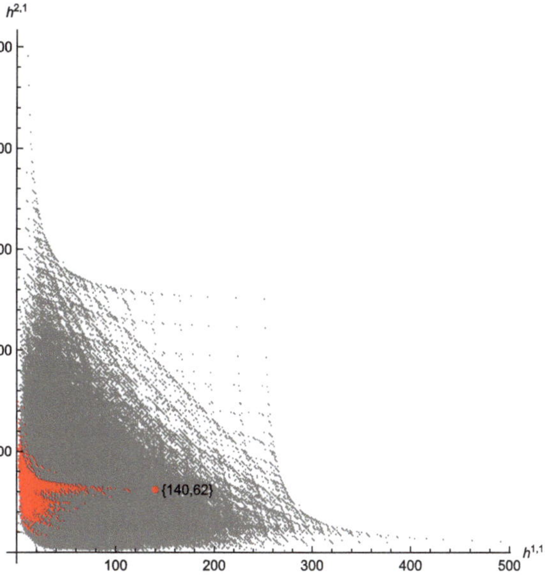

Fig. 1 Hodge numbers of the 473.8 million families of Calabi–Yau threefolds described by Kreuzer and Skarke through toric hypersurface constructions from 4D reflexive polytopes. Hodge numbers for the less than 30,000 cases without obvious elliptic fibers are shown in red; even many of these may have more subtle elliptic fibration structure in some phase. Since the number of topologically distinct elliptic Calabi–Yau threefolds is finite, this suggests finiteness of the full Calabi–Yau landscape

2 Triple Intersection Numbers and Topological Equivalence

As described above, an important open question is to identify a systematic algorithm for determining when two sets of discrete data that characterize a topological Calabi–Yau threefold are equivalent under a basis change. In particular, we focus on the specific part of that problem associated with the triple intersection numbers

$$C_{ijk} \in \mathbb{Z}\,, \quad i, j, k \leq N\,. \tag{1}$$

These intersection numbers form a 3-index symmetric tensor so that $C_{ijk} = C_{jik} = C_{ikj}$.

We wish to address the following mathematical question:

Problem: *Given two 3-index symmetric tensors C, C' of degree N associated with Calabi–Yau threefold geometries, does there exist an integer linear change of basis $\Lambda_{ij} \in \mathrm{SL}(N, \mathbb{Z})$ such that*

$$C'_{ijk} = \Lambda_{il}\Lambda_{jm}\Lambda_{kn}C_{lmn}\,? \tag{2}$$

Because the set of possible linear transformations Λ is infinite, there is no simple way to check the equivalence in finite time. In fact, there is no known algorithm to check this equivalence in finite time, whether or not it is known that the triple intersection numbers correspond to allowed Calabi–Yau threefold geometries.

Some partial results in this direction have been found. In particular, there are certain functions of the triple intersection numbers that are invariant under a basis change; thus, if the invariants differ between the intersection numbers C and C', there clearly cannot be an equivalence. Some of these invariants are summarized in [2]; for example, $\gcd(\{C_{ijk}\}_{1\le i,j,k\le N})$ is clearly invariant under any transformation $\Lambda_{ij} \in \mathrm{SL}(N, \mathbb{Z})$. Recently, some new invariants have been identified using the methodology of limiting mixed Hodge structures [17].

Because there is no known algorithm for checking this equivalence, this is a natural candidate for the application of machine learning (ML) methodology. In recent years, as elaborated in some detail in the recent text [3], there has been a great deal of work on using ML methods to solve discrete problems related to Calabi–Yau geometries, with varying degrees of success [18–33]. On one hand, one might hope that by presenting an ML system with a variety of data of triple intersection numbers with families of equivalence known from construction it may be possible to find some approximate method for checking equivalence. More ambitiously, one might hope to identify through this approach some systematic structure in the triple intersection numbers such as new invariants that may lead to an analytic solution of the problem.

Direct application of standard ML techniques to this problem does not immediately give insight, so we turn to a related problem to understand how machine learning can address discrete transformation problems of this type.

3 Machine Learning and a Simpler Matrix Problem

As an analogue of the problem of identifying when two 3-index tensors are equivalent under a basis change, we can consider the simpler problem of using machine learning to figure out when two matrices (i.e., 2-index tensors) are similarly equivalent. To relate this to some known results, we also drop for now the condition that the entries are integer. We can then consider the following simpler problem:

Problem: *Given complex matrices A_{ij}, B_{ij}, does there exist a unitary transformation Λ so that*

$$B = \Lambda A \Lambda^{\dagger}\ ? \tag{3}$$

There is a known set of criteria that determines when two such matrices are equivalent in this way. *Specht's theorem* [34] states that $A \sim B$ iff

$$\begin{aligned} \mathrm{Tr}\, A &= \mathrm{Tr}\, B\,, \\ \mathrm{Tr}\, A^2 &= \mathrm{Tr}\, B^2\,, \\ \mathrm{Tr}\, AA^{\dagger} &= \mathrm{Tr}\, BB^{\dagger}\,, \\ &\vdots \end{aligned} \tag{4}$$

i.e., $\mathrm{Tr}\, W(A, A^{\dagger}) = \mathrm{Tr}\, W(B, B^{\dagger})$ for all words W that are functions of the matrix and its adjoint.

In fact, for matrices of a given size N, it is sufficient to check the agreement between these invariants only up to a certain length. For $N = 2$, it is sufficient to check only up to words of degree 2 (i.e., 3 independent conditions). For $N = 3$, words up to degree 6 suffice (7 conditions). For general N it has been proven that it is sufficient to check up to degree $N\sqrt{2N^2/(N-1)+1/4}+N/2-2$, and it has been conjectured that the degree of sufficiency actually scales linearly in N [35, 36].

Thus, we know that for matrices (2-index tensors), this is a solvable problem even when the entries are non-integer. Thus, it is perhaps illuminating to see how machine learning can do at learning to check such equivalences, and we can see if a simple machine learning system can be designed that can essentially "discover" Specht's theorem.

Motivated by these considerations, we have implemented some simple ML algorithms to attempt to solve the matrix equivalence problem using the neural network functionality built in to Mathematica.

The simplest experiment is to try random *real* matrices A, B with a simple multi-layer ReLU network, providing data in which half of the cases given are equivalent matrices and the other half are inequivalent, with no further constraints. This kind of network quickly converges to near-100% accuracy. Some investigation, however, shows that this network is essentially just checking the first (linear) condition in (4), i.e., comparing the traces of the input matrices. It is not surprising that a simple ML system can rapidly detect when two matrices satisfy this simple linear condition.

The next challenge is to provide a data set in which we fix the condition $\text{Tr}\,A = \text{Tr}\,B$ by hand for all pairs. Using such a dataset of 3×3 matrices, and a basic multi-layer ReLU network, with some training the network accurately recognizes pairs that are equivalent about 75% of the time and pairs that are inequivalent about 60% of the time. This is not a very good success rate. By changing the network architecture to explicitly compute all products of terms in the first layer by hand (see Fig. 2), however, the success rates immediately increase to 99% for equivalent pairs and 90% for inequivalent pairs. The error rates during training for both network architectures are shown in Fig. 3. Thus, with this change of architecture the neural network "discovers" the quadratic term in Specht's theorem. It seems, however, that higher products must similarly be included by hand to efficiently converge on a system that knows about the additional constraints in (4). Similar results hold for complex matrices, where we must use a slightly more complicated version of the architecture shown in Fig. 2 to include $\text{Tr}\,AA^\dagger$ as well as $\text{Tr}\,A^2$.

4 Conclusions

In this brief extended abstract we have described a basic problem in identifying equivalence of symmetric 3-index tensors that is central to classifying Calabi–Yau threefolds by discrete information characterizing their topology. We have outlined the first steps of an effort to use machine learning to solve a simpler version of the

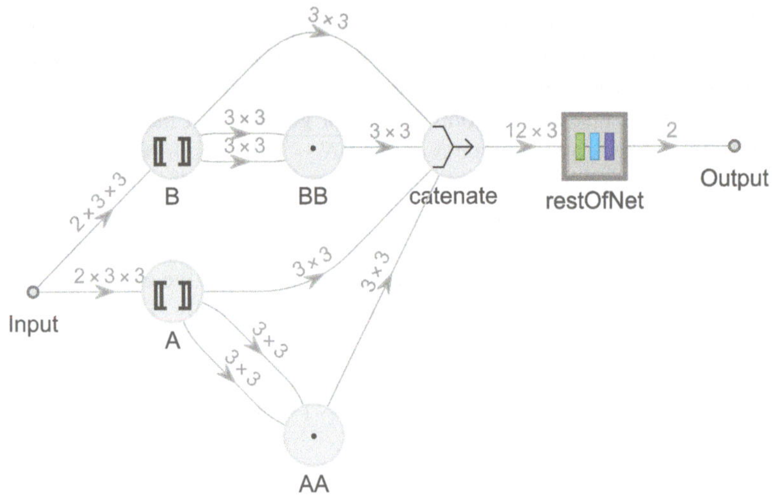

Fig. 2 Network architecture for learning orthogonal equivalence of 3 × 3 real matrices with fixed first-order invariant. The input matrices A and B are fed into dot layers that compute the matrix squares, and then A, A^2, B, and B^2 are fed into `restOfNet`, which is a multi-layer ReLU network

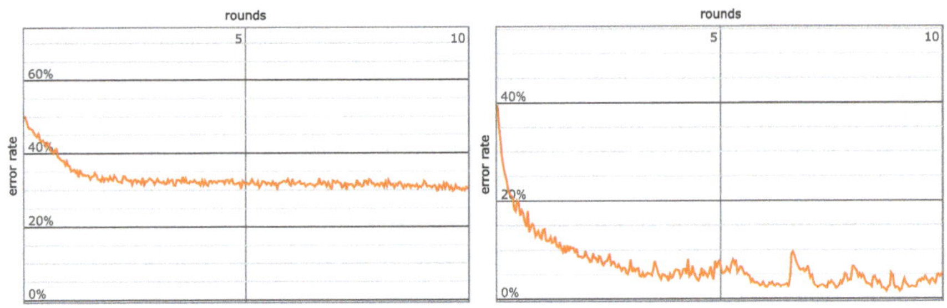

Fig. 3 Error rates during training for two different network architectures learning to identify orthogonal equivalence of 3 × 3 real matrices with fixed first-order invariant. On the left, a multi-layer ReLU network; on the right, the architecture shown in Fig. 2

problem with 2-index tensors where the analytic solution is known. Several lessons that can be taken from this are the following:

- Out of the box ML algorithms do not immediately yield insight in discrete problems of this kind.
- The fact that including product terms by hand in the ML architecture facilitates "discovery" by the neural network of nonlinear constraints suggests that networks with different basic functional units may perform dramatically differently on certain classes of problems; here we have an explicit understanding through Specht's theorem of this structure.
- It has been pointed out by, e.g., [37] that in principle a general nonlinear layer type in neural networks can learn the multiplication function, so in principle the dis-

tinction between these different functional units may be essentially some factor of overhead. It seems likely, however, that the time to learn such things may be exponentially large so that in practice network architecture and the form of nonlinear function used may be crucial in getting networks to learn certain structures.

The work described here is really only a first step at using ML methods for analyzing this problem. It would be interesting to extend the analysis further to reproduce higher order terms in the matrix case and to investigate whether sufficiently sophisticated network architectures could learn to recognize equivalent 3-index integer tensors for the Calabi–Yau equivalence problem, and if so whether it would be possible to extract new analytic information such as new Calabi–Yau threefold invariants from this analysis.

Acknowledgements We would like to thank Lara Anderson, James Gray, Harold Erbin, Tristan Hübsch, Max Tegmark, and Yinan Wang for helpful discussions. This work arose in the context of collaborative visits supported by a grant from the MISTI MIT-Africa-Imperial College seed fund initiative, and was supported in part by this grant. VJ is supported in part by the South African Research Chairs Initiative of the National Research Foundation, grant number 78554 and the Simons Foundation Mathematical and Physical Sciences Targeted Grants to Institutes, Award ID:509116. WT is also supported by DOE grant DE-SC00012567. APT is supported in part by the DOE (HEP) Award DE-SC0013528. This work was also supported in part by the National Science Foundation under Cooperative Agreement PHY-2019786.

References

1. P. Candelas, G. T. Horowitz, A. Strominger and E. Witten, “Vacuum configurations for superstrings,” Nucl. Phys. B **258**, 46 (1985).
2. T. Hübsch, “Calabi–Yau Manifolds: A Bestiary for Physicists,” World Scientific, 1992.
3. Y.-H. He, “The Calabi–Yau Landscape: From Geometry, to Physics, to Machine Learning,” Springer, Lecture Notes in Mathematics 2293, 2021.
4. S.-T. Yau, “Open problems in geometry,” Proc. Symp. Pure Math **54**, 1, 1-28 (1993).
5. C. T. C. Wall, “Classification problems in differential topology. V,” Inventiones mathematicae **1**, 355 (1966).
6. V. Batyrev, “Variations of the mixed Hodge structure of affine hypersurfaces in algebraic tori,” Duke Math. Journ. **69**, 349 (1993).
7. M. Kreuzer and H. Skarke, “Complete classification of reflexive polyhedra in four-dimensions,” Adv. Theor. Math. Phys. **4**, 1209 (2002) hep-th/0002240.
8. M. Kreuzer and H. Skarke, http://hep.itp.tuwien.ac.at/~kreuzer/CY.html.
9. P. Candelas, A. M. Dale, C. A. Lutken and R. Schimmrigk, “Complete intersection Calabi–Yau manifolds,” Nucl. Phys. B **298**, 493 (1988).
10. L. B. Anderson, F. Apruzzi, X. Gao, J. Gray and S. J. Lee, “A new construction of Calabi–Yau manifolds: Generalized CICYs,” Nucl. Phys. B **906**, 441 (2016) arXiv:1507.03235 [hep-th].
11. C. Vafa, “Evidence for F-Theory,” Nucl. Phys. B **469**, 403 (1996) arXiv:hep-th/9602022.
12. D. R. Morrison and C. Vafa, “Compactifications of F-Theory on Calabi–Yau threefolds – I,” Nucl. Phys. B **473**, 74 (1996) arXiv:hep-th/9602114; “Compactifications of F-Theory on Calabi–Yau threefolds – II,” Nucl. Phys. B **476**, 437 (1996) arXiv:hep-th/9603161.
13. A. Grassi, “On minimal models of elliptic threefolds,” Math. Ann. **290** (1991) 287–301.

14. M. Gross, "A finiteness theorem for elliptic Calabi–Yau threefolds," Duke Math. Jour. **74**, 271 (1994).
15. L. B. Anderson, X. Gao, J. Gray and S. J. Lee, "Multiple fibrations in Calabi–Yau geometry and string dualities," JHEP **1610**, 105 (2016) arXiv:1608.07555 [hep-th]; "Fibrations in CICY threefolds," JHEP **1710**, 077 (2017) arXiv:1708.07907 [hep-th].
16. Y. C. Huang and W. Taylor, "Fibration structure in toric hypersurface Calabi–Yau threefolds," JHEP **03** (2020), 172 arXiv:1907.09482 [hep-th].
17. T. W. Grimm, F. Ruehle and D. van de Heisteeg, "Classifying Calabi–Yau threefolds using infinite distance limits," Commun. Math. Phys. **382** (2021) no.1, 239-275 arXiv:1910.02963 [hep-th].
18. Y. H. He, "Deep-Learning the landscape," arXiv:1706.02714 [hep-th].
19. F. Ruehle, "Evolving neural networks with genetic algorithms to study the string landscape," JHEP **08** (2017), 038 arXiv:1706.07024 [hep-th].
20. J. Carifio, J. Halverson, D. Krioukov and B. D. Nelson, "Machine learning in the string landscape," JHEP **09** (2017), 157 arXiv:1707.00655 [hep-th].
21. Y. N. Wang and Z. Zhang, "Learning non-Higgsable gauge groups in 4D F-theory," JHEP **08** (2018), 009 arXiv:1804.07296 [hep-th].
22. K. Bull, Y. H. He, V. Jejjala and C. Mishra, "Machine learning CICY threefolds," Phys. Lett. B **785** (2018), 65-72 arXiv:1806.03121 [hep-th].
23. D. Klaewer and L. Schlechter, "Machine learning line Bundle cohomologies of hypersurfaces in toric varieties," Phys. Lett. B **789** (2019), 438-443 arXiv:1809.02547 [hep-th].
24. K. Bull, Y. H. He, V. Jejjala and C. Mishra, "Getting CICY high," Phys. Lett. B **795**, 700-706 (2019) arXiv:1903.03113 [hep-th].
25. C. R. Brodie, A. Constantin, R. Deen and A. Lukas, "Machine learning line bundle cohomology," Fortsch. Phys. **68** (2020) no.1, 1900087 arXiv:1906.08730 [hep-th].
26. M. Bies, M. Cvetič, R. Donagi, L. Lin, M. Liu and F. Ruehle, "Machine learning and algebraic approaches towards complete matter spectra in 4d F-theory," JHEP **01** (2021), 196 arXiv:2007.00009 [hep-th].
27. H. Erbin and R. Finotello, "Inception neural network for complete intersection Calabi–Yau 3-folds," Mach. Learn. Sci. Tech. **2** (2021) no.2, 02LT03 arXiv:2007.13379 [hep-th].
28. H. Erbin and R. Finotello, "Machine learning for complete intersection Calabi–Yau manifolds: A methodological study," Phys. Rev. D **103**, no.12, 126014 (2021) arXiv:2007.15706 [hep-th].
29. Y. H. He and A. Lukas, "Machine learning Calabi–Yau four-folds," Phys. Lett. B **815** (2021), 136139 arXiv:2009.02544 [hep-th].
30. H. Erbin, R. Finotello, R. Schneider and M. Tamaazousti, "Deep multi-task mining Calabi–Yau four-folds," Mach. Learn. Sci. Tech. **3** (2022) no.1, 015006 arXiv:2108.02221 [hep-th].
31. X. Gao and H. Zou, "Machine learning to the orientifold Calabi–Yau with string vacua," arXiv:2112.04950 [hep-th].
32. D. S. Berman, Y. H. He and E. Hirst, "Machine learning Calabi–Yau hypersurfaces," arXiv:2112.06350 [hep-th].
33. P. Berglund, B. Campbell and V. Jejjala, "Machine learning Kreuzer–Skarke Calabi–Yau threefolds," arXiv:2112.09117 [hep-th].
34. W. Specht, "Zur theorie der matrizen. II.," Jahresbericht der Deutschen Mathematiker-Vereinigung **50**, 19 (1940).
35. A. Paz, "An application of the Cayley–Hamilton theorem to matrix polynomials in several variables," Linear and Multilinear Algebra **15**, 161 (1984).
36. A. Freedman, R. Gupta and R. Guralnick, "Shirshov's theorem and representations of semigroups," Pacific J. Math. **181**, 159, (1997).
37. H. W. Lin, M. Tegmark and D. Rolnick, "Why does deep and cheap learning work so well?," J. Stat. Phys. **168**, 1223 (2017) arXiv:1608.08225 [cond-mat].

Mechanical Methods in Arithmetic Geometry?

Minhyong Kim

1991 Mathematics Subject Classification 11G05 · 11G30 · 11G40 · 81T13 · 81T45

1 Some Remarks on Physics, Mathematics, and Infinity

The problem of understanding infinite sums and the related difficulty of infinitesimals have bothered mathematicians since ancient times, starting at least from the 'paradox' of Zeno of Elea. Many of the infinities of modern physics can be viewed as versions of Zeno's paradox. For Zeno, there was a commonplace physical process, such as the flight of an arrow, that could be viewed as a composition of an infinite collection of subprocesses. The total time required for the process was evidently finite from experiment, but the mathematics of the period did not provide a framework for adding up the times for the subprocesses to obtain the observed finite answer. In our age, the situation is as frustrating for a class of basic finite numbers, e.g., scattering amplitudes in a quantum field theory. When the scattering process is broken into constituent subprocesses, our current mathematics still does not allow us to sum up the effects in a coherent manner consistent with reality.

The infinities that plague arithmetic geometry at present are of a nature different from those of quantum field theory, in that we have no external evidence that would lead us to believe them to be tameable. Belief in a refined notion of finitude (analytic

Supported by the New Horizons Grant EP/V046888/1 from the EPSRC of the United Kingdom.

M. Kim (✉)
International Centre for Mathematical Sciences, Edinburgh, UK
e-mail: minhyong.kim@icms.ac.uk

The Korea Institute for Advanced Study 85 Hoegiro, Dongdaemungu, Seoul, Republic of Korea

Y.-H. He et al. (eds.), *Nankai Symposium on Mathematical Dialogues*,
https://doi.org/10.1007/978-981-19-2328-9_24

Fig. 1 $V_{cyl} = (3/2)V_{sphere}$

continuation) is an uneasy consensus arising from a combination of tradition and authority based on a scattered collection of examples. This is one reason it might be useful to carry out a careful examination of analogies with problems of physics. A degree of effort might lead to satisfactory evidence for deep but mysterious conjectures, if not a roadmap for approaching the 'right answers' that are to replace the ostensibly infinite quantities.

A constructive ancient example is the famous theorem of Archimedes on the sphere and the cylinder (Fig. 1). It is probably of more than historical interest that Archimedes appears to have initially given a 'mechanical proof' involving infinitesimals starting from the law of levers as shown in Fig. 2 [9]. About this method, Archimedes wrote the following to Erathosthenes.

> I thought fit to write out for you and explain in detail in the same book the peculiarity of a certain method, by which it will be possible for you to get a start to enable you to investigate some of the problems in mathematics by means of mechanics. This procedure is, I am

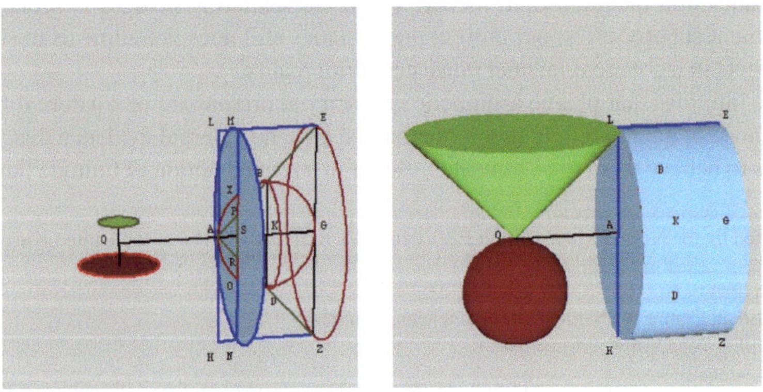

Fig. 2 $4V_{cyl} = 2[(4/3)V_{cyl} + V_{sphere}]$

persuaded, no less useful even for the proof of the theorems themselves; for certain things first became clear to me by a mechanical method, although they had to be demonstrated by geometry afterwards because their investigation by the said method did not furnish an actual demonstration. But it is of course easier, when we have previously acquired, by the method, some knowledge of the questions, to supply the proof than it is to find it without any previous knowledge.

We could do far worse than emulate his flexible outlook.

It might be worth remarking that there have also been ideologues since ancient times, who proclaimed an artificial division of mathematical (and maybe other) thinking into pure vs. applied, probably for a complicated set of personal and political reasons. Consider the case of Plutarch, a Greek intellectual of the Roman empire, writing around 100 CE. The chapter on the Roman general Marcellus in his influential book *Parallel Lives* contains a lengthy digression on Archimedes, in part because the great scientist was said to have been killed during the conquest of Syracuse led by Marcellus. After a discussion of Archimedes's achievements in engineering and physics, Plutarch writes:

Archimedes possessed so high a spirit, so profound a soul, and such treasures of scientific knowledge, that though these inventions had now obtained him the renown of more than human sagacity, he yet would not deign to leave behind him any commentary or writing on such subjects; but, repudiating as sordid and ignoble the whole trade of engineering, and every sort of art that lends itself to mere use and profit, he placed his whole affection and ambition in those purer speculations where there can be no reference to the vulgar needs of life; studies, the superiority of which to all others is unquestioned, and in which the only doubt can be whether the beauty and grandeur of the subjects examined, of the precision and cogency of the methods and means of proof, most deserve our admiration.

Even a casual look into the surviving works of Archimedes [1] will reveal this characterisation to be groundless.

2 Common and Complementary Concerns

The most obvious structure common to arithmetic geometry and quantum field theory is a ring. Number theory begins with the study of the ring $\mathbb{Z}$ of integers, while the duality between geometry and rings of functions is the starting point of scheme theory. Meanwhile, physics studies rings of observable quantities. In arithmetic geometry, it is important to consider sheaves and their sections more general than functions, both of which might be regarded as physical fields, such as electromagnetic fields or Higgs fields.

The divergence of views even in ring theory occurs at a very elementary level, since the building blocks of arithmetic geometry are not arbitrary rings, but those that are absolutely of finite type, e.g.,

$$\mathbb{Z}[i, \pi, e, \frac{1}{10}, \sqrt{2}, e^{2\pi i/7}, \zeta(3), \zeta(5)]$$

The square brackets here refer to closure under ring operations, i.e., addition, subtraction, and multiplication. More generally, they are rings of the form

$$\mathbb{Z}[x_1, x_2, \ldots, x_n]/I,$$

quotient rings over polynomial rings in finitely many variables over the integers. It seems to me that these should be of interest to physicists, in that most of actual physics requires only finitely many seed numbers on which one performs arithmetic operations. Uneasiness about restricting attention to such rings of finite type probably stems from the importance of geometry in physics. It is normally difficult to imagine geometric structures that make use only of such numbers excepts ones that are discrete in a quite superficial sense. However, arithmetic geometry is exactly concerned with sophisticated geometries over such number systems, starting from the construction of the *spectrum*:

$$\text{Spec}(\mathbb{Z}[x_1, x_2, \ldots, x_n]/I).$$

An *arithmetic scheme* is obtained by glueing a finite number of such spectra.

One might legitimately wonder if arithmetic schemes actually occur in physics. In this regard, note that any complex algebraic variety can be endowed with the structure of an underlying arithmetic schemes, since a choice of defining equations involves only finitely many coefficients. For example, the worldsheet of a closed string that starts and ends at the vacuum with finitely many insertions can be given the structure of an arithmetic scheme. Similarly, by way of the Nash-Tognoli embedding theorem, any compact smooth manifold can be given the structure of an arithmetic scheme.

In the other direction, arithmetic geometers should be made aware of the fact that quantum field theory has been developing algebraic structures that go far beyond the rings of usual mathematics. From the point of view of *extended operators*, rings correspond just to operators inserted at points. In physics at the moment, there are a plethora of higher categories (with ring like structures) corresponding to higher dimensional operators [6]. For example, operators inserted along lines will form a category, while 2-categories correspond to operators inserted along surfaces. These kind of refined structures should have an important role to play in arithmetic geometry. A number of other applications of higher category theory, such as derived spaces, have also been considered more carefully by mathematical physicists than by arithmetic geometers [3].

In the remaining sections, I will outline a specific difficult problem of infinity in arithmetic geometry for which ideas from physics are likely to be useful.

3 Structure of Finitely-Generated Rings

For the benefit of physicists, we give some brief preliminaries. In the following, an arrow is used to denote a ring homomorphism or a dual map of spaces.

3.1 The Spectrum

Given a commutative ring A with 1, the *spectrum* of A is the set of (equivalence classes of) pure states:

$$\mathrm{Spec}(A) := \{\rho : A \longrightarrow k(\rho)\}/\sim$$

where ρ is a ring homomorphism to a field $k(\rho)$.

The equivalence relation is given by

$$(\rho_1 : A \longrightarrow k(\rho_1)) \sim (\rho_2 : A \longrightarrow k(\rho_2))$$

if

$$\rho_2 = \phi \circ \rho_1,$$

where $\phi : k(\rho_1) \hookrightarrow k(\rho_2)$ is a homomorphism. In short, two states are equivalent if one is just the obtained from the other by an 'extension of values'.

The *point spectrum* $\mathrm{Spec}(A)_0 \subset \mathrm{Spec}(A)$ are the classes of surjective ρ. The following are some examples.

$$\mathrm{Spec}(\mathbb{Z})_0 = \{\mathbb{Z} \longrightarrow \mathbb{Z}/p\mathbb{Z} \mid p \text{ prime}\}$$

The larger set $\mathrm{Spec}(\mathbb{Z})$ also includes $\mathbb{Z} \hookrightarrow \mathbb{Q}$.

We have

$$\mathrm{Spec}(\mathbb{C}[x])_0 = \mathbb{C},$$

where $a \in \mathbb{C}$ corresponds to the homomorphism

$$\mathbb{C}[x] \longrightarrow \mathbb{C}$$

that evaluates at a:

$$f(x) \mapsto f(a).$$

The full spectrum $\mathrm{Spec}(\mathbb{C}[x])$ also includes $\mathbb{C}[x] \hookrightarrow \mathbb{C}(x)$.

Similarly,

$$\mathrm{Spec}(\mathbb{C}[x_1, x_2, \cdots, x_n])_0 = \mathbb{C}^n$$

where an n-tuple $\underline{a} \in \mathbb{C}^n$ determines an evaluation homomorphism on the polynomial ring. On the other hand, $\mathrm{Spec}(\mathbb{C}[x_1, x_2, \cdots, x_n])$ includes elements like

$$\mathbb{C}[x_1, x_2, \cdots, x_n] \hookrightarrow \mathbb{C}(x_1, x_2, \cdots, x_n),$$

but also an element

$$\rho_V : \mathbb{C}[x_1, x_2, \cdots, x_n] \twoheadrightarrow \mathbb{C}[x_1, x_2, \cdots, x_n]/I(V) \hookrightarrow k(V)$$

for each irreducible algebraic subset $V \subset \mathbb{C}^n$.

The classical theory of algebraic varieties would have regarded $\mathrm{Spec}(\mathbb{C}[x_1, x_2, \cdots, x_n])_0$ as all the points of affine space. Scheme theory, on the other hand, allows points that are 'higher-dimensional' in some sense.

To determine the point spectrum of

$$\mathbb{C}[x_1, x_2, \cdots, x_n]/(f_1, \ldots, f_k).$$

note that we can view it as a subset of $\mathrm{Spec}(\mathbb{C}[x_1, x_2, \cdots, x_n])_0$ via the composition

$$\mathbb{C}[x_1, x_2, \cdots, x_n] \longrightarrow \mathbb{C}[x_1, x_2, \cdots, x_n]/(f_1, \ldots, f_k) \xrightarrow{\rho} k.$$

On the other hand, you can see that an evaluation map at $\underline{a}$ on $\mathbb{C}[x_1, x_2, \cdots, x_n]$ factors through $\mathbb{C}[x_1, x_2, \cdots, x_n]/(f_1, \ldots, f_k)$ if and only if $f_i(\underline{a}) = 0$ for all i. That is,

$$\mathrm{Spec}(\mathbb{C}[x_1, x_2, \cdots, x_n]/(f_1, \ldots, f_k))_0 = \{\underline{a} \in \mathbb{C}^n \mid f_i(\underline{a}) = 0, \forall i\}.$$

3.2 *Maps of Spectra*

A ring homomorphism $A \longrightarrow B$ induces $\mathrm{Spec}(B) \longrightarrow \mathrm{Spec}(A)$.

For example, consider $\mathbb{Z} \hookrightarrow \mathbb{Z}[i]$. Here is a small list of prime decompositions inside the larger ring:

$$2 = (1-i)(1+i), \quad 5 = (2-i)(2+i), \quad 13 = (3-2i)(3+2i),$$

$$17 = (4-i)(4+i).$$

(Note, 3, 7, 11, 19... remain prime.)

At the level of spectra, these corresponds to commutative diagrams like

$$\begin{array}{ccc} \mathbb{Z}[i] & \longrightarrow & \mathbb{Z}[i]/(2-i) \\ \uparrow & & \uparrow \simeq \\ \mathbb{Z} & \longrightarrow & \mathbb{Z}/5 \end{array}$$

and

$$\begin{array}{ccc} \mathbb{Z}[i] & \longrightarrow & \mathbb{Z}[i]/(2+i) \\ \uparrow & & \uparrow \simeq \\ \mathbb{Z} & \longrightarrow & \mathbb{Z}/5 \end{array}$$

illustrating the points in the spectrum of $\mathbb{Z}[i]$ that lie above the same point of $\mathrm{Spec}(\mathbb{Z})$.

Once this formalism is in place, a topology on $\mathrm{Spec}(A)$ is defined by considering some maps $\mathrm{Spec}(B) \longrightarrow \mathrm{Spec}(A)$ as open sets, and

$$\mathrm{Spec}(B_1 \otimes_A B_2) \longrightarrow \mathrm{Spec}(A)$$

as the 'intersection' of the open sets $\mathrm{Spec}(B_i) \longrightarrow \mathrm{Spec}(A)$, and so on. Families of maps that have a reasonable collection of properties is called a *Grothendieck topology* on $\mathrm{Spec}(A)$ [10].

4 A Difficult Problem of Arithmetic Geometry

4.1 Zeta Functions

Given A finitely generated and $\rho : A \twoheadrightarrow k(\rho)$ an element of $\mathrm{Spec}(A)_0$, k is necessarily a finite field. Define the norm of the state ρ to be $\|\rho\| = |k(\rho)|$, the number of values taken. Define the **zeta function** of A by

$$\zeta_A(s) := \prod_{\rho \in \mathrm{Spec}(A)_0} \frac{1}{1 - \|\rho\|^s}.$$

Here is a collection of examples, some of them quite non-trivial.

$$\zeta_{\mathbb{Z}}(s) = \prod_p \frac{1}{1 - p^{-s}} = \sum_n \frac{1}{n^s}.$$

$$\zeta_{\mathbb{F}_p[x]}(s) = \prod_{f \in Irr^1(\mathbb{F}_p[x])} \frac{1}{1 - p^{-\deg(f)s}} = \frac{1}{1 - p^{1-s}}$$

$$\zeta_{\mathbb{F}_p[x_1, x_2, \ldots, x_n]}(s) = \frac{1}{1 - p^{n-s}}$$

$$\zeta_{\mathbb{Z}[x_1,x_2,\ldots,x_n]}(s) = \prod_p \zeta_{\mathbb{F}_p[x_1,x_2,\ldots,x_n]}(s) = \zeta_{\mathbb{Z}}(s-n)$$

$$\zeta_{\mathbb{F}_3[x,y]/(y^2+y-x^3+x^2+10x+20)}(s) = \frac{1+3^{-s}+3^{1-2s}}{1-3^{-s}}$$

In general, if $p \neq 11$, then

$$\zeta_{\mathbb{F}_p[x,y]/(y^2+y-x^3+x^2+10x+20)}(s) = \frac{1-a_p p^{-s}+p^{1-2s}}{1-p^{-s}},$$

where a_p varies in an interesting way with p.

Now consider the ring

$$\mathbb{Z}[x,y]/(y^2+y-x^3+x^2+10x+20).$$

Then

$$\zeta_{\mathbb{Z}[x,y]/(y^2+y-x^3+x^2+10x+20)} = \frac{1}{1-11^{-s}} \prod_{p \neq 11} \frac{1-a_p p^{-s}+p \cdot p^{-2s}}{1-p^{-s}}.$$

In fact,

$$\zeta_{\mathbb{Z}[x,y]/(y^2+y-x^3+x^2+10x+20)}(s) = \frac{\zeta_{\mathbb{Z}}(s)}{L(f,s)},$$

where

$$f(\tau) = e^{2\pi i\tau} \prod_n (1-e^{2n\pi i\tau})^2(1-e^{22n\pi i\tau})^2$$

is a modular form of weight 2 and

$$L(f,s) = \frac{1}{(2\pi)^s\Gamma(s)} \int_0^\infty f(iy)y^{s-1}dy$$

$$= \frac{1}{(2\pi)^s\Gamma(s)}[\int_{\sqrt{11}}^\infty f(iy)y^{s-1}dy + \int_{\sqrt{11}}^\infty f(iy)y^{1-s}dy].$$

4.2 *The Hasse-Weil Conjecture*

For any absolutely finitely-generated ring *A*,

$$\zeta_A(s) = \prod_{\rho \in \mathrm{Spec}(A)_0} \frac{1}{1-\|\rho\|^{-s}}$$

is a meromorphic function on $\mathbb{C}$ with at most finitely many poles, which occur at integers values of s.

A main motivation of the Langlands programme is to prove the Hasse-Weil conjecture by establishing a connection to automorphic forms as in the last example above [8].

5 L-Functions

Given a sheaf V on $\mathrm{Spec}(A)$, define

$$(*) \qquad L(A, V) = \prod_{x \in \mathrm{Spec}(A)_0} \frac{1}{\det([1 - F_x] | V_x^{I_x})}.$$

Kim [7] there are many things that need to be explained about this definition. Firstly, one might wish the right-hand side to be a complex number, but it's not even clear that the factors appearing in the product are complex. For now, we will assume that V is a sheaf of complex vector spaces. Another common situation is to have a sheaf of vector spaces over a field like $\bar{\mathbb{Q}}_\ell$. In that case, one can choose an isomorphism $\iota : \bar{\mathbb{Q}}_\ell \simeq \mathbb{C}$. One typically assumes that the determinants are algebraic numbers. Even after all this is done, the convergence is problematic. Partly for this reason, one considers

$$L(A, V(s)) := \prod_{x \in \mathrm{Spec}(A)_0} \frac{1}{\det([1 - \|x\|^{-s} F_x] | V_x^{I_x})}$$

for a complex variable s. The notation $V(s)$ requires explanation which we will not give here. There is a sense in which the sheaf itself is being twisted (inside a category of (collections of) Weil-Deligne representations). However, one can consider $V(s)$ simply as suggestive notation and focus on the right-hand side. With some assumptions on V, this is convergent for $Re(s) >> 0$. When V is a *motivic sheaf*, i.e., of geometric origin, this should be meromorphic in s. This is another version of the Hasse-Weil conjecture, and the one from the previous section can be reduced to this one, provided one allows sheaves with $\mathbb{Q}_\ell$-coefficients. (In fact, the factorisation of zeta functions into L-functions is often viewed as a 'spectral manifestation' of a category of motives.) Leaving the analyticity aside, one might view the key problem as being that of defining the infinite product (*) in general.

6 QFT Interpretation?

We have

$$\frac{1}{\det([1-F_x]|V_x^{I_x})} = Tr(F_x|Sym(V_x^{I_x})) =< Fr_x >,$$

where the bracket $< \cdot >$ refers to the expectation value of an operator. The superscript I_x is the local inertia subgroup. For sheaves that come from geometry, $V^{I_x} = V_x$ for all but finitely many x, so we will ignore it for simplicity of exposition. Formally,

$$< \prod_x F_x >= Tr(\prod_x F_x | \otimes_x \mathcal{H}_x) = \prod_x Tr(Fr_x|\mathcal{H}_x)$$

$$= \prod_x \frac{1}{\det([I - Fr_x]|V_x)}.$$

One might ask if this can itself be written as a path integral. As it stands, one has an analogue of an infinite number of Wilson loop functions. It is well-known since the classical paper of Witten that the Wilson loop function can be written as a path integral [11]. For this, one needs to consider the quantum mechanical path integral on a symplectic manifold whose quantisation is $Sym(V_x)$. It is natural in this context to view

$$\mathcal{H}_x := Sym(V_x)$$

as the formal quantisation of the symplectic vector space $V_x \times V_x^*$. That is, the space

$$\otimes_x \mathcal{H}_x$$

is formally the quantum Hilbert space associated to the space $\Gamma(V \times V^*)$ of discontinuous sections of the bundle $V \times V^*$. This suggests the existence of an 'arithmetic BF-theory' underlying the theory of L-functions, a connection that had already appeared in the paper [2].

One might legitimately ask if it is merely an idle exercise to attempt to write such quantities as path integrals. Possibly. On the other hand, the path integral formalism seems to be especially amenable to suitable 'deformations', which might render the problem of analytic continuation more tractable, either by way of the BV formalism or the method of changing the cycle of integration [3, 12]. That is, one might naively hope that the path integral approach would enable us to write the value

$$< \prod_x \|x\|^{-s} Fr_x >$$

as an integral over a cycle that converges for all values of s.

Acknowledgements I am very grateful to Yang-Hui He for the invitation to the stimulating online conference in summer 2021 hosted by the Chern Institute, which seemed to span the gamut of all possible topics in mathematics and physics. I'm also grateful to Magnus Carlson, Hee Joong Chung, Dohyeong Kim, Jeehoon Park, Hwajong Yoo for discussions related to the topic of this paper.

References

1. Archimedes *The Works of Archimedes*. T. Heath (trans.), Dover Publications (2002).
2. Carlson, Magnus; Kim, Minhyong A note on abelian arithmetic BF-theory. arXiv:1911.02236.
3. Costello, Kevin; Gwilliam, Owen Factorization algebras in quantum field theory. Vol. 2. New Mathematical Monographs, 41. Cambridge University Press, Cambridge, 2021.
4. Gaitsgory, Dennis; Rozenblyum, Nick A study in derived algebraic geometry. Vol. I, II. Mathematical Surveys and Monographs, 221. American Mathematical Society, Providence, RI, 2017.
5. Hartshorne, Robin Algebraic geometry. Graduate Texts in Mathematics, No. 52. Springer-Verlag, New York-Heidelberg, 1977.
6. Kapustin, Anton Topological field theory, higher categories, and their applications. Proceedings of the International Congress of Mathematicians. Volume III, 2021–2043, Hindustan Book Agency, New Delhi, 2010.
7. Kim, Minhyong Classical motives and motivic L-functions. Autour des motifs–École d'été Franco-Asiatique de Géométrie Algébrique et de Théorie des Nombres. Volume I, 1–25, Panor. Synthèses, 29, Soc. Math. France, Paris, 2009.
8. Langlands, R. P. L-functions and automorphic representations. Proceedings of the International Congress of Mathematicians (Helsinki, 1978), pp. 165–175, Acad. Sci. Fennica, Helsinki, 1980.
9. Mendel, Henry Archimedes Mechanical Method with Indivisibles, The Method, Prop. 2. https://web.calstatela.edu/faculty/hmendel/.
10. Milne, James S. Étale cohomology. Princeton Mathematical Series, No. 33. Princeton University Press, Princeton, N.J., 1980.
11. Witten, Edward Quantum field theory and the Jones polynomial. Comm. Math. Phys. 121 (1989), no. 3, 351–399.
12. Witten, Edward A new look at the path integral of quantum mechanics. Surveys in differential geometry. Volume XV. Perspectives in mathematics and physics, 345–419, Surv. Differ. Geom., 15, Int. Press, Somerville, MA, 2011.

Progress on BPS Spectra of 5d/6d $\mathcal{N} = 1$ Supersymmetric Field Theories

Sung-Soo Kim

Abstract We discuss recent progress on computing BPS spectra of higher dimensional supersymmetric gauge theories of eight supercharges. We generalize the blowup equation so that it allows one to compute the Nekrasov partition function on the Omega background for any 5d/6d supersymmetric quantum field theories which have either geometric or gauge theoretic description.

Supersymmetric field theories of eight supercharges in five or six dimensions (5d/6d) are a fruitful subject for physics and geometry. Lots of progress have been made along the classification of 6d superconformal theories [1–3] and little string theories [4], based on F-theory compactified on local elliptic Calabi-Yau (CY) threefolds. A large class of 5d superconformal theories has been also classified through gauge theoretic constructions [5–7] and M-theory compactified on local CY threefolds [6, 8–16]. These higher dimensionsional superconformal theories reveal very fascinating aspects which include tensionless strings, various dualities, and global symmetry enhancements.

Supersymmetric partition function provides a way of studying such rich features. In particular, the partition functions on $\mathbb{R}^4 \times S^1$ in 5d and $\mathbb{R}^4 \times T^2$ in 6d compute the number of the BPS states of the theory weighted by their electric charges and angular momenta. Many computational tools for computing the BPS spectrum of these higher dimensional theories have been developed over the years. For instance, the Nekrasov instanton partition function on the Ω-background and its generalization to various gauge groups with matter have been carried out based on the ADHM constructions [17–19]. Topological vertex method [20–22] also provides another systematic tool as long as a 5-brane web configuration for the theory of interest is known. See also [23, 24] for analytic method based on the Ding-Iohara-Miki algebra. However, each computational tool has some limitations. For example, the ADHM construction for exceptional gauge groups or matter in higher dimensional representations is still limited. For the topological vertex, there are many theories whose 5-brane webs are

S.-S. Kim (✉)
School of Physics, University of Electronic Science and Technology of China, Chengdu, Sichuan, China
e-mail: Sungsoo.kim@uestc.edu.cn

Y.-H. He et al. (eds.), *Nankai Symposium on Mathematical Dialogues*,
https://doi.org/10.1007/978-981-19-2328-9_25

unknown. The refined partition function with two Ω-background parameters ϵ_1, ϵ_2 for various 5-brane systems of orientifold planes is still challenging.

In this talk, we discuss yet another powerful tool, known as the blowup equation, developed by Nakajima-Yoshioka [25–27]. We then discuss recent progress which enables one to compute the BPS spectrum of higher dimensional supersymmetric field theories of any kind of gauge group as well matter [28–30]. See also the related works [31, 32]. Therefore, one can even compute the partition functions for those theories for which the AHDM construction and topological vertex are not yet applicable.

The blowup equation is a functional equation arising from identifying two partition functions on different backgrounds, one is the Ω-deformed $\mathbb{C}^2$ and the other is the one-point blow-up $\hat{\mathbb{C}}^2$ where the origin of the $\mathbb{C}^2$ is replaced by a compact 2-cycle $\mathbb{P}^1$. The $\hat{\mathbb{C}}^2$ can be parameterized by the projective coordinates $(z_0, z_1, z_2) \sim (\lambda^{-1} z_0, \lambda z_1, \lambda z_2)$ for $\lambda \in \mathbb{C} \setminus \{0\}$. The Lorentz generators $J_{1,2}$ act on the $\hat{\mathbb{C}}^2$ by

$$(z_0, z_1, z_2) \mapsto (z_0, e^{\epsilon_1} z_1, e^{\epsilon_2} z_2) , \tag{1}$$

with $\epsilon_{1,2}$ associated with the Cartans of the $SO(4)$ rotations, which are also known as the Omega deformation parameters. There are hence two fixed points of the Lorentz rotations, the North pole and the South pole of the $\mathbb{P}^1$ where the local coordinates around these fixed points are given as $(z_0 z_1, z_2/z_1)$ and $(z_1/z_2, z_0 z_2)$ and their weights under $J_{1,2}$ actions hence can be represented respectively as $(\epsilon_1, \epsilon_2 - \epsilon_1)$ and $(\epsilon_1 - \epsilon_2, \epsilon_2)$ at two poles. By localization, the partition function are given as a sum over magnetic fluxes $\vec{n}$ on the $\mathbb{P}^1$ associated with he maximal torus of the gauge group G. The background magnetic fluxes $\vec{B}$ associated with global symmetries can be turned on, which is not summed over. At each flux background labeled by $\vec{n}$ and $\vec{B}$, the partition function is localized at two fixed points on the $\mathbb{P}^1$ and the path integral near each fixed point reduces to that of local Ω-deformed $\hat{\mathbb{C}}^2$ with shifted chemical potentials due to the magnetic fluxes. As a consequence, the partition function $\hat{Z}$ can be written as [25–27],

$$\begin{aligned} &\Lambda(m_j; \epsilon_1, \epsilon_2) \hat{Z}(\phi_i, m_j, B_j; \epsilon_1, \epsilon_2) \\ &= \sum_{\vec{n}} (-1)^{|\vec{n}|} \hat{Z}^{(N)}(\phi_i + n_i \epsilon_1, m_j + B_j \epsilon_1; \epsilon_1, \epsilon_2 - \epsilon_1) \cdot \hat{Z}^{(S)}(\phi_i + n_i \epsilon_2, m_j + B_j \epsilon_2; \epsilon_1 - \epsilon_2, \epsilon_2) , \end{aligned} \tag{2}$$

where $|\vec{n}| = \sum_i n_i$ and the scalar ϕ_i in the vector multiplet get the expectation value in the Cartan subalgebra of the gauge group G. Here $\hat{Z}^{(N)}$ and $\hat{Z}^{(S)}$ are respectively the partition function $\hat{Z}$ with shifted chemical potentials evaluated near the North and South poles. The shifts in the chemical potentials ϕ_i and m_j reflect that magnetically charged states experience angular momentum shifts under the flux background. The prefactor $\Lambda(m_j; \epsilon_1, \epsilon_2)$ does not depend on dynamical parameters ϕ_i, but rather depends only on mass parameters m_j as well as $\epsilon_{1,2}$.

A 5d/6d field theory can have many blowup equations depending on background magnetic fluxes which reside on the blown-up $\mathbb{P}^1$ in $\hat{\mathbb{C}}^2$. We claim that such blowup equations can be solved recursively by expanding them in terms of Kähler parameters of the theory with two seeds: (i) *the effective prepotential* and (ii) *the consistent magnetic fluxes*. The effective prepotential is composed of the cubic prepotential and mixed Chern-Simons terms in the low energy theory on the Coulomb branch, induced by integrating out charged fermions. More explicitly, on the Coulomb branch of the moduli space of 5d $\mathcal{N}=1$ theory, and the low energy theory is described by the cubic prepotential [6] is given by

$$\mathcal{F}=\frac{m}{2}K_{ij}\phi_i\phi_j+\frac{\kappa}{6}d_{ijk}\phi_i\phi_j\phi_k+\frac{1}{12}\left(\sum_{e\in\mathbf{R}}|e\cdot\phi|^3-\sum_f\sum_{w\in\mathbf{w}_f}|w\cdot\phi+m_f|^3\right), \tag{3}$$

where $m=1/g^2$ is the inverse gauge couplings squared, κ is the classical Chern-Simons level which is non-zero only for $SU(N)$ with $N\geq 3$, $K_{ij}=\mathrm{Tr}(T_iT_j)$ of the gauge group G is the Killing form of G, and $d_{ijk}=\frac{1}{2}\mathrm{Tr}T_i\{T_j,T_k^a\}$ with the generator T_i in the fundamental representation of G. $\mathbf{R}$ and $\mathbf{w}_f$ are the roots and the weights for the f-th hypermultiplet with mass m_f of G, respectively. Together with the cubic prepotential, the effective prepotential $\mathcal{E}$ also includes the mixed gauge/gravitational Chern-Simons terms C_i^G [33–35] and the mixed gauge/$SU(2)_R$ Chern-Simons terms C_i^R given by [36],

$$C_i^G=-\partial_i\left(\sum_{e\in\mathbf{R}}|e\cdot\phi|-\sum_f\sum_{w\in\mathbf{w}_f}|w\cdot\phi+m_f|\right),\qquad C_i^R=\frac{1}{2}\partial_i\sum_{e\in\mathbf{R}}|e\cdot\phi|\,, \tag{4}$$

where $C_i^R=2$ for all i is in the Dynkin base. The effective potential $\mathcal{E}$ is then given by

$$\mathcal{E}(\phi,m;\epsilon_1,\epsilon_2)=\frac{1}{\epsilon_1\epsilon_2}\left[\mathcal{F}+\frac{1}{48}C_i^G\phi^i(\epsilon_1^2+\epsilon_2^2)+\frac{1}{2}C_i^R\phi^i\epsilon_+^2\right], \tag{5}$$

where $\epsilon_+\equiv\frac{\epsilon_1+\epsilon_2}{2}$.

The BPS partition function as a Witten index receives perturbative and non-perturbative instanton contributions which factorizes as

$$Z=e^{\mathcal{E}}\cdot\mathrm{PE}\left[-\frac{1+p_1p_2}{(1-p_1)(1-p_2)}\sum_{e\in\mathbf{R}^+}e^{-e\cdot\phi}+\frac{(p_1p_2)^{1/2}}{(1-p_1)(1-p_2)}\sum_f\sum_{w\in\mathbf{w}_f}e^{-|w\cdot\phi+m_f|}\right]\cdot Z_{\mathrm{inst}}, \tag{6}$$

where $p_{1,2}\equiv e^{-\epsilon_{1,2}}$ and $\mathbf{R}^+$ denotes the positive roots for the gauge group, and PE means the Plethystic exponential of a letter index $f(\mu)$ with a chemical potential μ defined as $\mathrm{PE}\left[f(\mu)\right]\equiv\exp\left(\sum_{n=1}^{\infty}\frac{1}{n}f(n\mu)\right)$.

After the transition from $\hat{\mathbb{C}}^2$ to $\mathbb{C}^2$, the partition function in (2) can be written as

$$\hat{Z}(\phi, m; \epsilon_1, \epsilon_2) = e^{\mathcal{E}(\phi,m;\epsilon_1,\epsilon_2)} \cdot Z_{GV}(\phi, m; \epsilon_1 + 2\pi i, \epsilon_2)\,, \tag{7}$$

where the shift of the fermion number operator in the presence of the magnetic fluxes is represented by the redefinition of angular momentum chemical potential $\epsilon_1 \to \epsilon_1 + 2\pi i$. Here Z_{GV} is the index part of the BPS partition function, which is actually the refined Gopakumar-Vafa (GV) invariant [37], defined as

$$Z_{GV}(\phi, m; \epsilon_1, \epsilon_2) = \text{PE}\left[\sum_{j_l, j_r, \mathbf{d}} (-1)^{2(j_l+j_r)} N^{\mathbf{d}}_{j_l, j_r} \frac{\chi^{SU(2)}_{j_l}(p_1/p_2)\,\chi^{SU(2)}_{j_r}(p_1 p_2)}{(p_1^{1/2} - p_1^{-1/2})(p_2^{1/2} - p_2^{-1/2})} e^{-\mathbf{d}\cdot\mathbf{m}}\right], \tag{8}$$

where $\mathbf{d}$ denotes the charge of a BPS state, $\mathbf{m}$ stands for the chemical potentials (or Kähler parameters) ϕ, m, and $N^{\mathbf{d}}_{j_l, j_r}$ is the degeneracy of a single-particle BPS state with spin (j_l, j_r) and charge $\mathbf{d}$, and $\chi^{SU(2)}_j$ is the $SU(2)$ character of spin j. Also, $j_l \equiv \frac{J_1 - J_2}{2}$ and $j_r \equiv \frac{J_1 + J_2}{2}$. The blowup equation (2) can be expressed in terms of power series in the fugacities and can be solved iteratively. We then rewrite the blowup equation as

$$\begin{aligned}\Lambda(m_j; \epsilon_1, \epsilon_2)\hat{Z}_{GV}(\phi_i, m_j; \epsilon_1, \epsilon_2) &= \sum_{\vec{n}} (-1)^{|\vec{n}|} e^{-V(\phi_i, m_j, \vec{n}, \vec{B}; \epsilon_1, \epsilon_2)} \\ &\times \hat{Z}^{(N)}_{GV}(\phi_i + n_i\epsilon_1, m_j + B_j\epsilon_1; \epsilon_1, \epsilon_2 - \epsilon_1) \cdot \hat{Z}^{(S)}_{GV}(\phi_i + n_i\epsilon_2, m_j + B_j\epsilon_2; \epsilon_1 - \epsilon_2, \epsilon_2)\,,\end{aligned} \tag{9}$$

where $V(\phi_i, m_j, \vec{n}, \vec{B}; \epsilon_1, \epsilon_2) \equiv \mathcal{E} - \mathcal{E}^{(N)} - \mathcal{E}^{(S)}$ with the appropriate shift of chemical potentials on each poles. We expand both sides of the blowup equation (9) and then try to find an iterative solution of $\hat{Z}_{GV}$. Importantly, we can use the fact that the GV-invariant should take a special form in (8). Also, spins of states at each order are bounded by the maximum spin $(j_l^{\max}, j_r^{\max})$ in the series expansion, and the characters $\chi^{SU(2)}_j$ with different spins are all orthogonal to each other. Plugging the ansatz of the GV-invariant with a finite number of trial states for a given charge $\mathbf{d}$ into the blowup equation and expanding it, we can recursively solve the equations to evaluate multiplicities $N^{\mathbf{d}}_{j_l, j_r}$ of BPS states.

Now let us discuss the consistent magnetic fluxes which are the second seed of the blowup equation and their choice is important when solving the blowup equation. As the magnetic fluxes on the $\mathbb{P}^1$ in the blowup $\hat{\mathbb{C}}^2$ cannot be arbitrary. They satisfy suitable quantization conditions. The consistent magnetic fluxes can be better understood from geometry. Namely, M-theory compactification on a smooth non-compact 3-fold X which gives rise to a 5d superconformal theory or a 6d superconformal theory on S^1 (possibly with twists) in a singular limit. See [11, 15] for more details. Many examples of 5d $\mathcal{N} = 1$ theories have been engineered in M-theory on local CY three-folds. A smooth 3-fold X can be locally described as a neighborhood of a collection of Kähler surfaces S_i. A Kähler surface S_i inside X is represented by either a local $\mathbb{P}^2$ or a Hirzebruch surface of degree n with b blowups, denoted by $\mathbb{F}_n^b$. The magnetic flux n_i (or B_j) can be turned on for each (non-)compact divisor

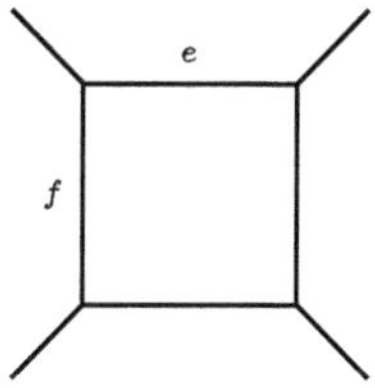

(a) $\mathbb{F}_0$

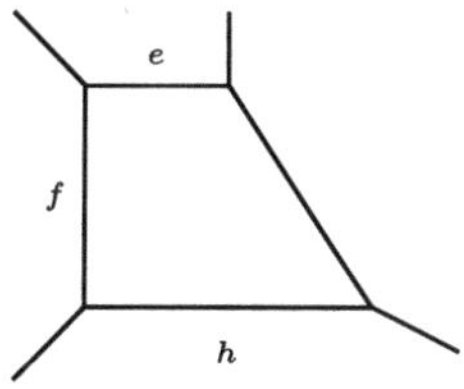

(b) $\mathbb{F}_1$

Fig. 1 Geometric constructions of **a** $SU(2)_0$ and **b** $SU(2)_\pi$. The fiber $\mathbb{P}^1$ in a Hirzebruch surface $\mathbb{F}_n$ is denoted by f, the base $\mathbb{P}^1$ is denoted by e, and $h = e + f$ for $\mathbb{F}_1$

in a CY three-fold. The curve C_i can consistently couple to a magnetic flux F if the following condition is satisfied [32]. The magnetic flux on a charged M2-brane state of spin (j_l, j_r) wrapping on C_i satisfies

$$\mathsf{F} \cdot C_i \text{ is integral/half-integral, when } 2(j_l + j_r) \text{ is odd/even.} \tag{10}$$

This then provides the consistent fluxes for obtaining the partition function from the blowup equation. Generically, it can be stated that the total fluxes on the W-bosons of the gauge group should be integral and those on the perturbative hypermultiplets should be half-integral. Namely,

$$\vec{n} \cdot e \in \mathbb{Z} \,, \qquad \vec{n} \cdot w_f + B_f \in \mathbb{Z} + \frac{1}{2} \,, \tag{11}$$

for all roots e, the weights w_f associated with all hypermultiplets f, and the fluxes B_f for flavor symmetries.

As an instructive example, consider 5d $SU(2)_{0,\pi}$ without flavor. The subscript denote the two allowed discrete theta parameters, arising from non-trivial 4th Homotopy $\pi_4(SU(2)) = \mathbb{Z}_2$. Their geometries are depicted in Fig. 1.

One readily finds that the effective prepotential for $SU(2)_{0,\pi}$ is given by

$$\mathcal{E} = \frac{1}{\epsilon_1 \epsilon_2}\left(m\phi^2 + \frac{8}{6}\phi^3 - \frac{\epsilon_1^2 + \epsilon_2^2}{12}\phi + \epsilon_+^2 \phi\right), \tag{12}$$

As discussed eariler, the GV-invariant takes the factorized form as

$$Z_{GV}(\phi, m; \epsilon_1, \epsilon_2) = \mathrm{PE}\left[-\frac{1 + p_1 p_2}{(1 - p_1)(1 - p_2)} e^{-2\phi}\right] \cdot \sum_{k=0}^{\infty} q^k Z_k(\phi; \epsilon_1, \epsilon_2) \,. \tag{13}$$

It follows from (9) that the k-th order of the instanton expansion of the blowup equation can be schematically written as

$$\Lambda_k(\epsilon_1,\epsilon_2)+\hat{Z}_k(\phi;\epsilon_1,\epsilon_2)=p_1^{kB_m}\,\hat{Z}_k(\phi;\epsilon_1,\epsilon_2-\epsilon_1)+p_2^{kB_m}\,\hat{Z}_k(\phi;\epsilon_1-\epsilon_2,\epsilon_2)$$
$$+\,(\text{terms with }\hat{Z}_{r<k}\text{ and }\Lambda_{r<k})\,. \tag{14}$$

Here, the k-instanton partition function $\hat{Z}_k$ at k-th order in the expansion is independent of the background magnetic flux B_m and appears only with a trivial gauge flux $n = 0$. The consistent magnetic fluxes for the $SU(2)_0$ and $SU(2)_\pi$ theories without containing negative volumes vol$(C) < 0$ on the Coulomb branch are given as

$$\begin{cases}\mathbb{F}_0: & SU(2)_0 \quad n\in\mathbb{Z}\,,\quad B_m=-2,\ -1,\ 0,\ 1,2\\ \mathbb{F}_1: & SU(2)_\pi \quad n\in\mathbb{Z}\,,\quad B_m=-\frac{3}{2},\ -\frac{1}{2},\ \frac{1}{2},\ \frac{3}{2}\,.\end{cases} \tag{15}$$

With three sets of consistent fluxes (n, B_m) for both cases, we can compute a closed expression of the partition function at each instanton order. For example, the 1-instanton partition functions are given as follows:

$$Z_1^{SU(2)_0}(\phi;\epsilon_1,\epsilon_2)=\frac{p_1p_2(1+p_1p_2)e^{-2\phi}}{(1-p_1)(1-p_2)(1-p_1p_2e^{-2\phi})(e^{-2\phi}-p_1p_2)}\,, \tag{16}$$

$$Z_1^{SU(2)_\pi}(\phi;\epsilon_1,\epsilon_2)=-\frac{p_1^{3/2}p_2^{3/2}(1+e^{-2\phi})e^{-\phi}}{(1-p_1)(1-p_2)(1-p_1p_2e^{-2\phi})(e^{-2\phi}-p_1p_2)}\,. \tag{17}$$

By repeating this procedure for the blowup equations to higher orders in q, it is straightforward to obtain the k-instanton partition functions Z_k.

Let us now discuss theories of higher dimensional Coulomb branches. A simple yet non-trivial example would be the $SU(3)$ gauge theory at the Chern-Simons level κ without any flavors. They are geometrically engineered as

$$\mathbb{F}_{\kappa+1}\big|_1 \xrightarrow[\quad e \qquad\qquad h+(\frac{\kappa}{2}-1)f\quad]{} \mathbb{F}_1\big|_2 \qquad \text{for } \kappa \text{ even}, \tag{18}$$

$$\mathbb{F}_{\kappa+1}\big|_1 \xrightarrow[\quad e \qquad\qquad h+\frac{\kappa-1}{2}f\quad]{} \mathbb{F}_0\big|_2 \qquad \text{for } \kappa \text{ odd}. \tag{19}$$

We take the Coulomb branch of the theory to be $2\phi_1-\phi_2\geq 0$ and $2\phi_2-\phi_1\geq 0$ where $\phi_1, \phi_2 > 0$. It follows that the effective prepotential $\mathcal{E}$ for the $SU(3)_{\kappa\leq 7}$ theory on the Coulomb branch takes the following form

$$\mathcal{E}=\frac{1}{\epsilon_1\epsilon_2}\left(\mathcal{F}-\frac{\epsilon_1^2+\epsilon_2^2}{12}(\phi_1+\phi_2)+\epsilon_+^2(\phi_1+\phi_2)\right), \tag{20}$$
$$6\mathcal{F}=6m(\phi_1^2-\phi_1\phi_2+\phi_2^2)+3\kappa\phi_1\phi_2(\phi_1-\phi_2)+8\phi_1^3-3\phi_1^2\phi_2-3\phi_1\phi_2^2+8\phi_2^3\,,$$

and the consistent fluxes are

$$n_1, n_2 \in \mathbb{Z} \quad \text{and} \quad B_m = \begin{cases} -\frac{1}{2}, \frac{1}{2}, \frac{3}{2} & \kappa = 0, 2, 4, 6 \\ -1, 0, 1 & \kappa = 1, 3, 5, 7 \, . \end{cases} \tag{21}$$

As the GV-invariant of this theory can be factorized as

$$Z_{GV} = \text{PE}\Big[-\frac{1 + p_1 p_2}{(1-p_1)(1-p_2)} \Big(e^{-(2\phi_1-\phi_2)} + e^{-(2\phi_2-\phi_1)} + e^{-(\phi_1+\phi_2)} \Big) \Big] \cdot \sum_{k=0}^{\infty} q^k Z_k(\phi; \epsilon_1, \epsilon_2) \, , \tag{22}$$

with the instanton fugacity $q = e^{-m}$ and $Z_0 = 1$, the k-th instanton part can be computed in a similar iterative way. The result is presented in [29]. It is interesting to see how the duality between the $SU(3)_5$ theory and the $Sp(2)_\pi$ theory works. One way is to check their BPS spectra explicitly via the blowup equation. This duality is, in fact, realized as the base-fiber duality swapping two curve classes e_2 and f_2 in $\mathbb{F}_0$. The map between the parameters in the $SU(3)_5$ theory and those in the $Sp(2)_\pi$ theory can be obtained from the geometric realization. The volumes of 2-cycles in the $SU(3)$ frame are given by

$$\text{vol}(f_1) = 2\phi_1 - \phi_2, \quad \text{vol}(f_2) = -\phi_1 + 2\phi_2, \quad \text{vol}(e_2) = -2\phi_1 + 2\phi_2 + m \, , \tag{23}$$

while those in the $Sp(2)$ frame after exchanging e_2 and f_2 are

$$\text{vol}(f_1) = 2\phi_1 - \phi_2, \quad \text{vol}(f_2) = -2\phi_1 + 2\phi_2, \quad \text{vol}(e_2) = -\phi_1 + 2\phi_2 + m \, . \tag{24}$$

It follows form these volumes that one finds a natural map between the parameters in two dual frames as [38–40]

$$\phi_1^{SU} = \phi_1^{Sp} + \frac{1}{3} m^{Sp} \, , \quad \phi_2^{SU} = \phi_2^{Sp} + \frac{2}{3} m^{Sp} \, , \quad m^{SU} = -\frac{2}{3} m^{Sp} \, . \tag{25}$$

The effective prepotential of the $Sp(2)_\pi$ theory on the Coulomb branch is given by

$$\begin{aligned} \mathcal{E} &= \frac{1}{\epsilon_1 \epsilon_2} \left(\mathcal{F} - \frac{\epsilon_1^2 + \epsilon_2^2}{12} (\phi_1 + \phi_2) + \epsilon_+^2 (\phi_1 + \phi_2) \right) , \\ 6\mathcal{F} &= 6 m_{Sp} (2\phi_1^2 - 2\phi_1\phi_2 + \phi_2^2) + 8\phi_1^3 + 12\phi_1^2\phi_2 - 18\phi_1\phi_2^2 + 8\phi_2^3 \, , \end{aligned} \tag{26}$$

where this $\mathcal{E}$ of the $Sp(2)_\pi$ theory coincides with that of the $SU(3)_5$ theory in (20) under (25). In the $Sp(2)$ frame, one can find that the following three sets of consistent fluxes respecting the $Sp(2)$ structure:

$$n_1, n_2 \in \mathbb{Z} \, , \quad B_m = -1, 0, 1 \, . \tag{27}$$

With these seeds for the $Sp(2)_\pi$ theory as well as $SU(3)_5$ theory, one can find apply the blowup equation and obtain the BPS spectrum of the theory. We list the BPS

Table 1 BPS spectrum of $SU(3)_5$ for $d_1 \leq 2$ and $d_1, d_2, d_3 \leq 2$. Here, $\mathbf{d} = (d_1, d_2, d_3)$ labels the BPS states with charge $d_1 e_2 + d_2 f_1 + d_3 f_2$

$\mathbf{d}$	$\oplus N^{\mathbf{d}}_{j_l, j_r}(j_l, j_r)$	$\mathbf{d}$	$\oplus N^{\mathbf{d}}_{j_l, j_r}(j_l, j_r)$
(1, 0, 0)	$(0, \frac{1}{2})$	(1, 0, 1)	$(0, \frac{3}{2})$
(1, 0, 2)	$(0, \frac{5}{2})$	(1, 1, 0)	$(0, \frac{1}{2})$
(1, 1, 1)	$(0, \frac{1}{2}) \oplus (0, \frac{3}{2})$	(1, 1, 2)	$(0, \frac{3}{2}) \oplus (0, \frac{5}{2})$
(1, 2, 0)	$(0, \frac{1}{2})$	(1, 2, 1)	$(0, \frac{1}{2}) \oplus (0, \frac{3}{2})$
(1, 2, 2)	$(0, \frac{1}{2}) \oplus (0, \frac{3}{2}) \oplus (0, \frac{5}{2})$	(2, 0, 1)	$(0, \frac{5}{2})$
(2, 0, 2)	$(0, \frac{5}{2}) \oplus (0, \frac{7}{2}) \oplus (\frac{1}{2}, 4)$	(2, 1, 1)	$(0, \frac{3}{2}) \oplus (0, \frac{5}{2})$
(2, 1, 2)	$(0, \frac{3}{2}) \oplus 3(0, \frac{5}{2}) \oplus 2(0, \frac{7}{2}) \oplus (\frac{1}{2}, 3) \oplus (\frac{1}{2}, 4)$	(2, 2, 1)	$(0, \frac{1}{2}) \oplus (0, \frac{3}{2}) \oplus (0, \frac{5}{2})$

spectrum up to 2-instanton order in Table 1. From Table 1, one finds that the instanton part of the partition function Z_k for the $SU(3)_5$ theory is obtained by collecting all the possible BPS states associated with the charge of d_1, while that for the $Sp(2)_\pi$ theory is obtained from the charge f_2. Both completely agree with the known partition function for each theory, as expected.

Another interesting theory is the $SU(3)_{\kappa=8}$ theory which is expected from a flavor decoupling of a KK theory $SU(3)_{15/2} + 1\mathbf{F}$ [41]. It is interesting as there is no other way of computing the BPS spectrum for this theory, because so far neither 5-brane webs nor ADHM constructions for this theory are known. The corresponding BPS spectrum is listed in [29].

We close this talk by summarizing that we proposed two seed functions for the blowup equation for any 5d theories, the effective prepotential and the consistent magnetic fluxes, and how to determine them systematically. We also demonstrated that with these two seeds, one can obtain the BPS spectrum of the theory by solving the corresponding blowup equation recursively. Our proposal is applicable for non-Lagrangian theories which do not have any Lagrangian descriptions as well as for KK theories which are obtained by compactifying 6d theories on a circle with or without a twist. Moreover, co-dimension 4 defects of 5d/6d theories can be also considered in a similar fashion and such defect partition functions via the blowup equation are studied in [29].

Acknowledgements I would like to thank the organizers of the 2021 Nankai Symposium on Mathematical Dialogues (celebrating 110th anniversary of the birth of Prof. S.-S. Chern) and in particular, Yang-Hui He for kind invitation and correspondences. The author is grateful to Hee-Cheol Kim, Joonho Kim, Minsung Kim, Ki-Hong Lee, Kimyeong Lee, and Jaewon Song for collaborations on the projects which this article is based on.

References

1. J. J. Heckman, D. R. Morrison, and C. Vafa, *JHEP* **1405** (2014) 028.
2. J. J. Heckman, D. R. Morrison, T. Rudelius, and C. Vafa, arXiv:1502.05405.
3. L. Bhardwaj, arXiv:1502.06594.
4. L. Bhardwaj, M. Del Zotto, J. J. Heckman, D. R. Morrison, T. Rudelius, and C. Vafa, *Phys. Rev. D* **93** (2016), no. 8 086002.
5. N. Seiberg, *Phys.Lett.* **B388** (1996) 753–760.
6. K. A. Intriligator, D. R. Morrison, and N. Seiberg, *Nucl. Phys. B* **497** (1997) 56–100.
7. P. Jefferson, H.-C. Kim, C. Vafa, and G. Zafrir, arXiv:1705.05836.
8. M. R. Douglas, S. H. Katz, and C. Vafa, *Nucl. Phys. B* **497** (1997) 155–172.
9. M. Del Zotto, J. J. Heckman, and D. R. Morrison, *JHEP* **09** (2017) 147.
10. D. Xie and S.-T. Yau, *JHEP* **06** (2017) 134.
11. P. Jefferson, S. Katz, H.-C. Kim, and C. Vafa, *JHEP* **04** (2018) 103.
12. L. Bhardwaj and P. Jefferson, *JHEP* **07** (2019) 178; *JHEP* **10** (2019) 282,
13. F. Apruzzi, L. Lin, and C. Mayrhofer, *JHEP* **05** (2019) 187.
14. F. Apruzzi, C. Lawrie, L. Lin, S. Schäfer-Nameki, and Y.-N. Wang, arXiv:1906.11820; *JHEP* **11** (2019) 068; arXiv:1909.09128.
15. L. Bhardwaj, P. Jefferson, H.-C. Kim, H.-C. Tarazi, and C. Vafa, arXiv:1909.11666.
16. F. Apruzzi, S. Schafer-Nameki, and Y.-N. Wang, arXiv:1912.04264.
17. M. Atiyah, N. J. Hitchin, V. Drinfeld, and Y. Manin, *Phys. Lett. A* **65** (1978) 185–187.
18. N. A. Nekrasov, *Adv.Theor.Math.Phys.* **7** (2004) 831–864.
19. N. Nekrasov and A. Okounkov, *Prog. Math.* **244** (2006) 525–596.
20. M. Aganagic, A. Klemm, M. Marino, and C. Vafa, *Commun.Math.Phys.* **254** (2005) 425–478.
21. A. Iqbal, C. Kozcaz, and C. Vafa, *JHEP* **0910** (2009) 069.
22. H. Awata and H. Kanno, *Int.J.Mod.Phys.* **A24** (2009) 2253–2306.
23. J.-t. Ding and K. Iohara, *Lett. Math. Phys.* **41** (1997) 181–193.
24. K. Miki, *Journal of Mathematical Physics* **48** (2007), no. 12 123520.
25. H. Nakajima and K. Yoshioka, *Invent. Math.* **162** (2005) 313–355;
26. H. Nakajima and K. Yoshioka, math/0505553.
27. L. Gottsche, H. Nakajima, and K. Yoshioka, *Pure Appl. Math. Quart.* **5** (2009) 1029–1111.
28. J. Kim, S.-S. Kim, K.-H. Lee, K. Lee, and J. Song, *JHEP* **11** (2019) 092.
29. H.-C. Kim, M. Kim, S.-S. Kim, and K.-H. Lee, *JHEP* **04** (2021) 161.
30. H.-C. Kim, M. Kim, and S.-S. Kim, *JHEP* **08** (2021) 131.
31. J. Gu, M.-x. Huang, A.-K. Kashani-Poor, and A. Klemm, *JHEP* **05** (2017) 130.
32. M.-x. Huang, K. Sun, and X. Wang, *JHEP* **10** (2018) 196.
33. F. Bonetti and T. W. Grimm, *JHEP* **05** (2012) 019.
34. F. Bonetti, T. W. Grimm, and S. Hohenegger, *JHEP* **07** (2013) 043.
35. T. W. Grimm and A. Kapfer, *JHEP* **05** (2016) 102,
36. P. Benetti Genolini, M. Honda, H.-C. Kim, D. Tong, and C. Vafa, *JHEP* **05** (2020) 058.
37. R. Gopakumar and C. Vafa, hep-th/9809187;hep-th/9812127.
38. D. Gaiotto and H.-C. Kim, *JHEP* **01** (2017) 019.
39. H. Hayashi, S.-S. Kim, K. Lee, and F. Yagi, *JHEP* **01** (2017) 093.
40. H. Hayashi, S.-S. Kim, K. Lee, and F. Yagi, *JHEP* **12** (2018) 016.
41. L. Bhardwaj, arXiv:1909.09635].

Acknowledgement: I would like to thank the organizers of the 2021 Nankai Symposium on Mathematical Dialogues celebrating 110th anniversary of the birth of Prof. S.-S. Chern, and in particular Yang-Hui He for kind invitation and correct guidance. The author is grateful to Hee-Cheol Kim, Kimyeong Lee, [illegible] [illegible]

Machine Learning Symmetry

Shailesh Lal

Abstract We review recent work in machine learning aspects of conformal field theory and Lie algebra representation theory using neural networks.

1 Introduction

One of the most striking examples of the success of machine learning and neural networks in analyzing physical systems has been in the context of the string landscape problem, initiated in [1–4], see [5, 6] for reviews. Neural networks have been used with great accuracy to predict various properties of compactification manifolds such as topological numbers [7–10], bundle cohomology [3, 11, 12], metrics [13, 14], and symmetry properties [15]. Machine Learning has also proved useful to find patterns in particle masses [16] and more generally to statistical predictions and model-building in string theory [17–21]. Interestingly, neural networks have been also used to even quantify the degree of likelihood that two string vacua will give rise to similar phenomenology [22]. At the heart of these successes is the robust pattern-finding ability of neural networks which allow us to directly infer the output given an input by learning heuristic association rules from previously available data.

If we, somewhat reductively, pose the landscape problem as the problem of scanning across a large parameter space to search for points where specific properties are realized, coupled with the computational complexity of probing even a single point, then it is apparent that the landscape problem is not unique to string theory but is quite generic to mathematical physics. For example, one might ask: 'among which in the set of representations of $\mathfrak{su}(5)$ may one embed the matter content of the standard model'? Or equally well, 'where in the set of conformal field theories does a given operator product expansion (OPE) coefficient take a particular value'? For an individual $\mathfrak{su}(5)$ representation or an individual conformal field theory, the answer may be obtained through a direct, albeit possibly involved, computation. We however, would like to address these questions across an in principle infinite set of

S. Lal (✉)
No. 544 Hefangkou Village, Huaibei Town, Huairou District, Beijing 101408, China
e-mail: shailesh.hri@gmail.com

Y.-H. He et al. (eds.), *Nankai Symposium on Mathematical Dialogues*,
https://doi.org/10.1007/978-981-19-2328-9_26

possibilities. Doing so might unveil new structures and organizing principles for the given parameter space, and eventually to new insight. In the following, we will report on some initial progress in using neural networks for these purposes [23, 24].

The works [23, 24] are in a sense part of a more general program of applying methods from data science and machine learning to study problems in mathematical physics, see [25, 26] for a pedagogical treatment. By now, machine learning and data science have been used to learn various aspects of number theory [27–29], quiver gauge theories and cluster algebras [30], knot theory [31–33] as well as graph theory [34] and commutative algebra [35]. Importantly, and in hindsight somewhat as a precursor of [24], support vector machines and neural networks were used to learn the algebraic structures of discrete groups and rings [36]. It was explicitly shown there that these machine learning frameworks are able to solve difficult problems in the representation theory of discrete groups without having to appeal to computationally expensive methods like Sylow theorems. In addition, neural network regressors have also been trained to identify Lie algebra symmetries in datasets in [37].

2 Machine Learning Conformal Field Theory

Consider d-dimensional Euclidean space with the inner product

$$\vec{x}^2 = x_1^2 + x_2^2 + \cdots + x_d^2 \tag{1}$$

and the symmetry transformations

$$\vec{x} \mapsto \vec{x} + \vec{c},\ \vec{x} \mapsto R \cdot \vec{x},\ \vec{x} \mapsto \lambda\vec{x},\ \vec{x} \mapsto \frac{\vec{x}}{|\vec{x}|^2}, \tag{2}$$

A conformal field theory, or CFT, is a quantum field theory which is invariant under these symmetries. These theories are in ubiquitous in physics, being the formalism to describe systems as diverse as phase transitions in a table-top experiment to quantum gravity. The past few years have seen a rapid spurt of new numerical and analytical results for CFT by means of the conformal bootstrap. In particular, one starts with a few minimal assumptions about a putative CFT, such as the spectrum of light operators. Then crossing symmetry of four-point functions, along with unitarity, strongly constrain the possible dynamics of such theories, see [38] for a review. These developments have opened up the possibility of exploring the landscape of these theories by robust numerical methods.

It further turns out that various defining properties of Conformal Field Theory are machine learnable to a near hundred percent accuracy [23]. We will focus here on the conformal block expansion of four point correlation functions of scalar conformal primaries. These are operators that transform under conformal transformations as

$$\mathcal{O}_\Delta(\lambda x) = \lambda^{-\Delta}\, \mathcal{O}_\Delta(\lambda x)\ . \tag{3}$$

Here Δ is known as the conformal dimension of the operator $\mathcal{O}_\Delta$. In the following, we will set the dimension d to 1. The 4-point function of the operator $\mathcal{O}_\Delta$ may then be expressed as [39]

$$\langle \mathcal{O}_\Delta(x_1)\,\mathcal{O}_\Delta(x_2)\,\mathcal{O}_\Delta(x_3)\,\mathcal{O}_\Delta(x_4)\rangle = \frac{\sum_{\{h\}} c_h{}^2 z^h\,{}_2F_1(h,h,2h;z)}{|x_{12}|^{2\Delta}|x_{34}|^{2\Delta}}\,. \tag{4}$$

Here z is the cross ratio

$$z = \frac{|x_{12}||x_{34}|}{|x_{13}||x_{24}|}\,, \tag{5}$$

and the sum runs over all operators $\mathcal{O}_h$ that appear in the OPE $\mathcal{O}_\Delta \times \mathcal{O}_\Delta$, and c_h is the corresponding OPE coefficient. Here the conformal dimention of $\mathcal{O}_h$ is h and the functions $z^h\,{}_2F_1(h,h,2h;z)$ are the conformal blocks.

In the following, we will demonstrate how neural networks distinguish between two classes of four-point functions, namely ones where a particular coefficient c_h is zero or non-zero respectively. To do so, it is useful to define an auxiliary function

$$f(z) = \sum_{\{h\}} c_h{}^2 z^h\,{}_2F_1(h,h,2h;z)\,. \tag{6}$$

The spectrum of conformal dimensions $\{h\}$ and the OPE coefficients c_h are not arbitrary and are constrained so that the overall four-point function is crossing symmetric. We will however consider a toy setting where the h are fixed to be positive integers and the c_h are generated randomly. We then define two classes of functions, *viz.*

$$f_1(z) = \sum_{n=0}^{10} c_n{}^2 z^n\,{}_2F_1(n,n,2n;z)\,,\quad f_2(z) = \sum_{n=0,\,n\neq 2}^{10} c_n{}^2 z^n\,{}_2F_1(n,n,2n;z)\,, \tag{7}$$

Some examples of these two classes of functions are visualized in Fig. 1a, b. We now machine learn the binary classification of functions for which $c_2 = 0$ and $c_2 \neq 0$. The input is prepared by discretizing z to lie on a grid of 100 equidistant points in $|z| \leq 0.8$ and generate 100000 instances each of functions f_1 and f_2 by randomly sampling the c_h in the interval (0, 1). Hence, each input datum is described by a 100-dimensional tuple, and the corresponding output label is a binary, valued in {0, 1}. A relatively simple `relu` activated MLP architecture trained on this data achieves near perfect performance [23]. In particular, when evaluating on 2000 test functions, we obtain the confusion matrix

$$\begin{pmatrix} 1000 & 0 \\ 67 & 933 \end{pmatrix}\,, \tag{8}$$

In sum, simple MLP architectures are able to distinguish between two classes of functions which look quite similar at first glance. In fact, it is possible to understand

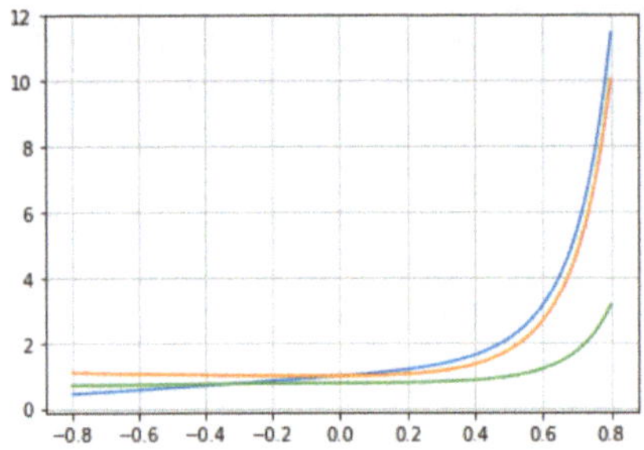

(a) Functions with $c_2 \neq 0$.

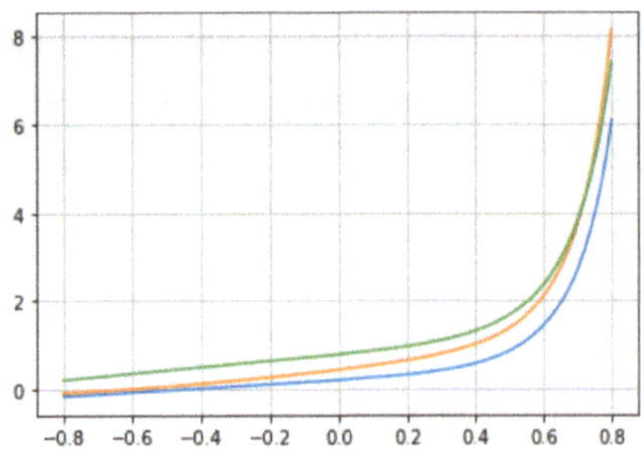

(b) Functions with $c_2 = 0$.

Fig. 1 Toy four-point functions for CFT_1

this difference quite explicitly. Principal component analysis applied to the randomly generated point clouds of Eq. 7 clearly visualizes the two classes of functions as being separated from each other.

3 Machine Learning Lie Algebras

One of the most fundamental problems in physics is the description of symmetries of a system, i.e. the set of operations that leaves the system invariant. This set has a natural group structure that may be imposed on it. The composition of two symmetry operations is quite clearly a symmetry operation, and any symmetry operation can also be inverted to restore the system to its original state. We will focus on the groups which have elements continuously connected to the identity, i.e. these elements may be parametrized as

$$U(\theta) = \exp\left(i\theta^a T_a\right), \theta^a \in \mathbb{R}. \tag{9}$$

These are the Lie groups, whose generators T^a form a Lie algebra, i.e. they are endowed with a composition rule $[\cdot, \cdot]$ known as the Lie bracket that obeys, among other properties,

$$[T_a, T_b] = i f_{ab}{}^c T_c. \tag{10}$$

Physicists usually focus on *unitary representations* of the group U over a vector space $\mathcal{V}$. The T_a then are represented by hermitian matrices on $\mathcal{V}$, and the Lie bracket is the usual matrix commutator.

A representation for which all the T_a cannot be simultaneously brought to block diagonal form is called *irreducible*. Representations which are not irreducible are called *reducible*, and may be expressed as a direct sum of irreducible representations. If a subset of the Lie algebra closes within itself under the same Lie bracket, it is known as a subalgebra. Further, irreducible representations of a Lie algebra decompose in general into a direct sum of irreducible representations of a given subalgebra. Carrying out such decompositions is in general a complicated analytical task [40, 41]. We refer the reader to [42] for more details regarding Lie algebras and their representation theory.

Given that particles carry unitary irreducible representations of the symmetry group, see e.g. [43], such computations are of key relevance to particle physics model building. As an example, consider the Lie algebras $\mathfrak{su}(N)$. These are the algebras corresponding to the group of unitary $N \times N$ matrices of determinant 1. The algebra $\mathfrak{su}(5)$ contains as a subalgebra $\mathfrak{su}(3) \times \mathfrak{su}(2) \times \mathfrak{u}(1)$, the symmetry algebra of the standard model. Classifying which representations of $\mathfrak{su}(5)$ contain the matter content of the standard model is an important problem for particle physics model building. Such computations are seen to run on exponential time scales [24] when carried out on `LieART`, a `Mathematica` software for computations in Lie algebra representation theory [41].

To machine-learn this decomposition, we note that unitary irreducible representations of $\mathfrak{su}(5)$ are labeled by an ordered 4-tuple of integers, see e.g. [42]. We restrict ourselves to those 4-tuples for which each individual entry ranges from 0 to 4. The dimensions of these representations range from 1, for $(0, 0, 0, 0)$ to 9765625 for $(4, 4, 4, 4)$. This data is used to train a `relu` activated MLP, to distinguish between representations of $\mathfrak{su}(5)$ that contain the standard model matter content and those that do not [24]. The training curves for this classification are shown in Fig. 3. The neural network trains to 90% test accuracy with a Matthews' phi-coefficient of 0.81. The confusion matrix obtained is

$$\begin{pmatrix} 43 & 0 \\ 8 & 29 \end{pmatrix} . \tag{11}$$

4 Outlook

In summary we see that neural networks are able to learn important facts about mathematical physics landscapes, both for conformal field theory and for the representation theory of Lie algebras. We hope that these promising results pave the way for a more exhaustive exploration of these landscapes using methods from data science and machine learning. We note especially the recent works on two dimensional CFTs using classifiers as well as autoencoders [44], along with the application of reinforcement learning to solve the equations of the conformal bootstrap itself [45]. It would also be interesting to examine if the methods of [22] could now be extended to these cases, and to glean interesting organizing principles by doing so. The pictorial representation of putative CFTs obtained by PCA, shown in Fig. 2 is already suggestive in this direction.

Finally, we note that symmetry in datasets was also characterized using neural networks in [37, 46–49]. It would be interesting to study how those methods apply to the examples studied here. More generally, we also note that ideas and methods from mathematical physics, in particular quantum field theory, have been used to explicate

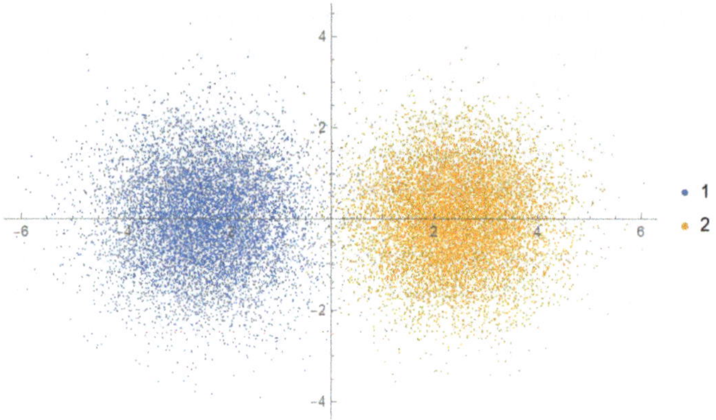

Fig. 2 The point clouds corresponding to f_1, shown as blue, and f_2, shown as yellow, represented in two dimensions by principal component analysis. Each point corresponds to a particular function f

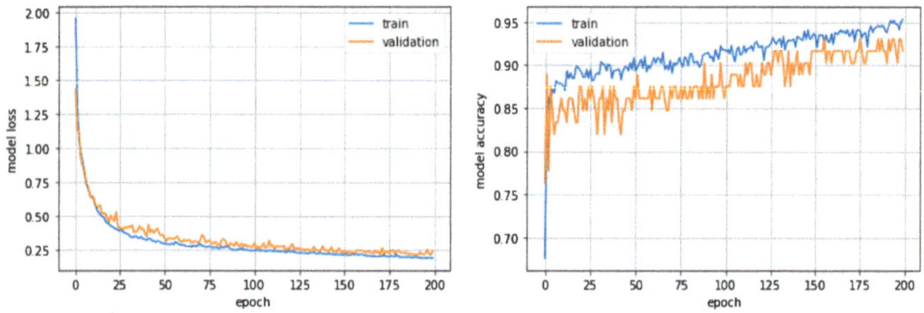

(a) Accuracy (b) Loss

Fig. 3 Training curves for $\mathfrak{su}(5)$ decomposition

the dynamics of neural networks (see e.g. [50–53]). Indeed, the very structures of quantum field theory and holography appear to be closely connected to certain neural networks [54]. We therefore expect that the interplay between mathematics, physics and data science will yield many new insights of mutual interest and benefit over the long term as well.

Acknowledgements Figures 1 and 2 are reproduced from the originals in [23]. SL's research was supported by the Simons Collaboration on the Non-perturbative Bootstrap and the Faculty of Sciences, University of Porto while [23, 24] were carried out.

References

1. Y. H. He, [arXiv:1706.02714 [hep-th].]
2. D. Krefl and R. K. Seong, Phys. Rev. D **96** (2017) no.6, 066014 [arXiv:1706.03346 [hep-th]].
3. F. Ruehle, JHEP **08** (2017), 038 [arXiv:1706.07024 [hep-th]].
4. J. Carifio, J. Halverson, D. Krioukov and B. D. Nelson, JHEP **09** (2017), 157 [arXiv:1707.00655 [hep-th]].
5. Y. H. He, Int. J. Mod. Phys. A **36** (2021) no.29, 2130017 [arXiv:2011.14442 [hep-th]].
6. F. Ruehle, Phys. Rept. **839** (2020), 1–117.
7. Y. H. He, Phys. Lett. B **774** (2017), 564–568.
8. H. Erbin and R. Finotello, Mach. Learn. Sci. Tech. **2** (2021) no.2, 02LT03 [arXiv:2007.13379 [hep-th]].
9. H. Erbin and R. Finotello, Phys. Rev. D **103** (2021) no. 12, 126014 [arXiv:2007.15706 [hep-th]].
10. H. Erbin, R. Finotello, R. Schneider and M. Tamaazousti, Mach. Learn. Sci. Tech. **3** (2022) no. 1, 015006 [arXiv:2108.02221 [hep-th]].
11. D. Klaewer and L. Schlechter, Phys. Lett. B **789** (2019), 438–443 [arXiv:1809.02547 [hep-th]].
12. C. R. Brodie, A. Constantin, R. Deen and A. Lukas, Fortsch. Phys. **68** (2020) no. 1, 1900087 [arXiv:1906.08730 [hep-th]].
13. A. Ashmore, Y. H. He and B. A. Ovrut, Fortsch. Phys. **68** (2020) no. 9, 2000068 [arXiv:1910.08605 [hep-th]].
14. M. Larfors, A. Lukas, F. Ruehle and R. Schneider, [arXiv:2111.01436 [hepth]].
15. X. Gao and H. Zou, [arXiv:2112.04950 [hep-th]].
16. Y. Gal, V. Jejjala, D. K. Mayorga Pena and C. Mishra,[arXiv:2003.10445 [hep-ph]].
17. R. Deen, Y. H. He, S. J. Lee and A. Lukas,[arXiv:2003.13339 [hep-th]].
18. J. Halverson, B. Nelson and F. Ruehle, JHEP **06** (2019), 003 [arXiv:1903.11616 [hep-th]].
19. J. Halverson and C. Long, Fortsch. Phys. **68** (2020) no. 5, 2000005 [arXiv:2001.00555 [hep-th]].
20. M. Larfors and R. Schneider, Fortsch. Phys. **68** (2020) no. 5, 2000034 [arXiv:2003.04817 [hep-th]].
21. H. Otsuka and K. Takemoto, JHEP **05** (2020), 047 [arXiv:2003.11880 [hepth]].
22. Y. H. He, S. Lal and M. Z. Zaz, [arXiv:2111.04761 [hep-th]].
23. H. Y. Chen, Y. H. He, S. Lal and M. Z. Zaz,[arXiv:2006.16114 [hep-th]].
24. H. Y. Chen, Y. H. He, S. Lal and S. Majumder, Phys. Lett. B **817** (2021), 136297 [arXiv:2011.00871 [hep-th]].
25. Y. H. He,[arXiv:1812.02893 [hep-th]].
26. Y. H. He,[arXiv:2101.06317 [cs.LG]].
27. L. Alessandretti, A. Baronchelli and Y. H. He, [arXiv:1911.02008 [math.NT]].
28. Y. H. He, E. Hirst and T. Peterken, J. Phys. A **54** (2021) no. 7, 075401 [arXiv:2004.05218 [hep-th]].
29. Y. H. He, K. H. Lee and T. Oliver, [arXiv:2010.01213 [math.NT]].
30. J. Bao, S. Franco, Y. H. He, E. Hirst, G. Musiker and Y. Xiao, Phys. Rev. D **102** (2020) no. 8, 086013 [arXiv:2006.10783 [hep-th]].
31. V. Jejjala, A. Kar and O. Parrikar, Phys. Lett. B **799** (2019), 135033 [arXiv:1902.05547 [hep-th]].
32. S. Gukov, J. Halverson, F. Ruehle and P. Sułkowski, Mach. Learn. Sci. Tech. **2** (2021) no. 2, 025035 [arXiv:2010.16263 [math.GT]].
33. A. Davies, P. Veličković, L. Buesing, *et. al.* Nature 600, 70–74 (2021).
34. Y. H. He and S. T. Yau,[arXiv:2006.16619 [math.CO]].
35. L. Amorós, O. Gasanova and L. Jakobsson, [arXiv:2109.11417 [math.AC]].
36. Y. H. He and M. Kim,[arXiv:1905.02263 [cs.LG]].
37. S. Craven, D. Croon, D. Cutting and R. Houtz,[arXiv:2112.07673 [hep-ph]].
38. D. Poland, S. Rychkov and A. Vichi, Rev. Mod. Phys. **91** (2019), 015002 [arXiv:1805.04405 [hep-th]].
39. M. Hogervorst and B. C. van Rees, JHEP **11** (2017), 193 [arXiv:1702.08471 [hep-th]].

40. R. Slansky, Phys. Rept. **79** (1981), 1–128.
41. R. Feger and T. W. Kephart, Comput. Phys. Commun. **192** (2015), 166–195 [arXiv:1206.6379 [math-ph]].
42. J. Fuchs and C. Schweigert, "Symmetries, Lie algebras and representations: A graduate course for physicists," Cambridge University Press.
43. S. Weinberg, "The Quantum Theory of Fields, Volume 1: Foundations," Cambridge University Press.
44. E. J. Kuo, A. Seif, R. Lundgren, S. Whitsitt and M. Hafezi,[arXiv:2106.13485 [cond-mat.str-el]].
45. G. Kántor, V. Niarchos and C. Papageorgakis, [arXiv:2108.09330 [hep-th]].
46. S. Krippendorf and M. Syvaeri, [arXiv:2003.13679 [physics.comp-ph]].
47. B. M. Dillon, G. Kasieczka, H. Olischlager, T. Plehn, P. Sorrenson and L. Vogel, [arXiv:2108.04253 [hep-ph]].
48. G. Barenboim, J. Hirn and V. Sanz, SciPost Phys. **11** (2021), 014 [arXiv:2103.06115 [cs.LG]].
49. K. Desai, B. Nachman and J. Thaler, [arXiv:2112.05722 [hep-ph]].
50. E. De Mello Koch, R. De Mello Koch and L Cheng, IEEE Access, vol. 8, pp. 106487–106505, 2020, [arXiv:1906.05212 [cs.LG]].
51. N. Halmagyi and S. Lal, [arXiv:2007.10181 [math-ph]].
52. J. Halverson, A. Maiti and K. Stoner, Mach. Learn. Sci. Tech. **2** (2021) no. 3, 035002 [arXiv:2008.08601 [cs.LG]].
53. H. Erbin, V. Lahoche and D. O. Samary,[arXiv:2108.01403 [hep-th]].
54. K. Hashimoto, Phys. Rev. D **99** (2019) no. 10, 106017 [arXiv:1903.04951 [hep-th]].

Kac–Moody Eisenstein Series

Kyu-Hwan Lee

Abstract In this survey, we will present recent results on convergence and holomorphy of Kac–Moody Eisenstein series.

2010 Mathematics Subject Classification Primary 20G44 · Secondary 11F70

1 Motivation and Past Developments

After being developed by Langlands [32, 33] in great generality, the theory of Eisenstein series has played fundamental roles in Langlands' Program. In particular, meromorphic continuation of Eisenstein series on reductive groups, established in the seminal work of Langlands [33], has been a foundational basis for the study of L-functions by means of the Langlands–Shahidi method (e.g. [29, 30]), which exploits analytic properties of Eisenstein series together with the fact that L-functions appear in the Fourier coefficients of Eisenstein series.

Eisenstein series also appear in many other places throughout number theory and representation theory. The scope of applications extends to geometry and mathematical physics. For example, Eisenstein series on exceptional Lie groups have been shown to occur explicitly as coefficients of correction terms in certain maximally supersymmetric string theories [24–26].

On the other hand, since we have seen many successful generalizations of finite dimensional constructions to infinite dimensional Kac–Moody algebras and groups [27, 31], it is a natural question to ask whether one can generalize the theory of Eisenstein series to Kac–Moody groups. Such an attempt is not merely for the sake of generalization. Even though it is conjectural at the moment, a satisfactory theory

This work was partially supported by a grant from the Simons Foundation (#712100).

K.-H. Lee (✉)
Department of Mathematics, University of Connecticut, Storrs, CT, USA
e-mail: khlee@math.uconn.edu

Y.-H. He et al. (eds.), *Nankai Symposium on Mathematical Dialogues*,
https://doi.org/10.1007/978-981-19-2328-9_27

of Eisenstein series on Kac–Moody groups would have huge impacts on some of the central problems in number theory related to Ramanujan Conjecture, Lindelöf Hypothesis and Langlands Functoriality [3, 38].

It has also been discovered that discrete subgroups of Kac–Moody groups and their automorphic forms appear as symmetries in high-energy theoretical physics. In particular, the group $E_{10}(\mathbb{Z})$ is conjectured to be the discrete invariance group for certain functions that arise in 11-dimensional supersymmetric string theory [8, 15]. Automorphic forms on E_{10} and E_{11} are conjectured to encode higher derivative corrections in string theory and M–theory [7, 9, 14, 39], and Eisenstein series on these Kac–Moody groups appear as central objects [12, 13].

All of these potential applications require establishing analytic properties of Kac-Moody Eisenstein series, including convergence and meromorphic continuation.

In his pioneering work [16–18], Garland extended the classical theory of Eisenstein series to arithmetic quotients $G_{\mathbb{Z}}\backslash G_{\mathbb{R}}/K$ of affine Kac–Moody groups G. In particular, he established absolute convergence for spectral parameters in a Godement range, and then proved a meromorphic continuation beyond it [19–22]. Absolute convergence has been generalized to affine Kac–Moody groups over number fields by Liu [35]. Garland, Miller, and Patnaik [23] showed that affine Eisenstein series induced from cusp forms are entire functions of the spectral parameter. In the function field case, analogous works were established in [1, 28, 34, 37].

Beyond the affine case, Carbone et al. [5] studied Eisenstein series on the rank 2 hyperbolic Kac–Moody groups with symmetric generalized Cartan matrices, and established almost-everywhere convergence of the series. They also defined and calculated the degenerate Fourier coefficients for the Eisenstein series and showed that the cuspidal Eisenstein series are entire, extending the results on the affine Kac–Moody groups.

2 Recent Results

In this section, we will review recent results in [4, 6] for Eisenstein series on more general Kac–Moody groups.

2.1 Convergence

In [4], Carbone, Garland, Lee, Liu and Miller studied the absolute convergence of Eisenstein series on general Kac–Moody groups. Almost-everywhere convergence is proven for an arbitrary symmetrizable Kac–Moody group, but everywhere convergence is established under an assumption on the root system (Property 1 below).

Let G be a representation-theoretic Kac–Moody group, and let $\mathfrak{g}$ be the corresponding real Kac–Moody algebra with a fixed Cartan subalgebra $\mathfrak{h}$. We assume that $\mathfrak{g}$ is infinite-dimensional and non-affine. Let $r = \dim(\mathfrak{h})$ denote the rank of G, I the index set $\{1, \ldots, r\}$, and Φ_+ (resp., Φ_-) the positive (resp., negative) roots of $\mathfrak{g}_{\mathbb{C}}$. Then $G_{\mathbb{R}}$ has the Iwasawa decomposition $G_{\mathbb{R}} = UA^+K$, where U is a maximal pro-unipotent subgroup, A^+ is the connected component of a maximal torus, and K is a subgroup of G playing the role of the maximal compact subgroup.

Define the Borel Kac–Moody Eisenstein series $E_\lambda(g)$ for $g \in G_{\mathbb{R}}$ and $\lambda \in \mathfrak{h}_{\mathbb{C}}^*$ by

$$E_\lambda(g) \;=\; \sum_{\gamma \in (\Gamma \cap B)\backslash \Gamma} a(\gamma g)^{\lambda+\rho}, \tag{1}$$

where $a(g)$ is the A^+–component of the Iwasawa decomposition of g, ρ is the Weyl vector, $\Gamma = G_{\mathbb{Z}}$ is the arithmetic subgroup, and $B \supset NA^+$ is a Borel subgroup. Let W be the Weyl group of $\mathfrak{g}$. We denote by $\ell(w)$ the length of $w \in W$ and define

$$\Phi_w \;:=\; \Phi_+ \cap w^{-1}\Phi_-, \;\; w \in W. \tag{2}$$

Consider the following property.

Property 1 *Every nontrivial $w \in W$ can be written as $w = vw_\beta$ for some $v \in W$, where $\ell(v) < \ell(w)$, w_β is the reflection associated to a positive simple root β, and $\alpha - \beta$ is never a real root for any $\alpha \in \Phi_v$.*

The main result in [4] is as follows.

Theorem 2 ([4]) *Assume that $\lambda \in \mathfrak{h}_{\mathbb{C}}^*$ satisfies* $\mathrm{Re}(\langle \lambda, \alpha_i^\vee \rangle) > 1$ *for each simple coroot $\alpha_i^\vee$, $i \in I$, and that Property 1 holds. Then the Kac–Moody Eisenstein series $E_\lambda(g)$ converges absolutely for $g \in \Gamma U A_{\mathfrak{C}} K$, where $A_{\mathfrak{C}} \subset A^+$ is the image of the Tits cone $\mathfrak{C}$ under the exponential map* $\exp : \mathfrak{h} \to A^+$.

The condition on λ is precisely the Godement range, and appears in the classical theory [33]. Property 1 holds when G is of rank 2 or when the Cartan matrix is symmetric and has sufficiently large entries. However, Property 1 is not true for all Kac–Moody root systems–it even fails in finite-dimensional groups.

In order to prove Theorem 2 we consider the constant term $E_\lambda^\sharp(g)$ of the series $E_\lambda(g)$, which is computed by the Gindikin–Karpelevich formula, and establish its absolute convergence. Unlike Theorem 2, which assumes Property 1, convergence of the constant term $E_\lambda^\sharp(g)$ holds for *all* symmetrizable G. As a consequence, almost-everywhere convergence of $E_\lambda(g)$ is also true for all symmetrizable G. When we compare $E_\lambda^\sharp(g)$ with $E_\lambda(g)$ to prove everywhere convergence, we need Property 1.

Though Theorem 2 is stated only for Borel Eisenstein series, its conclusions hold for *cuspidally* induced Eisenstein series as well, at least for parabolics with finite-dimensional Levi components. This fact was used in [6].

2.2 Entirety

In the process of these developments, a striking difference between the finite-dimensional case and the affine case was observed in [1, 23], where the affine Eisenstein series induced from cusp forms on finite-dimensional Levi subgroups were shown to be *entire* and not just meromorphic as they are in the finite-dimensional case. This phenomenon is not restricted to the affine case, and it was shown in [5] that cuspidal Eisenstein series on rank 2 symmetric hyperbolic Kac–Moody groups are also entire.

To understand the analytic properties of cuspidal Eisenstein series on Kac–Moody groups, one may naturally ask which parabolic subgroups of a Kac–Moody group make cuspidal Eisenstein series entire. To answer this question, a natural condition on parabolic subgroups (Property RD) is introduced in [6], and it was shown that cuspidal Eisenstein series attached to a parabolic subgroup satisfying Property RD is holomorphic on the full complex plane.

More precisely, let P be a maximal parabolic subgroup of $G_{\mathbb{R}}$ with fixed Levi decomposition and associated finite-dimensional Levi subgroup M, which is associated with a subset $\theta \subseteq I$. We denote by α_P the simple root associated to the one element index in $I \backslash \theta$. Let L be the derived subgroup of M. For a cusp form f on $(L \cap \Gamma)\backslash L$, we recall that f is unramified if f is right invariant under the action of $L \cap K$.

For such an f, we define the Eisenstein series $E_f(s, g), s \in \mathbb{C}, g \in G_{\mathbb{R}}$, in analogy with the classical case. We use the reduction mechanism of [2, 36] to obtain absolute convergence of the cuspidal Eisenstein series $E_f(s, g)$ from that of Borel Eisenstein series established in [4].

Denote by ϖ_P the fundamental weight associated to α_P, and by ρ_M the Weyl vector of M. Let

$$W^\theta = \{w \in W : w^{-1}\alpha_i > 0, i \in \theta\}.$$

Property 3 *A parabolic group P is said to satisfy* Property RD *if there exists a constant $D > 0$, such that for every nontrivial element $w \in W^\theta$ we have*

$$\langle D\varpi_P + \rho_M, \alpha^\vee \rangle \leq 0$$

for any positive root α such that $w^{-1}\alpha < 0$.

Property RD states that the coefficient of the simple root α_P grows faster than the coefficients of the simple roots in the subset θ. This property allows us to make use of the rapid decay of cusp forms on parabolic subgroups.

The main theorem in [6] can be stated as follows.

Theorem 4 ([6]) *Let f be an unramified cusp form on $(L \cap \Gamma)\backslash L$. If the maximal parabolic subgroup P satisfies Property RD, then for any compact subset $\mathfrak{S}$ of A_C, there exists a measure zero subset S_0 of $(\Gamma \cap U)\backslash U\mathfrak{S}$ such that $E_f(s, g)$ is an entire function of $s \in \mathbb{C}$ for $g \in (\Gamma \cap U)\backslash U\mathfrak{S}K - S_0K$.*

Here a measure zero set appears because absolute convergence is only established almost everywhere for the Eisenstein series in general. In the setting of everywhere convergence established in [4], the measure zero set is not needed.

The main idea of the proof is to exploit rapid decay of a cusp form on the maximal parabolic subgroup, guaranteed by Property RD. It turns out that a large class of Kac–Moody groups have parabolic subgroups satisfying Property RD, including the Kac–Moody group G associated with the Feingold–Frenkel rank 3 hyperbolic Kac–Moody algebra [10]. It would be an interesting question for future investigation to characterize the full class of Kac–Moody groups that admit parabolic subgroups satisfying Property RD.

References

1. A. Braverman and D. Kazhdan, *Representations of affine Kac–Moody groups over local and global fields: a suvery of some recent results,* European Congress of Mathematics, 91–117, Eur. Math. Soc., Zürich, 2013.
2. A. Borel, *Automorphic forms on reductive groups.* Automorphic forms and applications, 7–39, IAS/Park City Math. Ser., 12, Amer. Math. Soc., Providence, RI, 2007.
3. D. Bump, S. Friedberg and J. Hoffstein, *On some applications of automorphic forms to number theory*, Bull. Amer. Math. Soc. (N.S.) **33** (1996), no. 2, 157–175.
4. L. Carbone, H. Garland, K.-H. Lee, D. Liu and S. D. Miller, *On the convergence of Kac–Moody Eisenstein series*, preprint, arXiv:2005.13636.
5. L. Carbone, K.-H. Lee and D. Liu, *Eisenstein series on rank 2 hyperbolic Kac–Moody groups*, Math. Ann. **367** (2017), 1173–1197.
6. L. Carbone, K.-H. Lee and D. Liu, *Entirety of cuspidal Eisenstein series on Kac-Moody groups*, to appear in Algebra Number Theory, arXiv:2008.11559.
7. T. Damour, A. Hanany, M. Henneaux, A. Kleinschmidt and H. Nicolai, *Curvature corrections and Kac–Moody compatibility conditions*, Gen. Rel. Grav. 38:1507–1528 (2006).
8. T. Damour, A. Kleinschmidt and H. Nicolai, *Constraints and the E_{10} coset model*, Class. Quant. Grav. 24:6097–6120, (2007).
9. T. Damour and H. Nicolai, *Higher order M–theory corrections and the Kac–Moody algebra E_{10}*, Class. Quant. Grav. **22** (2005) 2849–2880.
10. A. Feingold and I. Frenkel, *A hyperbolic Kac–Moody algebra and the theory of Siegel modular forms of genus* 2, Math. Ann. **263** (1983), no. 1, 87–144.

11. P. Fleig, *Kac–Moody Eisenstein series in string theory*, Dissertation, University of Berlin, (2013).
12. P. Fleig and A. Kleinschmidt, *Eisenstein series for infinite-dimensional U-duality groups*, J. High Energ. Phys. (2012) 2012: 54.
13. P. Fleig, A. Kleinschmidt and D. Persson, *Fourier expansions of Kac–Moody Eisenstein series and degenerate Whittaker vectors*, Commun. Num. Theor. Phys. **08** (2014), 41–100.
14. P. Fleig, H. Gustafsson, A. Kleinschmidt and D. Persson, *Eisenstein Series and Automorphic Representations With Applications in String Theory*, Cambridge Studies in Advanced Mathematics **176**, Cambridge University Press, 2018.
15. O. J. Ganor, *Two conjectures on gauge theories, gravity and infinite dimensional Kac–Moody groups*, arXiv:hep-th/9903110
16. H. Garland, *Eisenstein series on arithmetic quotients of loop groups*, Math. Res. Lett. **6** (1999), no. 5–6, 723–733.
17. H. Garland, *Certain Eisenstein series on loop groups: convergence and the constant term*, Algebraic Groups and Arithmetic, Tata Inst. Fund. Res., Mumbai, (2004), 275–319.
18. H. Garland, *Absolute convergence of Eisenstein series on loop groups*, Duke Math. J. **135** (2006), no. 2, 203–260.
19. H. Garland, *Eisenstein series on loop groups: Maass-Selberg relations. I*, Algebraic groups and homogeneous spaces, 275–300, Tata Inst. Fund. Res. Stud. Math., Tata Inst. Fund. Res., Mumbai, 2007.
20. H. Garland, *Eisenstein series on loop groups: Maass-Selberg relations. II,* Amer. J. Math. **129** (2007), no. 3, 723–784.
21. H. Garland, *Eisenstein series on loop groups: Maass-Selberg relations. III,* Amer. J. Math. **129** (2007), no. 5, 1277–1353.
22. H. Garland, *Eisenstein series on loop groups: Maass-Selberg relations. IV*, Lie algebras, vertex operator algebras and their applications, 115–158, Contemp. Math. **442**, Amer. Math. Soc., Providence, RI, 2007.
23. H. Garland, S. D. Miller and M. M. Patnaik, *Entirety of cuspidal Eisenstein series on loop groups*, Amer. Jour. of Math. **139** (2017), 461–512.
24. M. Green, S. D. Miller, J. Russo and P. Vanhove, *Eisenstein series for higher-rank groups and string theory amplitudes*, Commun. Number Theory Phys. **4** (2010), no. 3, 551–596.
25. M. Green, S. D. Miller and P. Vanhove, *Small representations, string instantons, and Fourier modes of Eisenstein series*, with an appendix by D. Ciubotaru and P. E. Trapa, J. Number Theor. **146** (2015), 187–309.
26. M. B. Green, J. G. Russo and P. Vanhove, *Automorphic properties of low energy string amplitudes in various dimensions*, Phys. Rev. D **81** (2010) 086008, arXiv:1001.2535.
27. V. G. Kac, *Infinite dimensional Lie algebras*, Cambridge University Press, 1990.
28. M. Kapranov, *The elliptic curve in the S-duality theory and Eisenstein series for Kac–Moody groups*, math.AG/0001005.
29. H. H. Kim, *Functoriality for the exterior square of* GL_4 *and the symmetric fourth of* GL_2, with appendix 1 by D. Ramakrishnan and appendix 2 by Kim and P. Sarnak, J. Amer. Math. Soc. **16** (2003), no. 1, 139–183.
30. H. H. Kim and F. Shahidi, *Functorial products for* $\mathrm{GL}_2 \times \mathrm{GL}_3$ *and the symmetric cube for* GL_2, with an appendix by C. J. Bushnell and G. Henniart, Ann. of Math. (2) **155** (2002), no. 3, 837–893.
31. S. Kumar, *Kac–Moody groups, their flag varieties and representation theory*, Progress in Mathematics, **204**, Birkhäuser Boston, Inc., Boston, MA, 2002.
32. R. P. Langlands, *Euler products*, Yale Mathematical Monographs **1**, Yale University Press, New Haven, Conn.-London, 1971.

33. R. P. Langlands, *On the functional equations satisfied by Eisenstein series*, Lecture Notes in Mathematics, Vol. **544**, Springer-Verlag, Berlin-New York, 1976.
34. K.-H. Lee and P. Lombardo, *Eisenstein series on affine Kac–Moody groups over function fields*, Trans. Amer. Math. Soc. **366** (2014), no. 4, 2121–2165.
35. D. Liu, *Eisenstein series on loop groups*, Tran. Amer. Math. Soc. **367** (2015), no. 3, 2079–2135.
36. C. Moeglin, J.-L. Waldspurger, *Spectral decomposition and Eisenstein series,* Cambridge Tracts in Mathematics, 113. Cambridge University Press, Cambridge, 1995.
37. M. M. Patnaik, *Geometry of loop Eisenstein series*, Ph.D. thesis, Yale University, 2008.
38. F. Shahidi, *Infinite dimensional groups and automorphic L-functions*, Pure Appl. Math. Q. **1** (2005), no. 3, part 2, 683–699.
39. P. C. West, *E(11) and M theory*, Class. Quant. Grav. **18** (2001) 4443, hep-th/0104081.

On Towers of Light States at Infinite Distance

Seung-Joo Lee

Abstract Upon investigating asymptotic regimes of the F-theory moduli space, we verify that a tower of light states arises as predicted by the Distance Conjecture. Specifically, we provide a geometric classification of the infinite distance limits and comprehensively analyze the light states of the associated effective theories. We thereby find for every infinite distance limit that the effective theory either reduces to a weakly-coupled (dual) string theory or decompactifies to a higher-dimensional theory. This is in full agreement with the Emergent String Conjecture, which clarifies the physical nature of the light particle tower either as the excitation modes of an emergent weakly-coupled string or as the Kaluza-Klein modes associated with a decompactification of the spacetime. The results reported encompass both the Kähler and the complex structure limits of F-theory, respectively, in 4 and 8 dimensions, and hence are indicative of the overarching microscopic origins of the light towers.

1 Introduction

The notion of the *Landscape*, as the set of all consistent low-energy effective theories of quantum gravity, has served for decades as a central idea behind much research activity in high energy physics. Perhaps an equally important notion is that of the *Swampland*, which is the complementary set to the Landscape, hence collecting all the effective theories that cannot incorporate quantum gravity at high energies. Whether it is for intellectual guidance towards new physics ideas in model building, or for deeper understanding of the mysterious nature of quantum gravity *per se*, it is important to come up with concrete criteria by which theories in the landscape can be distinguished from the ones in the swampland. This is one of the major goals of the so-called Swampland Program [1]. In this context, many general properties

Based on a talk at the "Nankai Symposium on Mathematical Dialogues," 2021.

S.-J. Lee (✉)
Department of Physics, Yonsei University, Seoul, Republic of Korea
e-mail: seungjoolee@ibs.re.kr

Center for Theoretical Physics of the Universe, Institute for Basic Science, Daejeon, South Korea

Y.-H. He et al. (eds.), *Nankai Symposium on Mathematical Dialogues*,
https://doi.org/10.1007/978-981-19-2328-9_28

have thus far been proposed as universal features that every consistent theory in the landscape is believed to exhibit, which go under the name of *swampland conjectures*.

Interestingly, while the individual conjectures have been put forward based on their own physical rationale, accumulating pieces of evidence continue to be revealed for the exciting interconnections amongst them, leading to an intricate web of conjectures. Arguably at the heart of this web lies the Swampland Distance Conjecture [2], which asserts that a tower of states is bound to become exponentially light at infinite distance in moduli space of quantum gravity; it is these light states that oftentimes serve in turn as the physical objects of which presence is predicted by other conjectures,[1] thereby providing fruitful insights and inspirations for the latter.[2]

Despite the distinguished role played by the Swampland Distance Conjecture, the physical nature and the microscopic, dynamical origin of the predicted light states remain elusive. Provided that string theory is an ideal framework to investigate the quantum nature of gravity in a controlled and computable setup, one can thus hope to better understand and gain intuitions on the conjecture by clarifying whether and how it is realized in string theory. In this article we therefore address its string theoretic aspect by invoking the recent proposal of the Emergent String Conjecture [7], which identifies the light tower of states either as the excitation modes of a light string (emergent string phase) or as the light Kaluza-Klein (KK) modes (decompactification phase).

The aim of this article is to explain how these two phases are universally realized by the asymptotic forms of the compactification geometry in the general setting of F-theory [8], which provides a non-perturbative description of Type IIB string theory in presence of 7-branes. To this end we will first classify infinite distance limits in the moduli space of elliptic Calabi-Yau $(n+1)$-folds X_{n+1},

$$\pi : X_{n+1} \to B_n \,, \tag{1}$$

over the Type IIB internal n-folds B_n. Based on such a geometric classification, we will then proceed to characterize the associated $(10-2n)$-dimensional effective theories in the boundary of moduli space. Specifically, in Sects. 2 and 3, we will discuss the geometry and the physics at infinite distance in the Kähler and the complex structure moduli of F-theory, respectively, in 4 and 8 dimensions, *i.e.*, with $n = 3$ and $n = 1$ in (1); the reported results can thus be viewed as providing intuitions on most general asymptotic regimes of F-theory moduli space.[3] We will conclude in Sect. 4 with summary and outlook.

[1] We refer the readers e.g. to the reviews [3–6] for general overview of the swampland program as well as for detailed account of many important and exciting conjectures.

[2] For an illustration, we briefly describe at the end of Sect. 2 how this light tower can be argued to include the Weak Gravity sublattice/tower.

[3] For recent progress towards the physics of infinite distance limits in various perturbative string setups, see e.g. [9–21], which, respectively, analyze the Kähler and the complex structure limits. See also [22] for related discussions on the M-theoretic front.

2 Kähler Deformations

While the systematic study of the Kähler limits in F-theory originally started in a $6d$ setting [23, 24], in this section we will discuss more general results obtained for $4d$ $N = 1$ F-theory, as analyzed more recently in [25, 26]. The Kähler moduli of $4d$ F-theory control the volumes of various cycles in the Type IIB 3-fold B_3, and hence, the strengths of the couplings in the effective theory in particular. At the same time the Kähler geometry also governs many other physical quantities, most importantly, the tensions of various solitonic strings arising from the D3 branes wrapped on complex curves. An intriguing link may thus arise between the coupling constants and the string tensions via the common Kähler geometry, resulting, e.g., in a stringy realization of the Weak Gravity Conjecture in the weak gauge coupling regime. We will come back to this aspect towards the end of the section after presenting more general results applicable to all the Kähler limits.

We start by noting that a most general Kähler deformation can be thought of as the combination of an overall and a relative scalings. Infinite distance limits are therefore naturally characterized by a pair of parameters, μ and λ, which are individually taken to infinity for the asymptotic scalings of the two kinds,

$$\mu \to \infty \quad \text{(overall scaling limit)}\,; \qquad \lambda \to \infty \quad \text{(relative scaling limit)}\,. \tag{2}$$

Firstly, the former, overall limit is responsible for the linear scaling of the Kähler class $\mathcal{J}_{B_3}$,

$$\mathcal{J}_{B_3} \sim \mu J_{B_3}\,. \tag{3}$$

Here, by definition, the rescaled Kähler class J_{B_3} leads to a finite rescaled volume of B_3, i.e.,

$$V_{B_3} = \frac{1}{3!}\int_{B_3} J_{B_3}^3 \sim 1\,, \tag{4}$$

while the actual volume of B_3 diverges in the limit $\mu \to \infty$ as

$$\mathcal{V}_{B_3} = \frac{1}{3!}\int_{B_3} \mathcal{J}_{B_3}^3 \sim \mu^3 \to \infty\,. \tag{5}$$

On the other hand, the relative limit governs the growth of the largest Kähler parameter appearing in the rescaled Kähler class J_{B_3},

$$J_{B_3} = \lambda J_0 + \sum_i t^i J_i\,. \tag{6}$$

Here, the classes J_0 and J_i are the Kähler cone generators of B_3 and the associated parameters $\lambda =: t^0$ and t^i are subject to the parametric relations,

$$t^i \ll t^0 \quad \text{or} \quad t^i \sim t^0\,, \qquad \forall i\,, \tag{7}$$

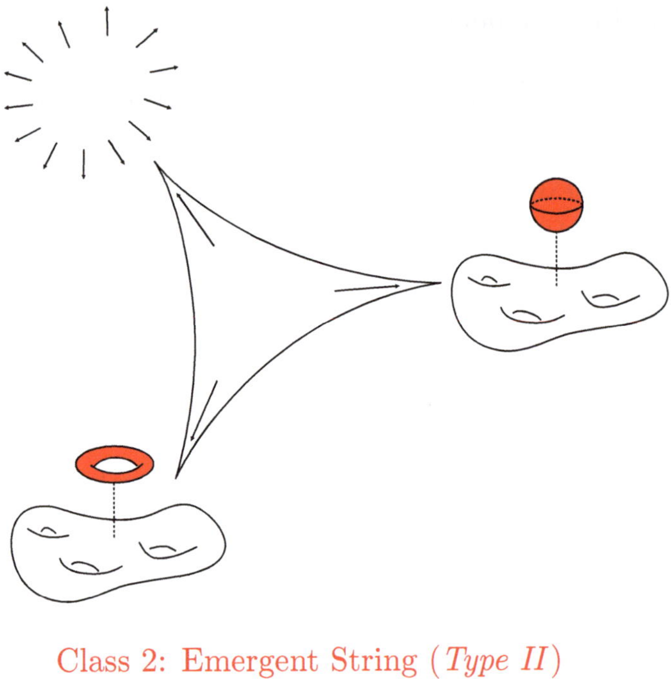

Fig. 1 Classification of infinite distance limits in the Kähler moduli space of 4*d* F-theory. Depicted in the top-left and the top-right corners are the limits corresponding to an emergent string phase, governed by the asymptotically weakly-coupled heterotic and Type II strings, respectively, and in the bottom corner are those corresponding to a decompactification phase

because $t^i < \lambda$ by definition of the large parameter λ. Combined with the finiteness of the rescaled volume (4), the limit $\lambda \to \infty$ thus enforces vanishing of some t^i, and hence, (relative) shrinking of certain holomorphic curves in turn.

As it turns out, depending largely on topology and geometry of such shrinking curves, we can classify the infinite distance Kähler limits into three qualitatively distinct classes as follows [26] (see Fig. 1 for a pictorial summary).

1. B_3 admits a $\mathbb{P}^1$-fibration of which fiber shrinks at a rate parametrically faster than any other non-contractable curves. D3 brane wrapping this fiber leads to an asymptotically tensionless *heterotic* string.
2. B_3 admits a T^2-fibration of which fiber shrinks at a rate parametrically faster than any other non-contractable curves. D3 brane wrapping this fiber leads to an asymptotitcally tensionless *Type II* string.
3. No unique $\mathbb{P}^1$ or T^2 fiber shrinks at a parametrically fastest rate (and hence, complementary to the previous two classes). However, expanding curves and surfaces have KK scale that wins against the tension of the solitonic critical strings, if any, indicating a decompactification.

For a more detailed description of the above three classes, let us denote by $C_0 \in H_2(B_3, \mathbb{Z})$ the homology class of the described fastest-shrinking fiber with genus $g(C_0) = 0$ or 1, and similarly, by $C_\infty \in H_2(B_3, \mathbb{Z})$ and $S_\infty \in H_4(B_3, \mathbb{Z})$ the homology classes of non-contractable curves and surfaces, respectively, that expand at a fastest parametric rate.

In the first two classes, the tension M_{str} of the relevant critical string turns out to scale as

$$\frac{M_{\text{str}}^2}{M_{\text{Pl}}^2} \sim \frac{\mathcal{V}_{C_0}}{\mathcal{V}_{B_3}} \sim \mu^{-2}\lambda^{-2}\,, \tag{8}$$

and the leading KK scale M_{KK} as

$$\frac{M_{\text{KK}}^2}{M_{\text{Pl}}^2} \sim \min\left(\frac{\mathcal{V}_{C_\infty}^{-1}}{\mathcal{V}_{B_3}}, \frac{\mathcal{V}_{S_\infty}^{-1/2}}{\mathcal{V}_{B_3}}, \frac{\mathcal{V}_{B_3}^{-1/3}}{\mathcal{V}_{B_3}}\right) \sim \mu^{-4}\lambda^{-1}\,. \tag{9}$$

The two scales thus compare as

$$\frac{M_{\text{str}}^2}{M_{\text{KK}}^2} \sim \mu^2\lambda^{-1}\,, \tag{10}$$

according to which we can naturally subdivide the space of large parameters λ and μ into three regimes, each exhibiting a characteristic phase of the effective physics as follows:

$$\lambda \ll \mu^2 \quad \longleftrightarrow \quad \text{decompactification} \tag{11}$$

$$\lambda \sim \mu^2 \quad \longleftrightarrow \quad \text{emergent string} \tag{12}$$

$$\lambda \gg \mu^2 \quad \longleftrightarrow \quad \text{potential pathology} \tag{13}$$

Note that the third regime (12) is described as being potentially pathological since the KK tower is decoupled from the string tower, contradicting that the critical dimension for a superstring is 10. In the first class of limits where the relevant light string is heterotic, it turns out that the day is saved by quantum corrections which are generically present with only 4 real supercharges constraining the effective theory: careful investigation of both perturbative α' and non-perturbative corrections teaches us that the limits in the regime (12) is obstructed. In the second class where the relevant light string is Type II, on the other hand, T-duality turns out to save the day.

In the last remaining class of limits, the fastest-shrinking fiber class C_0 is not uniquely defined. This indicates the presence of multiple critical strings which are equally relevant for physics. As already stated, however, the KK scale turns out to always win against the string tension, evading the pathology by entering into a decompactification phase instead.

Finally, let us briefly discuss how the addressed tower of light states at infinite distance can connect to other swampland conjectures, most prominently to the Weak

Gravity Conjecture [27] in its strongest, sublattice version [28–31]. For a quantitative control, we restrict our discussions to a special type of infinite distance limits where the $U(1)$ vector multiplets are asymptotically weakly coupled. As it turns out, of the three classes of limits discussed so far in full generality, only the first can arise, and hence, the heterotic string is the relevant asymptotically tensionless critical string. Then, not only do we know that there will arise a light tower but we can also be more precise about the spectrum. Specifically, a subset of the full spectrum is captured by the elliptic genus of the $N = (0, 2)$ heterotic worldsheet theory, of which (quasi-) Jacobi nature significantly constrains its functional behavior [32, 33]. This enables us to identify a sublattice[4] amount of charged particles for which the Coulomb repulsion is marginally stronger than the combined attraction of the gravitational and the Yukawa forces,

$$|F_{\text{Coulomb}}| \geq |F_{\text{Grav}}| + |F_{\text{Yuk}}|, \tag{14}$$

as predicted by the Sublattice Weak Gravity Conjecture in presence of scalar fields [34]. One can in fact verify that not only do these sublattice particles obey (14) at the classical level [25], but also upon incorporating quantum corrections [26].

3 Complex Structure Deformations

Having observed that the infinite distance limits in the Kähler moduli space always lead to a light particle tower of either string or KK excitation modes, we now change tack and consider the limits in the complex structure moduli space, which, together with the Kähler counterpart, comprise all deformations of F-theory. The complex structure deformations of F-theory encompass in particular the brane moduli in the open-string sector, which has largely been ignored until recently in study of infinite distance limits. In this section we will thus focus on the infinite complex structure limits of F-theory, in an arguably simplest non-trivial context of $8d$ effective theories.

To this end we investigate the complex structure deformations of elliptic K3 surfaces X_u, where $u \in D := \{u \in \mathbb{C} : |u| < 1\}$ is the deformation parameter. Viewed as a fibration over the unit disk D, the degeneration is thus described by a 3-fold family $\mathcal{X}$, whose generic member $X_{u\neq 0}$ is a smooth K3 surface while the central element $X_{u=0}$ degenerates into component surfaces

$$X_0 = \bigcup_{p=0}^{P} \bigcup_{i_p=1}^{N_p} X^p_{i_p} \quad \longrightarrow \quad B_0 = \bigcup_{p=0}^{P} B^p\,, \tag{15}$$

[4] The index for the embedding of this sublattice into the full charge lattice is determined as the intersection of the shrinking curve C_0 with the expanding gauge divisor S_∞ in the Type IIB internal space.

where $X^p_{i_p}$ are fibered over the rational base components B^p. The family $\mathcal{X}$ is called a Kulikov model only if a set of extra conditions are satisfied.[5] However, it is known that any degeneration can be put into such a Kulikov form by an appropriate combination of base changes and birational transformations [35–37], which, importantly, do not affect the effective physics.[6] Without loss of generality we may thus restrict only to the Kulikov models in analyzing the physics of complex structure degenerations.

Kulikov models are classified into Type I, II and III, where the latter two are known to be at infinite distance [36, 38, 39]. Our focus in this section thus lies in Kulikov models of Type II and III. Amongst many established properties of theirs, we point out that the surface components $X^p_{i_p}$ intersect at an elliptic and a rational curve, respectively, in Type II and III Kulikov models. Moreover, Type II models exhibit a pair of shrinking transcendental 2-tori, (γ_1, γ_2), and Type III models a single such 2-torus, γ_1. As in the Kähler limits of the previous sections, these 2-tori, due to their shirking nature in the complex structure limits, turn out to play a pivotal role in generating a light particle tower, which we will address momentarily.

Importantly, it turns out that elliptic Kulikov models of Type II and III can each be refined into two subtypes, so that the effective physics associated to each of the total of four subtypes exhibits a distinguished feature. In this context, the results that we will describe in this section are summarized in Fig. 2. To begin with, the refinement of elliptic Type II Kulikov models into Type II.a and II.b was already established in [40] as follows.[7]

- **II.a:** The central element X_0 takes the form (15) with $P = 1$ and $N_0 = N_1 = 1$: the two components X^0_1 and X^1_1 are rational elliptic surfaces and meet along the common smooth elliptic fiber sitting at $B^0 \cap B^1$ in the base.
- **II.b:** The central element X_0 takes the form (15) with $P = 0$ and $N_0 = 2$: the two components X^0_1 and X^0_2 are $\mathbb{P}^1$-fibrations over the common rational base B^0 and meet along its double cover ramified over four base points, which is thus an elliptic curve.

See Fig. 3 for an illustration of Type II.a and II.b Kulikov degenerations.

The physics of the associated effective theory had also been well-known. F-theory on a Type II.a Kulikov model is dual to heterotic string on a torus with the Kähler parameter taken to infinity ($T_{\text{het}} \to \infty$) [41–43], and hence, the spacetime fully decompactifies to 10d. On the other hand a Type II.b model realizes the so-called Sen's limit [43–45], where the effective theory is described as a weakly-coupled

[5] A degeneration is semi-stable if the components of the degenerate element X_0 are reduced and X_0 only involves normal-crossing singularities. Kulikov model is in turn defined as a semi-stable degeneration of K3 surfaces X_u of which 3-fold family $\mathcal{X}$ is Ricci-flat.

[6] Base change is a map $u \to u^k$ for $k \in \mathbb{Z}$ and hence does not alter the surface defined precisely at $u = 0$. The relevant birational transformations turn out to be base blowups, which correspond to zooming in towards the brane collision.

[7] The described canonical forms of X_0 can be achieved by appropriate base changes and birational transformations.

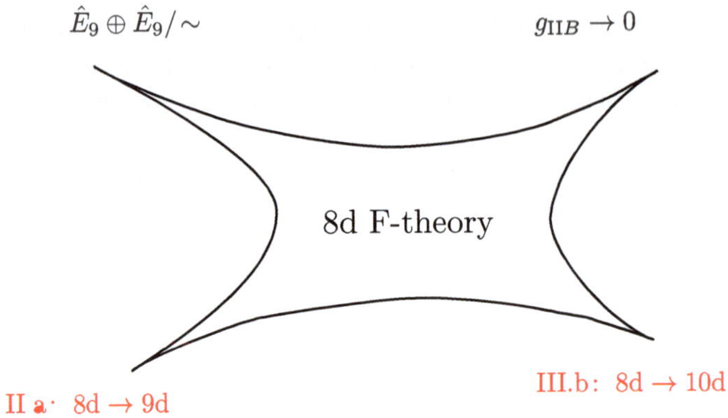

Fig. 2 Classification of infinite distance limits in the complex structure moduli space of 8*d* F-theory via refinement of elliptic Kulikov models. The limits in the top-right corner (models of Type II.b) correspond to an emergent string phase and those in the other three corners lead to a decompactification phase either to 9*d* (Type III.a) or to 10*d* (Type II.a and III.b)

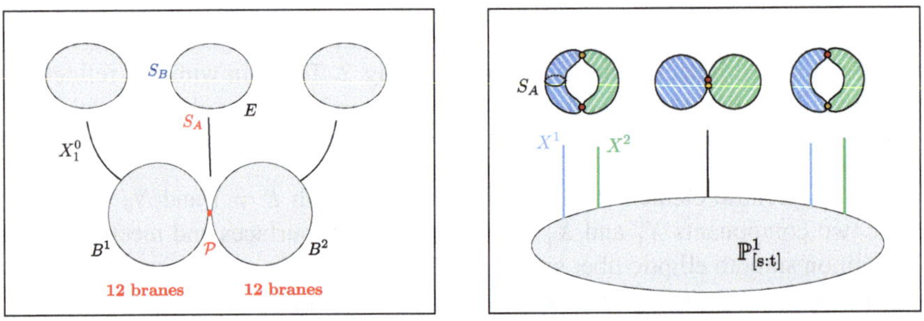

Fig. 3 Refinement of elliptic Kulikov models of Type II. Models of Type II.a (left) exhibit a degeneration of K3 surface into two rational elliptic components X_1^0 and X_1^1 meeting along the common smooth elliptic fiber, which sits at the intersection $\mathcal{P}$ of their bases B^0 and B^1. In models of Type II.b (right) the K3 degenerates into X_1^0 and X_2^0, each of which is a $\mathbb{P}^1$-fibration over the common base B^0; these two components meet along the elliptic curve realized as the double cover of B^0

Type IIB string in 8*d*. In fact, the characteristic structure for the geometry of these two subtypes enables us to immediately identify the described phases of the effective physics.

- For models of Type II.a, decompactification of two circular directions is indicated in M-theory by the pair of light KK-like towers, which result from M2 branes wrapping the shrinking transcendental 2-tori, (γ_1, γ_2). The latter in turn arise from fibering the A- and the B-cycles of the elliptic fiber $X_1^0 \cap X_1^1$ along the vanishing 1-cycle encircling the point $B^0 \cap B^1$ in the base; see Fig. 3 (left). These states correspond in F-theory to the towers resulting from (1, 0)- and (0, 1)-strings wound around the same vanishing 1-cycle of the base.

- For models of Type II.b, the "elliptic" fiber over the base B^0 has its A-cycle shrinking, and hence, the shirking 2-tori, (γ_1, γ_2), arise from adjoining this vanishing A-cycle in the fiber over the two 1-cycles in the base, to be precise, in the double-cover elliptic curve $X_1^0 \cap X_2^0$; see Fig. 3 (right). M2 branes wrapping those 2-tori again lead to a pair of light KK-like towers. However, there also arises a tensionless $(1, 0)$-string from M2 brane wrapping only the vanishing A-cycle. As it turns out, the KK and the string scales are parametrically the same, and hence, the spacetime does *not* decompactify.

More recently, elliptic Kulikov models of Type III have also been refined in the same vein into Type III.a and III.b models [46].[8] For a concise description of this refinement, let us first consider the *Kulikov Weierstrass model* $\mathcal{Y}$ by blowing down an elliptic Kulikov model $\mathcal{X}$ along the fiber, i.e., by contracting all the exceptional fibral curves. The central element Y_0 of $\mathcal{Y}$ thus decomposes as

$$Y_0 = \bigcup_{p=0}^{P} Y^p \quad \longrightarrow \quad B_0 = \bigcup_{p=0}^{P} B^p\,, \tag{16}$$

of which base B_0 is unchanged from (15). One can then blow $\mathcal{Y}$ further down along the base to an even more degenerate model $\widehat{\mathcal{Y}}$ by contracting all but one of the base components B^p. The central element $\hat{Y}_0$ of $\widehat{\mathcal{Y}}$ would then consist of a single surface component, elliptically fibered over a rational curve base $\hat{B}_0$,

$$\hat{Y}_0 \quad \longrightarrow \quad \hat{B}_0 =: \mathbb{P}^1_{[s:t]}\,, \tag{17}$$

where $[s : t]$ are the homogeneous coordinates of the base. In terms of the Weierstrass description of the F-theory K3 surface,

$$\hat{Y}_u : \qquad y^2 = x^3 + f(s, t;\, u)xz^4 + g(s, t;\, u)z^6\,, \tag{18}$$

it is in fact most natural to view the infinite complex structure limit as being initially realized in $\hat{Y}_0$. The involved birational transformations are schematically summarized as

$$\mathcal{X}ss \underset{\text{fibral blowdown}}{\overset{\text{fibral blowup}}{\longleftrightarrow}} \mathcal{Y} \underset{\text{base blowdown}}{\overset{\text{base blowup}}{\longleftrightarrow}} \widehat{\mathcal{Y}}\,. \tag{19}$$

As is well-known, the codimension-one singular fibers of a Weierstrass model $\hat{Y}_0$ can be characterized by the vanishing orders of the sections f and g, as well as the discriminant $\Delta := 4f^3 + 27g^2$, via the Kodaira-Néron classification. Here, the so-called *minimal* fiber in codimension-one locus of the base $\hat{B}_0 = \mathbb{P}^1_{[s:t]}$, e.g. at $s = 0$, is subject to

$$\mathrm{ord}_{\hat{Y}_0}(f)|_{s=0} < 4 \quad \text{or} \quad \mathrm{ord}_{\hat{Y}_0}(g)|_{s=0} < 6\,, \tag{20}$$

[8] See also the related independent analysis of [47–51] regarding compactification of the moduli space of elliptic K3s.

and represents a finite non-Abelian enhancement of the gauge algebra of the effective theory. Potential infinite distance limits must therefore exhibit either a codimension-one *non-minimal* fiber with

$$\mathrm{ord}_{\hat{Y}_0}(f, g, \Delta)|_{s=0} = (4+\alpha, 6+\beta, 12+\gamma)\,, \quad \text{for} \quad \alpha, \beta, \gamma \geq 0\,, \tag{21}$$

or else generic singular fibers in *codimension zero* (in absence of non-minimal fibers in codimension one). It has indeed been shown [46] that the former can lead to degenerations of Type II.a (if $\gamma = 0$) or of Type III (if $\gamma > 0$),[9] and the latter, to those of Type II.b.

As it turns out, however, the refinement of elliptic Type III Kulikov models is more naturally seen from the central element Y_0 (as opposed to $\hat{Y}_0$), of which decomposition (16) can be proven to take a chain form. Furthermore, $P \geq 2$ can be assumed without loss of generalities,[10] leading to a natural division of the components $Y^p \to B^p$ into the middle and the end surfaces, respectively, for $0 < p < P$ and for $p = 0$ or P. The refined classification is then motivated by the types of the generic fibers over the base components B^p, which are necessarily of Kodaira types I_{n_p} for $\mathcal{X}$ to be a Kulikov model.[11] One can then show for any Type III Kulikov Weierstrass models that $n_p > 0$ for the middle components, and hence, elliptic Type III models are naturally refined by the degeneration types of the end components Y^0 and Y^P as follows [46].

- **III.a:** The generic fibers of at least one end component are smooth, i.e., $n_0 = 0$ or $n_P = 0$.
- **III.b:** The generic fibers of both end components are singular, i.e., $n_0 > 0$ and $n_P > 0$.

Furthermore, special fibers of the degenerate components with $n_p > 0$ are restricted to only take one of the two forms,

$$\text{A-type} : \mathrm{ord}_{Y^p}(f, g, \Delta)|_{\mathrm{pt}\in B^p} = (0, 0, k) \quad \text{or} \quad \text{D-type} : \mathrm{ord}_{Y^p}(f, g, \Delta)|_{\mathrm{pt}\in B^p} = (2, 3, k)\,, \tag{22}$$

where D-type fibers are only allowed in the (degenerate) *end* components, and precisely two of them in each such end. See Fig. 4 for an illustration of this refinement.

As it turns out, the effective physics associated with the two subtypes of elliptic Type III models parallel their Type II counterparts to some extent and exhibit important distinguishing features at the same time. F-theory on a Type III.a model is again dual to heterotic string on a torus, but with both the Kähler and the complex structure parameters taken to infinity ($T_{\mathrm{het}} \sim U_{\mathrm{het}} \to \infty$), and hence, the spacetime partially decompactifies to $9d$. Similarly, a Type III.b model also renders the Type IIB string weakly-coupled, but on top of that decompactifies the spacetime to $10d$.

---

[9] To be precise, $\gamma > 0$ together with $\alpha = \beta = 0$ guarantee that the associated Kulikov model is of Type III. If α, β, and γ are all strictly positive, even a Type I Kulikov model might arise, as exemplified in [46].

[10] This is because $P \geq 2$ can always be achieved by a base change.

[11] Note that n_p coincide with N_p in (15) if the generic fibers are singular, i.e., if $n_p > 0$.

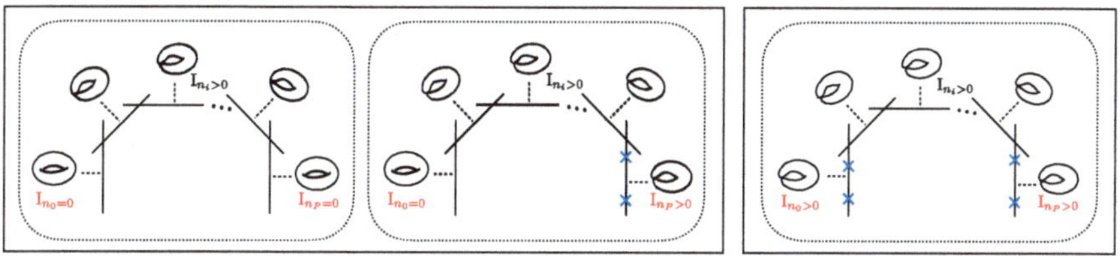

Fig. 4 Refinement of elliptic Type III Kulikov models in terms of the Weierstrass models $\mathcal{Y}$. Depicted are the possible configurations for the central element Y_0 of $\mathcal{Y}$. The degenerate surface Y_0 decomposes into a chain of components Y^p with generic fibers of Kodaira types I_{n_p} for $0 \le p \le P$, with $P \ge 2$. The components Y^p for $0 < p < P$ have generic fibers of Kodaira type I_{n_p} with $n_p > 0$. Elliptic Type III models are thus refined into Type III.a models where none/one of the end components degenerates, respectively, as in the first/second configuration (left), and Type III.b models where both ends degenerate as in the third configuration (right). Blue crosses on the end components denote D-type fibers

Such phases of the effective physics are easily read off from the geometric structure as follows [52].

- For models of Type III.a, decompactification of one circular direction is indicated in M-theory by the light KK-like tower, which results from M2 branes wrapping the shrinking transcendental 2-torus γ_1. The latter in turn arises from fibering the A-cycle of the elliptic fiber along the vanishing 1-cycle encircling the base point $B^p \cap B^{p+1}$ for any p[12]; see Fig. 5 for a schematic description of the intersection with $p = 0$. These states correspond in F-theory to the tower resulting from $(1, 0)$-strings wound around the same vanishing 1-cycle of the base.
- For models of Type III.b, there arises a tensionless $(1, 0)$-string from M2 brane wrapping the fibral A-cycle, which vanishes *globally* over the entire base B_0 unlike in models of Type III.a. However, two (or more) D-type fibers, and hence also the associated O-planes, are bound to collide in $\hat{B}_0$ of the degenerate K3 surface (17), since otherwise the base blowups would not have occurred to begin with. This indicates a complex structure degeneration of the Type IIB torus and hence, in turn, a decompactification of the spacetime to $10d$. Note that the 2-torus γ_1 is defined in the same way as in Type III.a case and leads to a light KK-like tower once wrapped by M2 branes. The other light tower, however, needs to be inferred indirectly.

Let us close this section by providing an alternative F-theoretic interpretation of the light towers resulting from Type III.a degenerations, with the central notion being the affine extension,[13]

$$E_{9-n} \longrightarrow \hat{E}_{9-n}\,, \quad n \ge 1\,, \tag{23}$$

[12] The generic singular fibers of Kodaira type I_{n_p} turn out to be mutually local in a global sense [52] so that a common 2-torus class γ_1 is defined for all p. Furthermore, the B-cycle cannot be fibered along the base 1-cycle due to the I_n monodromy around the point $B^p \cap B^{p+1}$.

[13] A similar interpretation can be given to Type II.a models, where the involved extension is of rank-2: $E_8 \to \hat{E}_8 \to \hat{E}_9$.

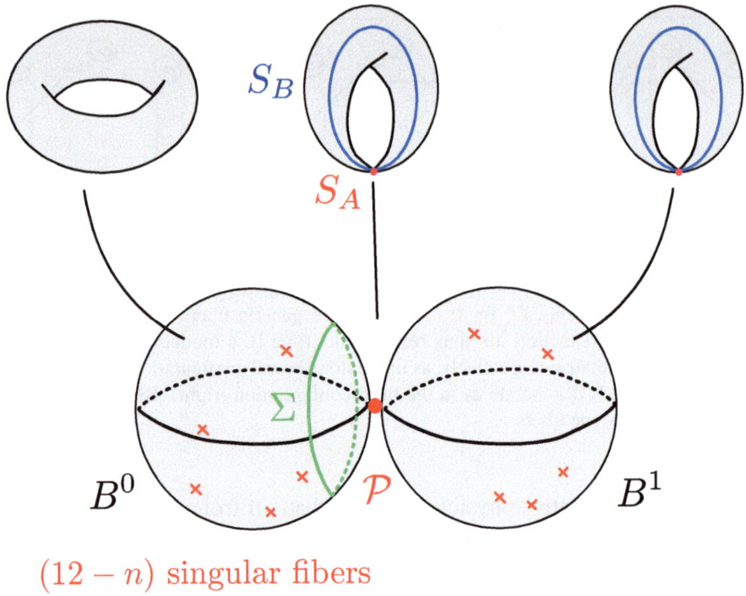

Fig. 5 The characteristic topology of a Type III.a Kulikov Weierstrass model at its rational elliptic end component, say, Y_0, adjoining the degenerate component Y^1 with generic I_n fibers. Red crosses on the base components denote the locations of singular fibers, $12 - n$ of which sit on B^0

associated with the $SL(2, \mathbb{Z})$ monodromy of a brane stack formed at infinite distance. Specifically, as depicted in Fig. 5, at the intersection of a rational elliptic end component, say, Y^0, and its adjacent degenerate component Y^1 lies an I_n fiber, where $n := n_1 \geq 1$.

In order for the combined monodromy of the 12 singular fibers on B^0 to be trivial, a total of $12 - n$ branes on B^0 away from $\mathcal{P} := B^0 \cap B^1$ must then generate the affine algebra $\hat{E}_{9-n}$, of which monodromy is the inverse of the A_{n-1} monodromy. In the described affinization (23) occurring at infinite distance,[14] the imaginary root leads to a BPS winding tower via (1, 0)-string. This thus serves as a light KK tower, indicating that the spacetime partially decompactifies to $9d$.

Apparently, the imaginary root only provides a KK tower and is ignored in the gauge algebra of the decompactified theory. We thus propose that the $9d$ gauge algebra acquires an E_{9-n} factor from the branes on B^0. As an immediate consequence of this interpretation, we learn that the maximal gauge algebras[15] in "$9d$ F-theory"

[14] A comprehensive analysis of monodromy actions was given in [53], which is suggestive of potential formation of $\hat{E}_{9-n}$ brane stacks. Kulikov models of Type III.a explicitly realize such affine extensions at infinite distance in the complex structure moduli.

[15] Strictly speaking, these may only constitute the non-abelian part of the gauge algebra. However, the claimed algebras already saturate the rank bound 17. We thus believe that additional $u(1)$ factors cannot be present, which is natural also in that all of the 24 branes have already been used up.

are $A_{m+n-1} \oplus E_{9-n} \oplus E_{9-m}$ (if both end components are rational elliptic) or $D_{8+n} \oplus E_{9-n}$ (if only one end is rational elliptic), where the constraints (22) on the allowed brane types have been used. Such gauge algebras have indeed been constructed in terms of explicit Kulikov Weierstrass models at infinite distance, practically by engineering appropriate non-minimal fibers in codimension one [52]. It is assuring that our F-theoretic results are in complete agreement with the recent classification from the heterotic side [54],[16] where the $9d$ algebras were analyzed via Wilson lines on S^1. This non-trivial match serves as an independent piece of evidence for our decompactification proposal.

4 Conclusions and Prospects

In this article we have reported on recent progress towards identifying towers of states expected to become light at infinite distance in moduli space of quantum gravity. Specifically, in the context of F-theory compactifications, the physical nature of such light towers has been clarified either as the excitation modes of an emergent weakly-coupled string or as the Kaluza-Klein modes associated with a decompactification, in agreement with the Emergent String Conjecture.

We started by geometrically classifying the infinite distance limits, both in the Kähler and in the complex structure moduli spaces of F-theory, respectively, in 4 and 8 dimensions. We have then proceeded to investigate the effective theories resulting from such asymptotic forms of the compactification geometry, and have thus confirmed, to begin with, that a tower of states does become light. This calls for an alternative way of describing the effective physics in the boundary of F-theory moduli space, notably, in terms of perturbative *heterotic* or *Type II* strings. The addressed light tower of states thus clarifies precisely how one should switch between different duality frames of string theory.

Both in the Kähler and in the complex structure sides of the story, we have seen that a pivotal role is played by the vanishing of non-contractible cycles, which we identified via our geometric classification. In all infinite distance limits, those shrinking cycles, once wrapped by various appropriate branes in the theory, lead either to a light KK-like tower of particles or to an asymptotically tensionless critical string, which in turn generates a light tower of excitation particles. In the latter situation, the geometry can in principle allow for multiple different species of lightest critical strings, which may look pathological at first sight. However, we have seen that in all such cases certain cycles in the geometry expand fast in such a way that the associated KK scale is parametrically lighter than the string scale, clarifying the fate of the limit as a decompactification phase instead.

In principle, effective tensionless branes with more than one spatial extensions may also arise in the same vein, again as a solitonic object in the effective theory [56]. In many cases, however, their tensions do not lead to a lightest scale of the effective

[16] See also the previous analysis of [55].

theory. Furthermore, in other cases where classical geometry alone does allow for higher-dimensional branes of which tension determines the lightest scale, quantum corrections in the moduli space turn out to obstruct those classical limits [57], thereby supporting the Emergent String Conjecture. This is similar in spirit to how quantum corrections obstruct seemingly pathological limits where the KK scale is lighter than the string scale, as discussed in Sect. 2. It remains to be seen if and how analogous quantum effects may arise for infinite complex structure limits.

Admittedly, our analysis in this article has been strictly in the context of string theory. The appearance of light towers has nonetheless proven to be by far non-trivial and is thus suggestive of a potential effective-theoretic interpretation of the Emergent String Conjecture.[17] Progress along this line would even more strengthen our understanding of light towers at infinite distance and could possibly serve as strong evidence of string theory as *the* framework of quantum gravity.

Acknowledgements The author is grateful to Daniel Klaewer, Wolfgang Lerche, Guglielmo Lockhart, Timo Weigand and Max Wiesner for collaborations on the projects which this article is based upon.This work is supported in part by IBS under the project code, IBS-R018-D1.

References

1. C. Vafa, hep-th/0509212.
2. H. Ooguri and C. Vafa, *Nucl. Phys. B***766** (2007) 21–33, [hep-th/0605264].
3. T. D. Brennan, F. Carta, and C. Vafa, *PoS***TASI2017** (2017) 015, [arXiv:1711.00864].
4. E. Palti, *Fortsch. Phys.***67** (2019), no. 6 1900037, [arXiv:1903.06239].
5. M. van Beest, J. Calderón-Infante, D. Mirfendereski, and I. Valenzuela, arXiv:2102.01111.
6. M. Graña and A. Herráez, *Universe***7** (2021), no. 8 273, [arXiv:2107.00087].
7. S.-J. Lee, W. Lerche, and T. Weigand, arXiv:1910.01135.
8. C. Vafa, *Nucl. Phys. B***469** (1996) 403–418, [hep-th/9602022].
9. P. Corvilain, T. W. Grimm, and I. Valenzuela, *JHEP***08** (2019) 075, [arXiv:1812.07548].
10. F. Baume, F. Marchesano, and M. Wiesner, *JHEP***04** (2020) 174, [arXiv:1912.02218].
11. R. Blumenhagen, D. Kläwer, L. Schlechter, and F. Wolf, *JHEP***06** (2018) 052, [arXiv:1803.04989].
12. T. W. Grimm, E. Palti, and I. Valenzuela, *JHEP***08** (2018) 143, [arXiv:1802.08264].
13. A. Joshi and A. Klemm, *JHEP***08** (2019) 086, [arXiv:1903.00596].
14. T. W. Grimm, C. Li, and I. Valenzuela, *JHEP***06** (2020) 009, [arXiv:1910.09549]. [Erratum: JHEP 01, 007 (2021)].
15. N. Gendler and I. Valenzuela, *JHEP***01** (2021) 176, [arXiv:2004.10768].
16. T. W. Grimm, *JHEP***10** (2021) 153, [arXiv:2010.15838].
17. T. W. Grimm, J. Monnee, and D. van de Heisteeg, arXiv:2103.12746.
18. B. Bastian, T. W. Grimm, and D. van de Heisteeg, arXiv:2105.02232.
19. E. Palti, arXiv:2107.01539.
20. B. Bastian, T. W. Grimm, and D. van de Heisteeg, arXiv:2108.11962.
21. T. W. Grimm, E. Plauschinn, and D. van de Heisteeg, arXiv:2110.05511
22. F. Xu, *JHEP***10** (2020) 045, [arXiv:2006.02350].
23. S.-J. Lee, W. Lerche, and T. Weigand, *JHEP***10** (2018) 164, [arXiv:1808.05958].

[17] See e.g. [58, 59] for an EFT string point of view on infinite distance limits.

24. S.-J. Lee, W. Lerche, and T. Weigand, *Nucl. Phys. B*938 (2019) 321–350, [arXiv:1810.05169].
25. S.-J. Lee, W. Lerche, and T. Weigand, *JHEP*08 (2019) 104, [arXiv:1901.08065].
26. D. Klaewer, S.-J. Lee, T. Weigand, and M. Wiesner, *JHEP*03 (2021) 252, [arXiv:2011.00024].
27. N. Arkani-Hamed, L. Motl, A. Nicolis, and C. Vafa, *JHEP*06 (2007) 060, [hep-th/0601001].
28. C. Cheung and G. N. Remmen, *Phys. Rev. Lett.*113 (2014) 051601, [arXiv:1402.2287].
29. B. Heidenreich, M. Reece, and T. Rudelius, *JHEP*02 (2016) 140, [arXiv:1509.06374].
30. B. Heidenreich, M. Reece, and T. Rudelius, *JHEP*08 (2017) 025, [arXiv:1606.08437].
31. M. Montero, G. Shiu, and P. Soler, *JHEP*10 (2016) 159, [arXiv:1606.08438].
32. S.-J. Lee, W. Lerche, G. Lockhart, and T. Weigand, *JHEP*01 (2021) 162, [arXiv:2005.10837].
33. S.-J. Lee, W. Lerche, G. Lockhart, and T. Weigand, arXiv:2012.00766.
34. E. Palti, *JHEP*08 (2017) 034, [arXiv:1705.04328].
35. V. Kulikov, *Math. USSR Izv. 11*05 (1977) 957–989.
36. V. Kulikov, *Math. USSR Izv. 17*02 (1981) 339–342.
37. U. Persson and H. Pinkham, *Ann of Math. (2) 113*01 (1981) 45–66.
38. R. Friedman and D. Morrison, *Progr Math.*29 (1983) 1–32.
39. U. Persson, *Mem. Amer. Math. Soc. 11*189 (1977).
40. A. Clingher and J. W. Morgan, *Commun. Math. Phys.*254 (2005) 513–563, [math/0308106].
41. D. R. Morrison and C. Vafa, *Nucl. Phys.*B473 (1996) 74–92, [hep-th/9602114].
42. D. R. Morrison and C. Vafa, *Nucl. Phys.*B476 (1996) 437–469, [hep-th/9603161].
43. P. S. Aspinwall and D. R. Morrison, *Nucl. Phys. B*503 (1997) 533–564, [hep-th/9705104].
44. A. Sen, *Nucl. Phys. B*475 (1996) 562–578, [hep-th/9605150].
45. A. Clingher, R. Donagi, and M. Wijnholt, *Adv. Theor. Math. Phys.*18 (2014), no. 3 613–658, [arXiv:1212.4505].
46. S.-J. Lee and T. Weigand, arXiv:2112.07682.
47. V. Alexeev, A. Brunyate, and P. Engel, arXiv:2002.07127.
48. A. Brunyate, *PhD Thesis at University of Georgia* (2015).
49. Y. Odaka and Y. Oshima, arXiv:1810.07685.
50. Y. Odaka, arXiv:2010.00416.
51. K. Ascher and D. Bejleri, arXiv:1902.10686.
52. S.-J. Lee, W. Lerche, and T. Weigand, arXiv:2112.08385.
53. O. DeWolfe and B. Zwiebach, *Nucl. Phys. B*541 (1999) 509–565, [hep-th/9804210].
54. A. Font, B. Fraiman, M. Graña, C. A. Núñez, and H. P. De Freitas, *JHEP*10 (2020) 194, [arXiv:2007.10358].
55. F. A. Cachazo and C. Vafa, hep-th/0001029.
56. A. Font, A. Herráez, and L. E. Ibáñez, *JHEP*08 (2019) 044, [arXiv:1904.05379].
57. R. Álvarez-García, D. Kläwer, and T. Weigand, arXiv:2112.09136.
58. S. Lanza, F. Marchesano, L. Martucci, and I. Valenzuela, *JHEP*02 (2021) 006, [arXiv:2006.15154].
59. S. Lanza, F. Marchesano, L. Martucci, and I. Valenzuela, *JHEP*09 (2021) 197, [arXiv:2104.05726].

Cubic Action in Double Field Theory

Chen-Te Ma

Abstract We study Double Field Theory on a $2d$-dimensional doubled torus with $N_L + N_R = 2$, where N_L and N_R are the numbers of left- and right-moving oscillators. The massive states, $N_L \neq N_R$, provide the momentum and winding numbers simultaneously. The fields of Double Field Theory need to satisfy a constraint imposed on the string states. In the target space, we provide a unique constraint up to the cubic order compatible with the integration by part. We make a correspondence of fields between the massless and massive cases. We then use the gauge symmetry to build the action. For the quadratic order, the mass term at the order of $1/\alpha'$ appears when $N_L \neq N_R$. We can also introduce the additional interacting term to construct the gauge-invariant cubic action. Since the massive states do not follow a consistent truncation, a consistent theory possibly cannot appear from the states of $N_L + N_R = 2$. We show that the expectation is wrong up to the cubic order.

1 Introduction

String Theory on a d-dimensional torus has a T-duality group O(d, d; $\mathbb{Z}$). This duality exchanges the momentum number n_a and the winding number m_a, where $a = 1, 2, \ldots, d$. *Double Field Theory* introduces the coordinates (X^a and $\tilde{X}_a$ on a torus and the dual one to make T-duality *manifest* [1].

In String Theory, an on-shell state has to satisfy the *level matching* and the free string on-shell conditions. The level matching condition constrains the difference between the number of left- and right-moving oscillators (N_L and N_R). In Double Field Theory, the level matching condition gives the following constraint [1]:

$$\partial^a \tilde{\partial}_a f = -\frac{N_L - N_R}{\alpha'} f \equiv -\frac{\lambda}{2} f,$$

C.-T. Ma (✉)
Asia Pacific Center for Theoretical Physics, Pohang University of Science and Technology, Pohang, Gyeongsangbuk-do, South Korea
e-mail: yefgst@gmail.com

Y.-H. He et al. (eds.), *Nankai Symposium on Mathematical Dialogues*,
https://doi.org/10.1007/978-981-19-2328-9_29

$$\partial_a \equiv \frac{\partial}{\partial x^a}, \; \tilde{\partial}^a \equiv \frac{\partial}{\partial \tilde{x}_a}, \tag{1}$$

where f is a fluctuation field or a gauge parameter. The x^a ($\tilde{x}_a$) is the ordinary (dual) coordinate in the target space theory. It is necessary to impose this condition to f for obtaining a *gauge symmetry*. For the massless state ($N_L = N_R$), this condition is called a weak constraint. The simplest solution is the strong constraint (imposing the weak constraint on all products), which annihilates half of the degrees of freedom. The non-trivial feature in Double Field Theory is the interaction between the *target space* and *dual space*. The existence of massive state ($N_L \neq N_R$) relies on a solution *beyond* strong constraint. The solution following a strong constraint cannot generate a massive state. Hence we are interested in studying massive mode for exploring Double Field Theory.

In this conference proceeding, we show the cubic action in Double Field Theory with a state of $N_L + N_R = 2$ [2, 3]. We generalize the constraint to cubic products with $\lambda \neq 0$ [2, 3]. The generalization is unique by requiring compatibility with the integration by parts [2–5]. Because the product is commutative, the gauge transformation is closed [2, 3]. We observe the spin-2 Fierz-Pauli fields from the massive states $(N_L, N_R) = (2, 0); (0, 2)$ [2, 3].

2 Cubic Action

For $(N_L, N_R) = (1, 1)$, the fluctuation fields are: graviton field $h_{jk}(x^\mu, x^a, \tilde{x}_a)$; Kalb-Ramond field $b_{jk}(x^\mu, x^a, \tilde{x}_a)$; scalar dilaton field $\Phi(x^\mu, x^a, \tilde{x}_a)$, where $j, k = 1, 2, \ldots, D$. We label the non-compact directions by $\mu = 0, 1, \ldots, 25 - d$. Here D is the number of compact and non-compact directions. We consider the bosonic string. Therefore, the dimensions of a target space are 26. In the massive mode $(N_L, N_R) = (2, 0); (0, 2)$, we replace the Kalb-Ramond field by the field strength associated with A_j [2, 3]. We show the constraint and the cubic action [2, 3].

2.1 Constraint

The constraint is [1] $\partial_M \partial^M f = -\lambda f$, where $\partial_M \equiv \left(\tilde{\partial}^a \; \partial_a\right)^T$. We raise or lower the doubled indices by the O(d, d) metric [1]

$$\eta = \begin{pmatrix} 0 & I \\ I & 0 \end{pmatrix}. \tag{2}$$

The implementation of the constraint on the momentum space is convenient. The star product provides the convenient notation

$$A * 1 \equiv \sum_K A_K \exp(iKX)\delta_{KK,\lambda}, \tag{3}$$

where

$$X^M \equiv (\tilde{x}_a, x^a), \qquad K^M \equiv (p_a, \omega^a); \tag{4}$$

$$KK \equiv K^M K_M, \qquad KX \equiv K^M X_M. \tag{5}$$

We will assume that all doubled fields satisfy the constraints without writing the star product when showing the action. The compact coordinates and the duals are x^a and $\tilde{x}_a$, respectively. The momenta and windings along with the compact directions are p_a and ω^a, respectively. The x^a's and $\tilde{x}_a$'s are the compact coordinates and the duals, respectively. The p_a's and ω^a's are the momenta and windings along with the compact directions, respectively. The $*$-product imposes a constraint $\partial^M\partial_M(A*1) = -\lambda(A*1)$, only one required in showing the quadratic action [2, 3].

The integration by parts gives the constraints $K_A + K_B = 0$ and $K_A + K_B + K_C = 0$ to the quadratic and cubic terms, respectively [2–5]. Therefore, we obtain the commutative quadratic products $A * B = B * A$ and the associative triple products $A * (B * C) = (A * B) * C$ [2–5]. Hence the gauge transformation is closed without modifying it [2–5]. The non-vanishing λ still shows information beyond the strong constraint [2, 3].

2.2 Action

The massless cubic action is given by [1]

$$\begin{aligned}
&S^{(1)}_{DFT}\\
&= \frac{1}{16\pi G_N}\int [dxd\tilde{x}]\left[\frac{1}{4}e_{jk}\Box e^{jk} + \frac{1}{4}(\bar{D}^k e_{jk})^2 + \frac{1}{4}(D^j e_{jk})^2 - 2\Phi D^j\bar{D}^k e_{jk} - 4\Phi\Box\Phi\right.\\
&+\frac{1}{4}e_{jk}\Big((D^j e_{lm})(\bar{D}^k e^{lm}) - (D^j e_{lm})(\bar{D}^m e^{lk}) - (D^l e^{jm})(\bar{D}^k e_{lm})\Big)\\
&+\frac{1}{2}\Phi\Big((D^j e_{jk})^2 + (\bar{D}^k e_{jk})^2\\
&+\frac{1}{2}(D_l e_{jk})^2 + \frac{1}{2}(\bar{D}_l e_{jk})^2 + 2e^{jk}(D_j D^l e_{lk} + \bar{D}_k\bar{D}^l e_{jl})\Big)\\
&\left.+4e_{jk}\Phi D^j\bar{D}^k\Phi + 4\Phi^2\Box\Phi\right],
\end{aligned} \tag{6}$$

where $\Box \equiv (D^2 + \bar{D}^2)/2$, G_N is a gravitational constant, and e_{jk} is a non-linear combination of h_{jk} and b_{jk} [1]. The massive cubic action is given by [2, 3]

$$
\begin{aligned}
& S_{DFT}^{(2)} \\
= & \frac{\lambda}{16\pi G_N}\int [dxd\tilde{x}]\bigg[-\frac{\lambda}{4}e_{jk}e^{jk}+4\lambda\Phi^2 \\
& +\frac{1}{4}\Phi e^{lk}e_{lk}-4\Phi\Phi\Phi \\
& -\frac{1}{2}A^m e_{mk}D_l e^{lk}-2A_l e^{lk}\bar{D}_k\Phi+A^m\bar{D}_k(A_m)D_l(e^{lk})+2\lambda A_l A^l\Phi\bigg]. \qquad (7)
\end{aligned}
$$

The total action is the sum of $S_{DFT}^{(1)}$ and $S_{DFT}^{(2)}$. For $(N_L, N_R) = (1, 1)$, the massive cubic action vanishes. We replace the Kalb-Ramond field with the field strength (associated with a one-form field) for the massive states.

3 Outlook

When $N_L \neq N_R$, we showed a heavy mass, proportional to $1/\alpha'$. The experimental result of graviton mass term provides a lower bound to α'. Because one can use the observation of the graviton mass to constrain the scale of α', the result is non-trivial.

Acknowledgements The author acknowledges the YST Program of the APCTP and would like to thank Nan-Peng Ma for his encouragement.

The author thanks the Chern Institute of Mathematics, Nankai University.

Discussion during the workshop, "Mathematical Dialogues (Nankai Symposium on Physics, Geometry and Number Theory)", was helpful to complete this work.

References

1. C. Hull and B. Zwiebach, "Double Field Theory," JHEP **09**, 099 (2009) https://doi.org/10.1088/1126-6708/2009/09/099arXiv:0904.4664 [hep-th].
2. C. T. Ma and F. Pezzella, "More Stringy Effects in Target Space from Double Field Theory," JHEP **08**, 113 (2020) doi:https://doi.org/10.1007/JHEP08(2020)113arXiv:1909.00411 [hep-th].
3. C. T. Ma, "Cubic action in double field theory," Nucl. Phys. B **971**, 115528 (2021) doi:https://doi.org/10.1016/j.nuclphysb.2021.115528arXiv:2104.09414 [hep-th].
4. C. T. Ma and F. Pezzella, "Supergravity with Doubled Spacetime Structure," Phys. Rev. D **95**, no.6, 066016 (2017) doi:https://doi.org/10.1103/PhysRevD.95.066016arXiv:1611.03690 [hep-th].
5. C. T. Ma and F. Pezzella, "Geometric Low-Energy Effective Action in a Doubled Spacetime," Nucl. Phys. B **930**, 135–154 (2018) https://doi.org/10.1016/j.nuclphysb.2018.03.004arXiv:1706.03365 [hep-th].

Two Kind of Dark Energy Models for Current Cosmology Tensions: Finslerian and Back-Reaction Effect Forms

Xin-He Meng

Abstract In this presentation, we introduce two kind of dark energy models: 1. the Finslerian extension gravity framework for dark energy models. To reconcile the current tensions between high and low redshift observations, we perform the first constraints on the Finslerian cosmological models including the effective dark matter and dark energy components and find that all the four Finslerian models could alleviate effectively the Hubble constant (H_0) tension as well as the amplitude of the root-mean-square density fluctuations (σ_8) tension between the Planck measurements and the local Universe observations at the 68% confidence level. The addition of a massless sterile neutrino and a varying total mass of active neutrinos to the base Finslerian (additionally) two-parameter model, respectively, reduces the H_0 tension from 3.4σ to 1.9σ and alleviates the σ_8 tension better than the other three Finslerian models. Computing the Bayesian evidence, with respect to ΛCDM model, our analysis shows a *weaker* preference for the base Finslerian model and *moderate* preferences for its three one-parameter extensions. Based on the model-independent Gaussian Processes, we also propose a new linear relation which can describe the current redshift space distortions data, and by using the most stringent constraints provided so far, we have also obtained the limits on typical model parameters for three one-parameter extension models. 2. Back-reaction as an effective dark energy model for the H_o tension problem. The Universe accelerated expansion phenomenon may be regarded as an averaged effect caused by small scale inhomogeneities of the universe matter-energy distribution with the use of the spatial averaged approach of Buchert, in which two models are considered, one of which assumes that the back-reaction term as $\mathcal{Q}_{\mathcal{D}}$ and the averaged spatial Ricci scalar $\langle \mathcal{R} \rangle_{\mathcal{D}}$ obeys the scaling laws of the volume scale factor $a_{\mathcal{D}}$ at adequately late times, and another adopts the ansatz that the back-reaction term $\mathcal{Q}_{\mathcal{D}}$ is simply a constant in the recent universe. Thanks to the effective geometry introduced by Larena et al. in their previous work, we confront these two back-reaction models with latest type Ia supernova and Hubble parameter observations, coming out with the results that the constant back-reaction model is slightly favored over the other model, and the best fitting back-reaction term

X.-H. Meng (✉)
Department of Modern Physics, School of Physics, Nankai University, Tianjin, P.R. China
e-mail: xhm@nankai.edu.cn

Y.-H. He et al. (eds.), *Nankai Symposium on Mathematical Dialogues*,
https://doi.org/10.1007/978-981-19-2328-9_30

in the scaling back-reaction model behaves almost like a constant. Also, the numerical results show that the constant back-reaction model predicts a smaller expansion rate and decelerated expansion rate than the other model does at redshifts higher than about ($z >$)1, and the both back-reaction terms begin to accelerate the universe expansions at a redshift z around 0.5.

Keywords Cosmological model · Dark energy models · Finsler geometry · Cosmological back-reaction effects · H_o tension

1 Presentation

In the past over two decades, a variety of cosmic probes such as supernova Type Ia (SNe Ia) [1, 2], baryonic acoustic oscillations (BAO) [3] and cosmic microwave background (CMB) anisotropies [4, 5] have confirmed the standard cosmological paradigm, namely the so-called inflationary Λ-cold-dark-matter (ΛCDM) model. Although recent measurements of CMB anisotropies by the Planck satellite [6] have verified, once again, the high efficiency of the ΛCDM model in explaining the cosmological phenomena, there are still two main tensions between high and low redshift cosmic probes. On the one hand, the Hubble expansion rate derived indirectly from the latest Planck data [7] under the assumption of the ΛCDM model is lower than the direct local measurement from Riess et al. over the 3σ confidence level (C.L.) by using improved SNIa calibration techniques [8] (hereafter H_0 tension). On the other hand, the derived σ_8 parameter, the amplitude of the rms density fluctuations today in linear regime, from Planck data is, nonetheless, higher than the same quantity measured by several low redshift surveys including lensing, cluster counts and redshift space distortions (RSD) [7, 9, 10] (hereafter σ_8 tension).

Up to date, it is still unclear whether these tensions originate from unknown systematic uncertainties in different methods used for measurements, or possibly small deviations from the ΛCDM model at all, indicating new physics is needed. In order to alleviate one or both of the above discrepancies, several solutions have been proposed, including dynamical dark energy [11–13], unified dark fluid [14], dark radiation [15, 16], modified gravities [17, 18], extended parameter space [19, 20], cosmic voids [21], decaying dark matter [22–24], ultralight axions [25], and so forth. Since the realistic nature of dark sector at the present time is still unknown, it is interesting and reasonable to give a possible explanation by exploring the connection between gravitation and new geometry. Modified Einstein's gravity may throw new light on the discrepancies.

Finsler geometry [26–28], which involves Riemann geometry as its special case, opens naturally a new prospect to understand the current cosmological puzzles. This new geometry keeps the elegant properties of Riemann geometry, i.e., the isometric group is a Lie group on a Finslerian manifold, while it admits less Killing vectors than a Riemannian spacetime does. In general, there are at most $n(n-1)/2+1$

independent Killing vectors in a n dimensional non-Riemannian Finslerian spacetime. Considering the simplest possible asymmetrical generalization of Riemannian metric, Randers [29] proposed the well-known Randers space, a subclass of Finslerian space. In the framework of Randers space, a generalized Friedmann-Robertson-Walker (FRW) cosmological model based on Randers-Finsler geometry has been discussed [30] and a modified dispersion relation of free particles has also been studied [31].

In history, the gravitational aspects of a Finslerian space were investigated for a long time [32–35]. The gravitational field equations (GFEs) derived from a Riemannian osculating metric were exhibited in Ref. [36]. For such a metric, the FRW-like cosmological scenario and the anisotropies of the Universe were also studied [30, 37]. Nonetheless, their GFEs are derived without considering the Bianchi identity and the general covariance principle of Einstein's gravity. It is interesting that, in Refs. [38–40], the authors have overcome these difficulties and derived the corresponding GFEs by constructing a Randers-Finsler space of approximate Berwald type (hereafter Berwald space), which is just an extension of a Riemannian space.

In this talk, we focus on exploring whether the cosmic evolutional model derived from the Berwald space may reconcile the current H_0 and σ_8 tensions between high and low redshift observations. Our main result is that with the observational datasets the Finslerian cosmological models considered in this analysis could alleviate effectively H_0 and σ_8 tensions in some instances.

Model.—For a Finsler-Berwald FRW Universe, the square of the effective dimensionless Hubble parameter $E(z)$ [38] is expressed as

$$\begin{aligned} E^2(z) =& \frac{9\alpha+\beta+12}{4(\alpha+1)(\beta+3)}\Omega_m(1+z)^{\frac{6(3\alpha+\beta+6)}{9\alpha+\beta+12}} \\ &+ \frac{9(1-\omega)\alpha+(1+3\omega)\beta+12}{4(\alpha+1)(\beta+3)}\Omega_{de} \\ &\times (1+z)^{\frac{6[3(1-\omega)\alpha+(1+3\omega)\beta+6(1+\omega)]}{9(1-\omega)\alpha+(1+3\omega)\beta+12}}, \end{aligned} \tag{1}$$

where z, ω, Ω_m, Ω_{de}, α and β denote the redshift, equation of state (EoS) of dark energy, effective matter and dark energy density parameters today, and effective time-component and space-component parameters, respectively. Here "*effective*" means that the physical quantities are derived based on non-Riemannian Berwald space. Note that this model reduces to ΛCDM when $\omega = -1$ and $\alpha = \beta = 0$. Since we have much interest in studying the evolution of the late Universe, we ignore the contribution from the radiation component. Due to the fact that the spatial curvature of our Universe is very close to zero [6], for simplicity, we also ignore the contribution of the effective "*curvature*" and just consider a flat Finsler-Berwald FRW Universe (see also Ref. [38]).

To investigate how the Finslerian universe reconcile the global with local measurements, we take five cosmological models into account. The base ΛCDM model with the usual set of cosmological parameters: $\{\Omega_b h^2, \Omega_c h^2, 100\theta_{MC}, \tau, \ln(10^{10}A_s), n_s\}$, where $\Omega_b h^2$ and $\Omega_c h^2$ are, respectively, the baryon and CDM densities today, θ_{MC}

is the ratio between sound horizon and angular diameter distance at the decoupling epoch, τ is the Thomson scattering optical depth due to reionization, $\ln(10^{10}A_s)$ and n_s are the amplitude and spectral index of primordial scalar perturbation power spectrum at the pivot scale $K_0 = 0.05$ Mpc^{-1}, respectively. Here $h \equiv H_0/100$ and H_0 is the Hubble constant. The simplest Finslerian cosmological model is a two-parameter extension of ΛCDM which corresponds to the case $\omega = -1$ in Eq. (1). Hereafter we call it FΛ model, which is the most important model considered in this analysis. In order to test overall the abilities of different Finslerian models to relieve tensions and verify the stabilities of our results with respect to (w.r.t.) assumptions on Finsler geometry, we also consider three one-parameter extensions to the FΛ model: (i) The Finslerian ωCDM (Fω) model with constant EoS of dark energy; (ii) The Finslerian active neutrino (Fν) model which allows the summed mass of three species of neutrinos Σm_ν to vary with a degenerate hierarchy; (iii) The Finslerian sterile neutrino (Fsν) model allowing the effective number of relativistic species N_{eff} to vary but $\Sigma m_\nu = 0.06$ eV fixed.

To perform constraints on the model parameters, we have modified carefully the publicly Markov Chain Monte Carlo code CosmoMC [41, 42] and Boltzmann code CAMB [43]. Meanwhile, we choose flat priors to the above model parameters and marginalize the foreground nuisance parameters provided by Planck. We use the temperature and polarization measurements from Planck (P) [6], RSD (R) data from the latest "Gold-2017" compilation in Ref. [44] and "other" (O) data sets such as SNIa data from the JLA compilation [45], BAO data [46–49], cosmic chronometer data from Ref. [50], HST data from Ref. [8], weak lensing data from CFHTLenS survey [51] and CMB lensing data [52]. For simplicity, we denote different data combinations as P, PR and PRO, respectively. A comprehensive analysis about the impacts of different data sets on the Finslerian scenarios are exhibited in a follow-up study [53]. We find that the models may relieve both the H_o and S_8 tensions to some extents with datasets fittings and constraints.

Back-reaction Models. Though the above Finslerian extended framework can relieve the tensions, the additional parameters introduced within the mathematical structure make the original and relatively concise six-parameter base LCDM cosmology seems too freely flexible. An alternative has been considered to explain the dark energy phenomenon as an averaged effect caused by small scale inhomogeneities of the universe, as in references therein [54, 55].

In order to consider cosmology model without assuming a FLRW background, it is necessary to answer a longstanding question that how to average a general inhomogeneous model. To date the macroscopic gravity (MG) approach is probably the most well known attempt at averaging in space-time. Although it is the only approach that gives a prescription for the correlation functions which emerge in an averaging of the Einstein's field equations, so far it required a number of assumptions about the correlation functions which make the theory less convictive. Therefore, in this presentation we adopt another averaged approach which is put forward by Buchert, as see references therein [55], in despite of its foliation dependent nature, such approach is quite simple and hence becomes the most well studied theoretical framework of averaged models. Since the averaged field equations in such approach do not form a closed set, one needs to make some assumptions about the back-reaction term appearing in the averaged equations. By taking the assumption that the back-reaction term $\mathcal{Q}_\mathcal{D}$ and the averaged spatial Ricci scalar $\langle \mathcal{R} \rangle_\mathcal{D}$ obey the scaling laws of the volume scale factor $a_\mathcal{D}$, Buchert has proposed a simple back-reaction model. To confront such model with observations, Larena et. al. present the effective geometry with the introduction of a template metric that is only compatible with homogeneity and isotropy on large scales of FLRW cosmology instead of on all scales. As was pointed out by Larena et. al., the scaling solution cannot be expected to fully represent the realistic back-reaction effect throughout the whole history of the universe since we expect that the realistic back-reaction term will change considerably at redshift $z_\mathcal{D} \sim 10$. However, since we only use the datasets of type Ia supernova and observational Hubble parameter in this paper, we merely concern the behavior of the back-reaction term at adequately late times, i.e. $z_\mathcal{D} \lesssim O(1)$, which means that although we assume $\mathcal{Q}_\mathcal{D}$ obeys scaling laws of $a_\mathcal{D}$ in such redshift range , it can behave very differently at higher redshifts, particularly, such term encounters rapid change when $z_\mathcal{D} \sim 10$ because of the structure formation effects, and becomes negligible when $z_\mathcal{D} \gtrsim 1000$, which is reasonable because of the consistence between perturbation theory predictions and CMB observations. Nevertheless, we still doubt that the scaling solution is a prime description of the late-time back-reaction term, so we propose another parameterization of $\mathcal{Q}_\mathcal{D}$ by simply setting it as a constant at late times, and it turn out that such model is preferred by observations. We find the models are favored by the current observational datasets.

The spatial averaged approach of Buchert is demonstrated with presentation of the averaged equations for the volume scale factor $a_\mathcal{D}$, with that we introduce the template metric, which is a necessary tool to test the theoretical predictions with observations, and computation of observables to astrophysics datasets. To compare the back-reaction models with observational datasets, we apply a simple likelihood analysis for the proposed two back-reaction models by confronting them with latest type Ia supernova and Hubble parameter observations datasets, and we find that the back-reaction models can relieve the H_o tensions for various parameter constraints. We should say the tensions are still open questions and may call for new (astro-)physics.

Acknowledgements The work is partially supported by the Natural Science Foundation of China and other various grants over past years. The contents presented here are mainly from collaborating works with my previous two Ph.D students for several years efforts. We thank the meeting organizers for a wonderful dialogue between Math and Physics. Also we would like to dedicate this presentation to memorize the late Prof. S. S. Chern.

References

1. A. G. Riess et al. (Supernova Search Team), Astron. J. **116**, 1009 (1998).
2. S. Perlmutter, et al. (Supernova Cosmology Project), Phys. Rev. Lett. **83**, 670 (1999).
3. D. H. Weinberg et al., Phys. Rep. **530**, 87 (2013).
4. C. L. Bennett et al. (WMAP Collaboration), Astrophys. J. Suppl. Ser. **208**, 20 (2013).
5. P. Ade et al. (Planck Collaboration), Astron. Astrophys. **571**, A16 (2014).
6. P. Ade et al. (Planck Collaboration), Astron. Astrophys. **594**, A14 (2016).
7. N. Aghanim et al. (Planck Collaboration), A & A, **596**, A107 (2016).
8. A. G. Riess et al., Astrophys. J. **826**, 1 (2016).
9. R. Battye, T. Charnock, and A. Moss, Phys. Rev. D **91**, 103508 (2015).
10. E. Macaulay, I. K. Wehus, and H. K. Eriksen, Phys. Rev. Lett. **111**, 161301 (2013).
11. S. Kumar and R. Nunes, Phys. Rev. D **94**, 123511 (2016).
12. M. Kunz, S. Nesseris, and I. Sawicki, Phys. Rev. D **92**, 063006 (2015).
13. E. Di Valentino et al., [arXiv: 1704.00762v1].
14. Deng Wang, Yang-Jie Yan, and Xin-He Meng, as in Eur. Phys. J. C77(2017)10,660, [arXiv: 1701.03362v1]; 2103.14788.
15. Mark Wyman et al., Phys. Rev. Lett. **112**, 051302 (2014).
16. B. Leistedt, H. V. Peiris, and L. Verde, Phys. Rev. Lett. **113**, 041301 (2014).
17. E. Di Valentino, A. Melchiorri, and J. Silk, Phys. Rev. D **93**, 023513 (2016).
18. A. De Felice and S. Mukohyama, Phys. Rev. Lett. **118**, 091104 (2017).
19. E. Di Valentino, A. Melchiorri, and J. Silk, Phys. Rev. D **92**, 121302 (2015).
20. E. Di Valentino, A. Melchiorri, and J. Silk, Phys. Lett. B **761**, 242 (2016).
21. K. Ichiki, C. M. Yoo, and M. Oguri, Phys. Rev. D 93, 023529 (2016).
22. K. Enqvist et al., J. Cosmol. Astropart. Phys. **09** (2015) 067.
23. Z. Berezhiani, A. D. Dolgov, and I. I. Tkachev, Phys. Rev. D **92**, 061303 (2015).
24. Peng Wang, Xin-He Meng, Class. Quant. Grav. **22,** 283-294 (2005).
25. R. Hlozek, D. Grin, D. J. E. Marsh, and P. G. Ferreira, Phys. Rev. D **91**, 103512 (2015).
26. D. Bao, S. S. Chern, and Z. Shen, *An Introduction to Riemann-Finsler Geometry*, Graduate Texts in Mathmatics **200**, Springer, New York, 2000.
27. S. S. Chern, *Finsler geometry is just Riemann geometry without the quadratic restrictions*, Notice of AMS, 959 (1996).
28. S. S. Chern, Sci. Rep. Nat. Tsing Hua Univ. Ser. **A 5**, 95 (1948); or Selected Papers, vol. **II**, **194**, Springer, 1989.
29. G. Randers, Phys. Rev. **59**, 195 (1941).
30. P. C. Stavrinos, A. P. Kouretsis, and M. Stathakopoulos, Gen. Relativ. Gravit. **40**, 1403 (2008).
31. Z. Chang and X. Li, Phys. Lett. B **663**, 103 (2008).
32. Y. Takano, Lett. Nuovo Cimento **10**, 747 (1974).
33. R. K. Tavakol and N. Van Den Bergh, Phys. Lett. A **112**, 23 (1985).
34. S. Ikeda, Ann. Phys. **44**, 558 (1987).
35. G. Yu. Bogoslovsky, Phys. Part. Nucl. **24**, 354 (1993).
36. G. Asanov, *Finsler Geometry, Relativity and Gauge Theories*, Reidel Pub.Com., Dordrecht, 1985.
37. A. P. Kouretsis, M. Stathakopoulos, and P. C. Stavrinos, Phys. Rev. D **79**, 104011, (2009).
38. Z. Chang, M. Li, and X. Li, Chinese Phys. C **36**, 710 (2012).
39. Z. Chang and X. Li, Phys. Lett. B **668**, 453 (2008).
40. X. Li and Z. Chang, Phys. Rev. D **90**, 064049 (2014).
41. A. Lewis and S. Bridle, Phys. Rev. D **66**, 103511 (2002).
42. A. Lewis, Phys. Rev. D **87**, 103529 (2013).
43. A. Lewis, A. Challinor, and A. Lasenby, Astrophys. J. **538**, 473 (2000).
44. S. Nesseris, G. Pantazis, and L. Perivolaropoulos, [arXiv: 1703.10538v1].
45. M. Betoule et al. (SDSS collaboration), Astron. Astrophys. **568**, A22 (2014).
46. F. Beutler et al., Mon. Not. R. Astron. Soc. **423**, 3430 (2012).

47. C. Blake et al., Mon. Not. R. Astron. Soc. **415**, 2876 (2011).
48. W. J. Percival et al. (SDSS Collaboration), Mon. Not. R. Astron. Soc. **401**, 2148 (2010).
49. L. Anderson et al., Mon. Not. R. Astron. Soc. **439**, 83 (2014).
50. M. Moresco et al., J. Cosmol. Astropart. Phys. **05**, (2016) 014.
51. C. Heymans et al., Mon. Not. Roy. Astron. Soc. **432**, 2433 (2013).
52. P. Ade et al. (Planck Collaboration), Astron. Astrophys. **594**, A15 (2016).
53. Deng Wang and Xin-He Meng, as in 1709.04141; ibid, PRD 96(2017)2,023538;ibid, DPU,21(2018)55 [2103.14790];ibid, EPJC 77(2017)11,725.
54. Y H Yao and X H Meng, Mod.Phys. Lett.A **34**(2019)21,1950167; ibid, PDU 27(2020)100462; ibid,30(2020)100729; ibid,33(2021)100852[2011.09160].
55. Y H Yao and X H Meng,CMP 71(2019)9,1109[1707.00111]

Confined Vortex Surfaces and Turbulent Statistics. Talk at Nankai Symposium, August 2021

Alexander Migdal

1 Brief Summary

We offer a new approach to turbulence that one can study analytically and numerically.

We are solving the Navier-Stokes equations for an infinite time evolution, given that the laminar flow is unstable. It is implied that energy is pumped in through remote boundary to compensate the energy dissipation taking place inside the flow on vorticity structures.

This unstable solution could cover the whole phase space of the fluid Hamiltonian mechanics; however, we conjecture that there are some low-dimensional attractors in that space. The solution would cover the subspace of these attractors like the Newtonian dynamics covers the energy surface in ergodic motion.

We also know the phenomenon of anomalous dissipation: the dissipation in extreme turbulence occurs in the regions of high vorticity, which compensates for the viscosity factor in front of the enstrophy integral.

Furthermore, we conjecture that these regions of high vorticity are vortex surfaces. There are several reasons to believe so, both theoretical and (numerical or real) experimental sides.

The Hamiltonian dynamics of vortex surfaces [1, 2] can be solved exactly in some important cases, and the resulting solutions have all nice properties, including anomalous dissipation.

We have found in our previous work [3, 4] that the steady-state of the vortex surface is a promising candidate for such an attractor in phase space. At first sight, it appears that these steady vortex surfaces have some local degrees of freedom,

A. Migdal (✉)
Abu Dhabi Investment Authority, Abu Dhabi, United Arab Emirates
e-mail: sasha.migdal@gmail.com

Department of Physics, New York University Abu Dhabi, Saadiyat Island, Abu Dhabi, United Arab Emirates

Y.-H. He et al. (eds.), *Nankai Symposium on Mathematical Dialogues*,
https://doi.org/10.1007/978-981-19-2328-9_31

corresponding to the arbitrary shape of the surface, as long as the gap $\Gamma = \Delta\Phi$ in velocity potential $\vec{v} = \vec{\nabla}\Phi$ satisfies certain integral equations, following from the minimization of the Hamiltonian. This minimization condition was shown to be equivalent to the vanishing normal velocity $\vec{\sigma} \cdot \vec{v} = 0$ at both sides of the vortex surface.

In particular, we suggested in [4] that the tails of PDF for the circulation are related to the instanton (classical solution of Euler equations) with certain topology of so called Alice string in string theory. Such vortex sheets were in fact observed in liquid helium [5] which is a particular case of zero viscosity incompressible fluid.

Using instanton solution we predicted the exponential distribution with pre-exponential power factor, perfectly matching observed PDF in DNS [6].

At that time, we speculated that these surface degrees of freedom were described by some version of a solvable string theory [4]. There are two kinds of surface degrees of freedom: the surface shape and the internal metric on this surface, described by the Liouville field.

In fluid dynamics terms, the fluctuations of this internal metric correspond to the tangent motions of the fluid, equivalent to the reparametrization of this surface.

The shape of the surface was assumed frozen in the turbulent flow in [4], and the remaining internal metric was assumed to be responsible for the multifractal scaling laws for the moments of velocity circulation.

The new recent understanding eliminated the shape of the surface as degrees of freedom while placing no restrictions on the internal metric on this surface.

Some surfaces are "more equal" than others because they are stable at the microscopic level. This stability follows from new exact solution [7, 8] of the Navier-Stokes equation for an infinite plane in a presence of generic background strain.

Abstracting from this solution, we conjectured in [9] that the local stability condition of the vortex surface with the local normal vector $\vec{\sigma}$ and local mean boundary value of the strain tensor $\hat{S}$ reduces to three equations (with $\vec{v}_{\pm}$ being the boundary values of velocity on each side)

$$\vec{v}_{\pm} \cdot \vec{\sigma} = 0; \tag{1}$$

$$\hat{S} \cdot (\vec{v}_{+} - \vec{v}_{-}) = 0; \tag{2}$$

$$S_{nn} = \vec{\sigma} \cdot \hat{S} \cdot \vec{\sigma} < 0 \tag{3}$$

We call these equations the Confined Vortex Surface or *CVS* equations. They are supposed to hold at every point of the vortex surface, thus imposing extra boundary conditions on the Euler equations.

The strain being the time-odd variable, the third stability condition provides the dynamical mechanism of spontaneous irreversibility of the Euler dynamics.

As the local strain tensor for the fixed vortex surface is uniquely determined by conventional Neumann boundary conditions for potential flow on each side, the second *CVS* boundary condition restricts allowed shapes of vortex surfaces.

It is truly remarkable that such a simple equation, involving just three letters, leads to so many specific consequences for the turbulent flow.

The negative normal strain presses vorticity into the *CVS* surface from both sides, leading to a narrow Gaussian profile of vorticity in the normal direction. The width of this boundary layer goes to zero as some negative power of Reynolds number in the turbulent limit.

The spontaneous stochasticity and the corresponding the turbulent statistics were studied in the next paper [10]. We assume that our flow and the surface are subject to a background strain, a random traceless matrix adding up from Biot-Savart tails of many other vortex structures located far away from this one. Such random strain is described by eigenvalues $a \leq b \leq c$ adding up to zero. The parameters a, b, c obey a Gaussian distribution multiplied by a famous Vandermonde determinant.

$$P_\sigma(a, b, c) = \sqrt{\frac{3}{\pi}}\theta(b-a)\theta(c-b)(b-a)(c-a)(c-b)$$
$$\exp -\frac{a^2+b^2+c^2}{2\sigma^2} \tag{4}$$

The normal strain is uniform all around our vortex surface $S_{nn} = -c$, so that we have one of the stability criteria fulfilled if the axis of the vortex tube points along the main eigenvector of the strain, with a positive eigenvalue c. The tangent strain is degenerate: there is zero strain along the tangent velocity gap.

The energy dissipation occurs at the surface, and we have demonstrated [9] that this surface integral is conserved in the Navier-Stokes dynamics in the turbulent limit.

$$\mathcal{E} = \frac{\sqrt{\nu}}{2\sqrt{\pi}} \sum_S \int_S dS \sqrt{-\hat{S}_{nn}(\Delta\vec{v})^2}; \tag{5}$$

where $\int dS$ refers to the surface integral and $\hat{S}_{nn}$ is the normal component of the local strain.

As we see it now, $\mathcal{E}$ is conserved because of the *CVS* stability of the vortex surfaces hidden behind the rough picture of the energy flow in Fourier space. One cannot describe the shape of the discontinuity surface by a finite (or even convergent) set of Fourier coefficients of the velocity field.

We describe the turbulence phenomenon as a fluctuating geometry instead of fluctuating velocity field. In the modern QFT, such a pair of complementary descriptions represents Duality.

In the ADS-CFT theory [11] the conformal field theory (fluctuating vector gauge field) is dual to the fluctuating geometry of curved space. The strong coupling of the conformal field theory corresponds to the (solvable) weak coupling limit of the fluctuating geometry.

A more direct analogy is the description of the Burgers' one-dimensional turbulence of compressible fluid as a statistics of shocks- one-dimensional analogs of our vortex surfaces. This shock statistics is also a dual theory of geometric objects, weakly fluctuating in the extreme turbulent limit when they reduce to step functions of the coordinate.

In our theory, the energy flows from large scales to small scales. It dissipates at small scales in thin boundary layers of the *CVS* surface, but this spatial picture is different from the hierarchical cascade of energy "from a big eddy to a smaller one."

The only way to reconcile these two pictures is to assume the Russian doll of nested tubes, passing the energy from the outer tube to the inner ones, with each tube dissipating energy as discussed above. Someone still has to find such a solution of the *CVS* equations.

The conservation of energy dissipation on the CVS surface makes this parameter $\mathcal{E}$ an appropriate scaling factor for the K41 theory viewed as dimensional analysis, which describes the medium Reynolds numbers so well.

We hope that there exist full 3D solutions to the *CVS* equations, with the finite area and finite volume inside. If found, such solutions would be the best candidates for turbulent statistics. Lacking that, we can analyze the cylindrical solutions as some approximation to reality or proof of concept.

Let us come back to our cylindrical solution. The cross-section of the tube has four cusps and is described by four branches of an algebraic hyperbola

$$|x||y|^{\mu} = \text{const}; \tag{6}$$

$$\mu = 1 - \frac{c}{a} \tag{7}$$

(see Fig. 1). This solution describes a vortex tube much larger than the viscous scale. All vorticity collapses to the boundary layer; this layer's width will decrease as a positive viscosity power in the turbulent limit.

The flow is, therefore, potential both inside and outside. The normal velocity vanishes on both sides of the surface of the tube, but the tangent velocity is finite, and there is a tangent gap.

Hollow vortex tubes were observed in the recent DNS [12], but it is hard to tell until some research would specifically hunt for hollow tubes in turbulent flow simulations.

The uniform strain tensor, distributed as a Gaussian random traceless matrix, is forcing our flow.

The origin of this random force is the accumulation of a large number of $1/R^3$ tails of the Biot-Savart laws for other remote vortex tubes with random locations and random orientations.

The stability condition $S_{nn} < 0$, coming from the microscopic analysis of the Navier-Stokes equation in a boundary layer, fixes the sign of the flow.

The same requirement follows from the Kelvin-Helmholtz stability of the vortex surface against the high wave-length perturbations. The low wave-length perturbations of the sharp vortex surface are always unstable, but presumably, viscous diffusion stabilizes those perturbations as it happens in the Burgers-Townsend solution.

Fig. 1 The stream plot in xy plane for $\mu = 1.7$. The green fluid is inside, the clear fluid is outside the vortex surface (red)

This stability condition is the origin of the flow's irreversibility. We choose the parameters so that the energy dissipation (and equal energy pumping) stays finite in the limit of vanishing viscosity.

We considered these closed vortex surfaces (vortex tubes) as a dilute gas, with a large mean distance between tubes, compared to their size.

In this dilute gas approximation, we derived the Gaussian random strain from an infinite number of remote vortex structures, adding to the Biot-Savart formula for the strain at large distances and averaging to the random constant stress tensor due to the Central Limit theorem.

This calculation leads to a self-consistency relation between the mean distance $\bar{R}$ between vortex tubes, and the variance σ of the random background strain:

$$\langle \mathcal{E} \rangle \propto \bar{R}^4 \sqrt{\nu \sigma^5}; \tag{8}$$

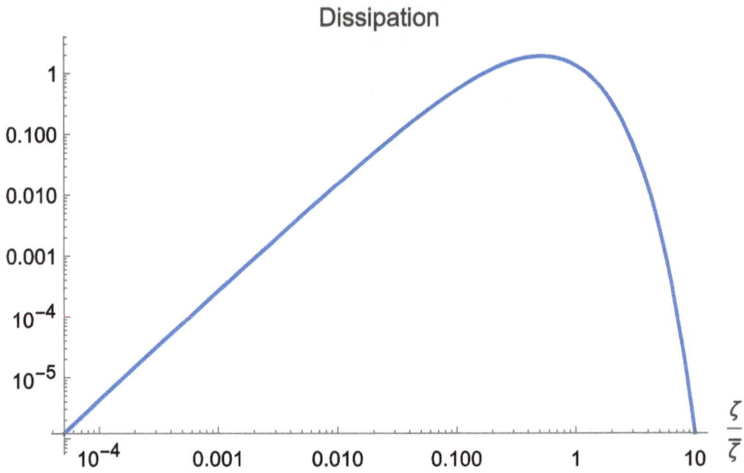

Fig. 2 The energy dissipation PDF (fixed perimeter) in log-log scale

The statistical distribution of the shape index μ, following from the Gaussian distribution of the background strain, offers the microscopic explanation for the "multifractal" scaling laws for the moments of velocity difference.

The moments of velocity difference are a more complex function of n, $|\Delta r|$ than just $|\Delta r|^{\zeta(n)}$, which could still be approximated by the power laws at specific integer values of n.

Using the dilute gas approximation, we computed the PDF for the energy dissipation normalized by the third power of the perimeter of the loop. It is plotted in log-log scale at Fig. 2. This curve is universal in our solution; there are no model approximations nor any dimensionless parameters.

In the same way, one can compute the distribution of the circulation and other observables, such as moments of velocity differences $\langle |v(r) - v(0)|^n \rangle$ or the Wilson loop $\left\langle \exp\left(\imath\gamma \oint_C v_\alpha dr_\alpha\right)\right\rangle$.

Since 2021, when this paper was submitted, the theory of turbulence has evolved significantly. In my recent papers (2023–2025), I have developed a microscopic theory of decaying turbulence based on the loop equation for the distribution of velocity circulation around an arbitrary closed loop as a functional of the loop's shape. This theory has received rigorous mathematical support (though not yet a complete proof) in the work of Camillo De Lellis and Elia Bruè. Moreover, the theory matches both classic and new experiments in pipes and wind tunnels, as well as recent large-scale numerical simulations of the Navier–Stokes equation by K. R. Sreenivasan and A. Rodhiya. These results are reviewed in my invited paper forthcoming in the "Philosophical Transactions of the Royal Society A".

Acknowledgements This work is supported by a Simons Foundation award ID 686282 at NYU. I am so grateful to Jim for his constant support and inspiration.

I am grateful to Ed Witten, Peter Sarnak, Thomas Spenser, Camillo De Lellis and other participants of the seminars at the Institute for Advance Study in Princeton for stimulating discussions.

Numerous talks at the Einstein Chair Seminar in Stony Brooks University with vigorous discussions with Dennis Sullivan and Theo Drivas also helped me develop and debug this theory.

The participants of the Max Planck Institute seminar also helped me with their questions and comments.

Finally, my friends Karim Shariff, Sasha Polyakov, Pasha Wiegman, Grisha Volovik, Nikita Nekrasov, Grisha Falkovich, Victor Yakhot and Ketapali Sreenivasan discussed my work with me several times, helping me understand the basic concepts and establish connections with conventional models of Turbulence.

References

1. A. A. Migdal, Random surfaces and turbulence, in *Proceedings of the International Workshop on Plasma Theory and Nonlinear and Turbulent Processes in Physics, Kiev, April 1987*, ed. V. G. Bar'yakhtar (World Scientific, 1988), p. 460.
2. M. E. Agishtein and A. A. Migdal, *Physica D: Nonlinear Phenomena*40, 91 (1989), https://doi.org/10.1016/0167-2789(89)90029-8.
3. A. Migdal, *International Journal of Modern Physics A*35, 2030018 (2020), arXiv:2007.12468v7 [hep-th], https://doi.org/10.1142/s0217751x20300185.
4. A. Migdal, *International Journal of Modern Physics A*36, 2150062 (2021),https://doi.org/10.1142/S0217751X21500627, https://doi.org/10.1142/S0217751X21500627.
5. G. E. Volovik, *UFN*185, 970 (09 2015), https://doi.org/10.3367/UFNr.0185.201509h.0970.
6. K. P. Iyer, K. R. Sreenivasan and P. K. Yeung, *Phys. Rev. X*9, 041006 (2019), https://doi.org/10.1103/PhysRevX.9.041006.
7. A. Migdal, *Physics of Fluids*33, 035127 (2021), https://doi.org/10.1063/5.0044724, https://doi.org/10.1063/5.0044724.
8. K. Shariff and G. E. Elsinga, *Physics of Fluids*33, 033611 (2021), https://doi.org/10.1063/5.0045243, https://doi.org/10.1063/5.0045243.
9. A. Migdal, Confined vortex surface and irreversibility. 1. properties of exact solution (2021), arXiv:2103.02065v10 [physics.flu-dyn].
10. A. Migdal, Confined vortex surface and irreversibility. 2. exact solutions and turbulent statistics (2021), arXiv:2105.12719v5 [physics.flu-dyn].
11. Wikipedia, AdS-CFT correspondence — Wikipedia, the free encyclopedia https://en.wikipedia.org/wiki/AdS/CFT_correspondence, (2019), [Online; accessed 28-October-2021].
12. D. Buaria, A. Pumir, E. Bodenschatz and P. K. Yeung, *New Journal of Physics*21, 043004 (2019), https://doi.org/10.1088/1367-2630/ab0756.

From Chern-Tenenblat to Jackiw-Teitelboim via Sine-Gordon

Jeff Murugan

Abstract The study of 2-dimensional surfaces of constant curvature constitutes a beautiful branch of geometry with well-documented ties to the mathematical physics of integrable systems. A lesser known, but equally fascinating, fact is its connection to 2-dimensional gravity; specifically Jackiw-Teitelboim (JT) gravity, where the connection manifests through a coordinate choice that roughly speaking re-casts the gravitational field equations as the sine-Gordon equation. In this language many well-known results, such as the JT-gravity black hole and its properties, were understood in terms of sine-Gordon solitons and their properties. In this talk, I will revisit some of these old ideas in the context of some of the recent exciting developments in JT-gravity and, more generally, low-dimensional quantum gravity and speculate on how some of these new ideas may be similarly understood.

1 Introduction

The marriage between geometry and physics has been a particularly fruitful one, and the name of S.S. Chern is instantly recognised as one of the principle architects at this interface. While his development of the theory of characteristic classes is familiar to most graduate students of theoretical physics by its application to topological quantum field theory, his contributions to nonlinear analysis are perhaps less so. This line of work is best exemplified in the seminal 1986 work of Chern and Tenenblat [1], in which they introduced and studied in painstaking detail, a class of partial differential equations describing pseudospherical surfaces of the type pictured in Fig. 1. To translate between these two descriptions, Chern and Tenenblat noted that the PDE

$$F\left(t, x, \phi, \frac{\partial\phi}{\partial t}, \frac{\partial\phi}{\partial x}, \ldots, \frac{\partial^l \partial^{k-l}\phi}{\partial t^l \partial x^{k-l}}\right) = 0\,, \tag{1}$$

J. Murugan (✉)
The Laboratory for Quantum Gravity and Strings, Department of Mathematics and Applied Mathematics, University of Cape Town, Cape Town, South Africa
e-mail: jeff@nassp.uct.ac.za

Y.-H. He et al. (eds.), *Nankai Symposium on Mathematical Dialogues*,
https://doi.org/10.1007/978-981-19-2328-9_32

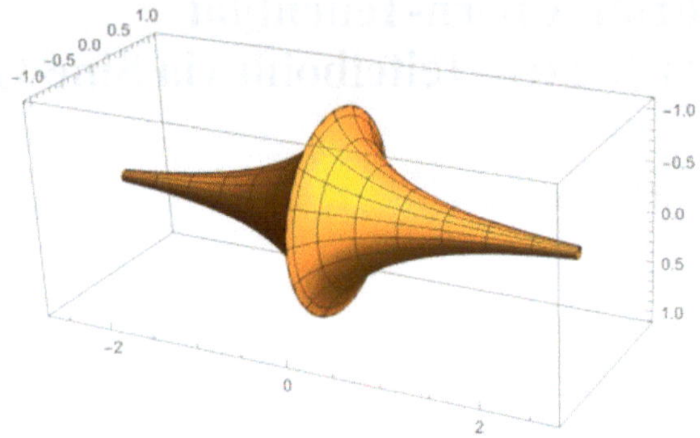

Fig. 1 A pseudospherical surface defined parametrically through $(x, y, z) = (\text{sech}(u)\cos(v), \text{sech}(u)\sin(v), u - \tanh(u))$

describes a pseudospherical surface $\mathscr{M}$ if 1-forms $\omega_i = f_{i1}\,\text{dx} + f_{i2}\,\text{dt}$ can be found, with $i = 1, 2, 3$ and smooth function $f_{ij}\,(t, x, \phi, \partial_t\phi, \partial_x\phi, \ldots)$, such that the associated *structure equations*

$$\text{d}\omega_i = \frac{1}{2}\epsilon_{ijk}\,\omega_j \wedge \omega_j \tag{2}$$

hold if ϕ is a solution of (1). Conversely, given a solution of (1) we can always construct a pseudosphere with metric $ds^2 = (\omega_1)^2 + (\omega_2)^2$, constant negative Gaussian curvature (which we can always normalize to -1), and for which we can identify ω_3 as the Levi-Civita connection 1-form.

Of course, not all differential equations are created equal. One of particular interest, both to Chern and Tenenblat in their original work, as well as to us in this article, is the *sine-Gordon* equation. To see how it arises in this construction, choose 1-forms

$$\omega_1 = \cos\frac{\phi}{2}\,(\text{dx} + \text{dt})\ ; \quad \omega_2 = \sin\frac{\phi}{2}\,(\text{dx} - \text{dt})\ ; \quad \omega_3 = \frac{1}{2}\,(\partial_x\phi\,\text{dx} - \partial_t\phi\,\text{dt}) \tag{3}$$

so that (2) holds iff the scalar satisfies

$$\Box\phi(t, x) = \sin\phi(t, x)\,. \tag{4}$$

This is by no means a unique choice of 1-forms that produce the sine-Gordon equation. In fact, Chern and Tenenblat themselves use

$$\omega_1 = \frac{1}{\eta}\sin\phi\,\text{dt}\,; \quad \omega_2 = \eta\,\text{dx} + \frac{1}{\eta}\cos\phi\,\text{dt}\,; \quad \omega_3 = \partial_x\phi\,\text{dx} \tag{5}$$

as a one-parameter family of 1-forms that all yield the sine-Gordon equation for $\phi(t, x)$. Different choices of ω_i yield, alternatively, the KdV, modified-KdV, Burgers and sinh-Gordon equations, among others. In addition to being prototypical examples of *integrable* equations, each of these 2-dimensional classical field theories also possess a rich spectrum of solitonic solutions.

The connection to 2-dimensional gravity is equally fascinating. Certainly, as everyone knows, Einstein gravity in $(2+1)$-dimensions is (dynamically) trivial. However it is a (slightly) lesser-known fact that Einstein gravity in $(1+1)$-dimensional is *not even trivial*. This is because in two spacetime dimensions the Einstein tensor $G_{\mu\nu} \equiv R_{\mu\nu} - \frac{1}{2}g_{\mu\nu}R$ vanishes *identically*. To circumvent this rather boring state of affairs, Jackiw [3] and, independently, Teitelboim [2] made the suggestion to base the dynamical equations for *linear* gravity[1] on the Reimann *scalar* instead, since in two dimensions, it is the simplest quantity that encodes all the local geometrical information about the spacetime. The resulting dilaton gravity theory, known as Jackiw-Teitelboim (JT) gravity, not only possesses a great deal of symmetry ($SO(2, 1)$ in $D = 2$), cosmological [4] and black hole solutions [5] and even a gauge-theoretic formulation [6, 7], but—and this is the crux of this note—in fact *every* generic solution of pseudospherical equations of the form (1) furnishes a classical solution of JT gravity [14]. In this sense, it would seem that JT gravity is deeply connected to the theory of integrable systems.

Spurred on by its relation to the Sachdev-Ye-Kitaev (SYK) model [8, 9], JT gravity is currently enjoying a dramatic renaissance; one that has produced a slew of results on, among others, gravity as an ensemble average [10], replica wormholes [11] and potential resolutions of the information loss paradox [13]. A key feature of these new developments is the inclusion of the *boundary data* which encodes the Schwarzian dynamics [12] and associated chaoticity properties of the theory. This is to be contrasted with the work carried out in the 1980s and 1990s which focused on the bulk dynamics. In this light then, it seems timeous to revisit the connection between JT gravity and integrable 2-dimensional models in the hope that there are lessons to be learnt from the geometry of pseudo-spherical surfaces that may yet prove valuable for quantum gravity. Alas this article will not contain them. An exposition of a lecture presented at the 2021 Nankai Symposium, it is instead two parts review (of the classical JT/sine-Gordon connection, following the treatment given in [15–17]), one part wild speculation of how this connection may be used to understand some of the more contemporary questions asked of the 2-dimensional physics of JT gravity. The devil, as they say, is in the details, and these are left to forthcoming work [19].

2 The Sine-Gordon/JT Connection

The sine-Gordon equation (4) is the paragon of a nonlinear PDE; integrable, possessing a rich spectrum of solitonic solutions and (it's quantum version is even) S-dual to the massive Thirring model via bosonization. Following the prescription above, corresponding to the sine-Gordon equation (4) is a pseudospherical Riemannian 2-manifold with metric

[1] Not to be confused with *linearized* gravity, which is something else entirely.

$$ds^2 = \sin^2\frac{\phi}{2}\,dt^2 + \cos^2\frac{\phi}{2}\,dx^2 \tag{6}$$

Similarly, the (perhaps less familiar) *Euclidean* sine-Gordon equation maps to a *Lorentzian* pseudosphere according to

$$\left(\partial_t^2 + \partial_x^2\right)\phi = \sin\phi \quad\longleftrightarrow\quad ds^2 = -\sin^2\frac{\phi}{2}\,dt^2 + \cos^2\frac{\phi}{2}\,dx^2\,. \tag{7}$$

Of course, which 2-manifold this is, depends on $\phi(t, x)$. In differential geometry, *Bäcklund transforms* map pseudospherical surfaces into each other via linear differential equations. Translated back into the language of PDEs, these act as solution-generating transforms that, starting with some (usually trivial) initial seed solution, generate arbitrarily many more. A familiar, albeit trivial, example of the Bäcklund transform is the first order *Cauchy-Riemann* equations,

$$\partial_x u = \partial_y v, \quad \partial_y u = -\partial_x v\,, \tag{8}$$

that relate the real and imaginary components of an analytic function $f(x, y) = u(x, y) + iv(x, y)$. Since each of these components themselves satisfy the second order Laplace equation $\Delta u(x, y) = \Delta v(x, y) = 0$, the Cauchy-Riemann equations act as an auto-Bäcklund transformation that maps the function $u(x, y)$ into its harmonic conjugate, $v(x, y)$. On the other hand, the Bäcklund transform

$$\partial_x v = \partial_x u + 2\kappa \exp\left(\frac{u+v}{2}\right), \quad \partial_y v = -\partial_y u - \frac{2}{\kappa}\exp\left(\frac{u-v}{2}\right), \tag{9}$$

maps a harmonic function $v(x, y)$ into a solution of the (nonlinear) Liouville equation $\Delta u(x, y) = \exp(u(x, y))$ [18]. Of course it is much simpler to solve the linear Laplace equation than the nonlinear Liouville equation. More to the point of this article, the linear system,

$$\partial_X v = \partial_X u + 2\kappa \sin\left(\frac{u+v}{2}\right), \quad \partial_Y v = -\partial_Y u + \frac{2}{\kappa}\sin\left(\frac{u-v}{2}\right), \tag{10}$$

furnishes an auto-Bäcklund transformation, B_κ for the sine-Gordon equation,[2] with parameter κ. To illustrate the power of (10) as a solution-generating transform, let's start with the trivial solution of (4), $v = 0$, as a seed. Substituting into (10), integrating, and converting back to (t, x) coordinates yields

$$u(t, x) = 4\tan^{-1}\left[\exp\left(\frac{x - x_0 - v(t - t_0)}{\sqrt{1 - v^2}}\right)\right]. \tag{11}$$

[2] In the sense that both u and v each satisfy the sine-Gordon equation, with X and Y regarded as lightcone coordinates for the d'Alembertian operator.

This is, of course just the sine-Gordon kink (or 1-soliton since it carries topological charge $Q = 1$). More generally, if ω is any seed solution, and $\omega_{12\cdots N} \equiv \mathsf{B}_{\kappa_N}(\cdots \mathsf{B}_{\kappa_2}(\mathsf{B}_{\kappa_1}(\omega)))$, then the *Bianchi permutability theorem* [18] gives an entirely algebraic construction of the N-soliton of the sine-Gordon equation, given the data $\{\omega, \omega_1, \omega_2, \ldots, \omega_{123\cdots N-1}\}$.

To appreciate the connection to 2-dimensional gravity, consider the JT-gravity action functional,

$$I_{JT}[\tau, g_{\mu\nu}] = \frac{1}{16\pi G_N} \int_{\mathcal{M}} d^2x \sqrt{-g}\, \tau \,(R + \Lambda) \,, \tag{12}$$

with (dimensionless) Newton's constant G_N, dilaton τ, and cosmological constant Λ, which we will set to 2 in what follows. Solutions of the associated equations of motion, $R + 2 = 0$ and $(\nabla_\mu \nabla_\nu - g_{\mu\nu})\tau = 0$, are manifestly constant curvature, and therefore maximally symmetric in the sense of admitting three independent Killing vectors $k^\mu_{(i)} = \epsilon^{\mu\nu}\partial_\nu \tau_{(i)}/\sqrt{-g}$, $i = 1, 2, 3$. All such solutions must be locally diffeomorphic to AdS_2 however, because a solution is determined by both the metric $g_{\mu\nu}$ *and* the dilaton τ, it may still be *globally* distinct from AdS_2. An example of this is the JT black hole,

$$ds^2 = -\left(r^2 - M\right) dt^2 + \left(r^2 - M\right)^{-1} dr^2 \,, \tag{13}$$

where M is constant and r is linearly related to the dilaton. This solution, which can also be arrived at by dimensionally truncating a 3-dimensional BTZ black hole, clearly has a horizon at $r = \sqrt{M}$. While the BTZ black hole is singular at $r = 0$, the same is not true of (13). However, there is a corresponding statement in JT gravity and that is that at $\tau = 0$, G_N blows up.

To make the connection to pseudospherical surfaces, let's take as an ansatz for the metric the line element given in (7). Substituting into the JT action functional reduces it to the form,

$$I_{JT}[\tau, \phi] = \frac{1}{16\pi G_N} \int_{\mathcal{M}} d^2x \, \tau \,(\Delta\phi - \sin\phi) \,. \tag{14}$$

Correspondingly, the equations of motion that arise from variation of I_{JT} with respect to the dilaton and the embedding coordinate, ϕ are the Euclidean sine-Gordon equation, $\Delta\phi = \sin\phi$ and, respectively, the second order linear equation, $(\Delta - \cos\phi)\tau = 0$. An interesting observation to be made here is that, even though (one of) the equations of motion reduce to the sine-Gordon equation in this gauge, the JT action does not reduce to the sine-Gordon action. In a remarkable series of articles [15–17], Gegenberg and Kunstatter showed that the theory is nevertheless quite rich. Two points in particular should be singled out. The first is that taking ϕ to be the sine-Gordon kink (11), reduces the ansatz to

$$ds^2 = -\mathrm{sech}^2\rho\, dt^2 + \tanh^2\rho\, dx^2 \,, \tag{15}$$

with $\rho \equiv (x - vt)/\sqrt{1 - v^2}$ and $x_0 = t_0 = 0$. Then, defining the new coordinates (t', r') through

$$dt'^2 \equiv dt^2 - v\frac{\sqrt{1-v^2}\,\sinh^2\rho}{1 - v^2\sinh^2\rho}d\rho\,, \quad r' \equiv \sqrt{1-v^2}\,\mathrm{sech}\rho\,, \tag{16}$$

yields precisely the JT black hole (13) if we identify the black hole mass M with the (square of the) spectral parameter, v^2. The resulting metric has ADM energy $E = v^2/2G_N$, is singular at the location of the kink, $\rho = 0$ and at $\rho = \infty$, and regular at the horizon $r' = v$. This construction is quite general; indeed the authors of [17] go on to show that an N-soliton solution of the sine-Gordon equation (of which the kink is the 1-soliton) can be mapped to a black-hole metric of JT gravity with a unique mass. The second point, while more speculative is no less intriguing, and asks if the N-soliton construction may provide some insight into the nature of the Bekenstein-Hawking entropy of the JT black hole. Ignoring much of the exotica[3] of the sine-Gordon spectrum, the total rest energy of an N-soliton is given in terms of the sine-Gordon coupling β by $E_0 = N/\beta$. As argued earlier, the N-soliton state can be constructed from lower topological charge states. This is more than just a mathematical statement encoded in the Bianchi permutability theorem; it is a physical fact. The degeneracy of this state is then computed as the number of ways of writing N as a sum of non-negative integers i.e. the *partition of* N, $p(N)$. While no exact closed form expression exists for $p(N)$, for large N it can be approximated by the Hardy-Ramanujan formula,

$$p(N) \sim \frac{1}{4N\sqrt{3}} \exp\left(\pi\sqrt{\frac{2N}{3}}\right)\,. \tag{17}$$

The corresponding entropy of the state $S \sim \log(p(N)) \sim \pi\sqrt{N}$, matching the Bekenstein-Hawking entropy of the JT black hole (13), up to (admittedly important) factors of order one. As the authors themselves remark, this argument is rather coarse and in need of refinement. Still, it is hardly likely to be merely coincidental.

3 Some Speculative Comments

Having summarised, perhaps overly succinctly, what was known about the Chern-Tenenblat/sine-Gordon/Jackiw-Teitelboim connection around the turn of the last century, we now proceed to speculate as to how this may fit into the present day context in which, at least in the holographic quantum gravity community, JT gravity is enjoying somewhat of a deluge of attention. Much of this interest can be traced back to Kitaev's remarkable observation that the low-energy limit of JT gravity matches that

[3] These include, soliton-anti-soliton bound states, breathers etc.

of the SYK model of N Majorana fermions interacting with quenched random q-body interactions [8, 9]. Key to this observation is the Gibbons-Hawking boundary term

$$I_{bdy} = \frac{1}{8\pi G_N} \int_{\partial \mathscr{M}} \tau_b K \,, \tag{18}$$

that needs to be added to I_{JT} in the holographic context, when the manifold $\mathscr{M}$ has a boundary $\partial\mathscr{M}$ in which the dual field theory resides. Here K is the trace of the extrinsic curvature tensor and $\tau_b = \tau|_{\partial\mathscr{M}}$, is the value of the dilaton on the boundary. If, as in this case, $\partial\mathscr{M}$ is a co-dimension-1 curve parameterized by some 'time' variable u then $\mathscr{M}$ is cut out along the trajectory $(t(u), z(u))$. If, in addition, T and n denote the tangent vector field and unit normal to the boundary curve respectively, then

$$K = -\frac{g(T, \nabla_T n)}{g(T,T)} = \frac{t'(t'^2 + z'^2 + zz'') - zz't''}{\left(t'^2 + z'^2\right)^{3/2}} \,. \tag{19}$$

Implementing boundary conditions on the induced metric and dilaton of $ds|_{\partial\mathscr{M}} = du/\epsilon$, $\tau_b = \tau_r/\epsilon$ and sending $\epsilon \to 0$ allows the extrinsic curvature to be written in terms of the Schwarzian derivative as $K = 1 + \epsilon^2 \mathsf{Sch}(t, u)$ to leading order in ϵ. This defines the so-called *Schwarzian action* that encodes the leading order corrections to the conformal limit and, important for the identification with the dual quantum mechanics on the boundary, controls the chaos properties of the theory as diagnosed by the out-of-time-order four-point correlators. Evidently, all the action is at the boundary.[4] To see how to understand these developments, particularly the Schwarzian action and its implications for chaos, in the sine-Gordon gauge, notice that the *pure* AdS_2 background maps to the sine-Gordon equation *on the infinite line*. The boundary conditions supplied at spatial infinity are famously integrable and precisely the reason we can construct its soliton spectrum via the Bäcklund transform. Cutting off the spacetime with some boundary curve has the effect of putting the sine-Gordon equation on a *finite, time-dependent* interval. Now, even on a static finite interval, boundary conditions that preserve integrablility are a set of measure zero, but when the endpoints of the interval are subject to external driving, the system is generically chaotic [20], exhibiting similar characteristics to the Sinai billiard problem one dimension higher (see Fig. 2).

However, the Schwarzian limit of JT-gravity is not just mildly chaotic; as the putative dual to the SYK model, it is *maximally chaotic* in the sense of saturating the MSS bound on the leading Lyapunov exponent [21]. Again, chaos in the bounded sine-Gordon equation should be anticipated, but why and how such a system exhibits maximal chaos remains unclear, especially because there is no such expectation of a classical system. On the topic of boundaries, while there is an extensive literature on pseudospheres without boundary, most of which can be faithfully mapped to

[4] Sorry, couldn't resist.

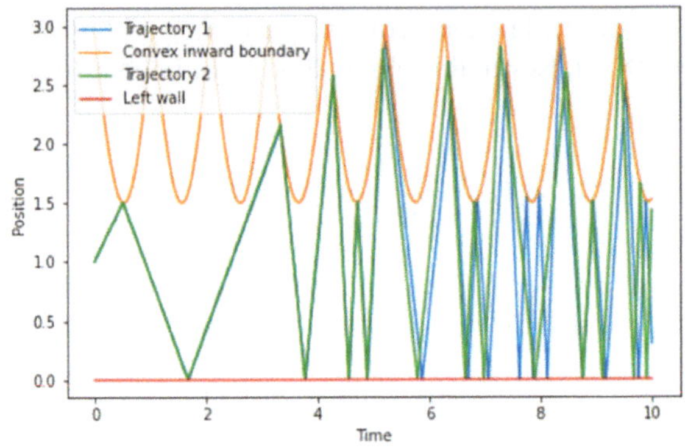

Fig. 2 A plot of the positions of two classical particles on a time-dependent finite interval, starting a distance of 10^{-3} away from each other with the left boundary remaining fixed and the right boundary's position given by $L(t) = 3\left(1 - \frac{1}{2}|\sin t|\right)$. The rapid divergence of trajectories is characteristic of classical chaotic behaviour

various solitons of the sine-Gordon equation, comparatively little is known about pseudospheres with boundaries, save that they are difficult to construct directly. On the other hand, with the map between sine-Gordon solutions and pseudospheres at hand, perhaps knowledge of sine-Gordon solitons on the finite interval, along with their various bound states may well facilitate such constructions. Our motivation for studying this problem is more utilitarian; one solution of nearly-AdS_2 gravity that has generated significant recent interest, in part because it offers an excellent laboratory for the controlled study of questions of information loss, the Hawking-Page transition, Page curves and emergent gravity, is a wormhole configuration that connects the two boundaries of the global spacetime [22]. This Maldacena-Qi wormhole is conjectured to be dual to two coupled SYK quantum dots and it is tempting to speculate that such a "double trumpet" solution may be constructed by gluing together two pseudospheres with boundary along the boundary, in which case the wormhole would map to a soliton bound state. Finally, there is the issue of quantization of JT gravity and the corresponding quantum sine-Gordon system, but perhaps that is a story for another Nankai symposium.

Acknowledgements I owe a debt of gratitude to Kayla Hopley who, in the course of her MSc research, found out about the connection between JT gravity and the sine-Gordon equation, and to Sebastian Zimper for producing Fig. 2. I am also indebted to Yang-Hui He for his kind invitation to present this work at the Nankai Symposium. I regret only that I, like the other participants, could not be there in person.

References

1. S.S. Chern and K. Tenenblat. Studies in Applied Mathematics **74**, 55–83 (1986)
2. C. Teitelboim, Phys. Lett. B **126**, 41–45 (1983) https://doi.org/10.1016/0370-2693(83)90012-6
3. R. Jackiw, Nucl. Phys. B **252**, 343–356 (1985) https://doi.org/10.1016/0550-3213(85)90448-1
4. R. B. Mann and S. F. Ross, Phys. Rev. D **47**, 3312–3318 (1993) https://doi.org/10.1103/PhysRevD.47.3312arXiv:hep-th/9206022 [hep-th].
5. R. B. Mann, A. Shiekh and L. Tarasov, Nucl. Phys. B **341**, 134–154 (1990) https://doi.org/10.1016/0550-3213(90)90265-F
6. K. Isler and C. A. Trugenberger, Phys. Rev. Lett. **63**, 834 (1989) https://doi.org/10.1103/PhysRevLett.63.834
7. A. H. Chamseddine and D. Wyler, Phys. Lett. B **228**, 75–78 (1989) https://doi.org/10.1016/0370-2693(89)90528-5
8. J. Maldacena and D. Stanford, Phys. Rev. D **94**, no.10, 106002 (2016) doi:https://doi.org/10.1103/PhysRevD.94.106002arXiv:1604.07818 [hep-th].
9. A. Kitaev and S. J. Suh, JHEP **05**, 183 (2018) https://doi.org/10.1007/JHEP05(2018)183arXiv:1711.08467 [hep-th].
10. D. Stanford and E. Witten, Adv. Theor. Math. Phys. **24**, no.6, 1475–1680 (2020) https://doi.org/10.4310/ATMP.2020.v24.n6.a4arXiv:1907.03363 [hep-th].
11. G. Penington, S. H. Shenker, D. Stanford and Z. Yang, arXiv:1911.11977 [hep-th].
12. T. G. Mertens, JHEP **05**, 036 (2018) https://doi.org/10.1007/JHEP05(2018)036arXiv:1801.09605 [hep-th].
13. T. J. Hollowood, S. Prem Kumar and A. Legramandi, J. Phys. A **53**, no.47, 475401 (2020) doi:https://doi.org/10.1088/1751-8121/abbc51arXiv:2007.04877 [hep-th].
14. E. G. Reyes, J. Phys. A **39**, L55–L60 (2006) https://doi.org/10.1088/0305-4470/39/2/L02
15. J. Gegenberg and G. Kunstatter, arXiv:hep-th/9709183 [hep-th].
16. J. Gegenberg and G. Kunstatter, Phys. Lett. B **413**, 274–280 (1997) https://doi.org/10.1016/S0370-2693(97)01118-0arXiv:hep-th/9707181 [hep-th].
17. J. Gegenberg and G. Kunstatter, Phys. Rev. D **58**, 124010 (1998) https://doi.org/10.1103/PhysRevD.58.124010arXiv:hep-th/9807042 [hep-th].
18. C. Rogers and W. F. Shadwick,
19. J. Murugan, "*The sine-Gordon/JT gravity connection revisited,*" In progress, (2022)
20. J. Murugan and S. Zimper, "*Chaos in the sine-Gordon equation with Floquet boundaries,*" In progress, (2022)
21. J. Maldacena, S. H. Shenker and D. Stanford, JHEP **08**, 106 (2016) https://doi.org/10.1007/JHEP08(2016)106arXiv:1503.01409 [hep-th].
22. J. Maldacena and X. L. Qi, arXiv:1804.00491 [hep-th].

Quantum Matter from Algebraic Geometry and Number Theory

Steven Rayan

Abstract The exciting and rapidly-growing field of topological materials has brought with it unexpected new connections between physics and pure mathematics. Algebraic topology in particular has played a significant role in classifying topological materials. In this brief note, I report on joint work with J. Maciejko and offer a brief look at another emerging chapter in this story in which algebraic geometry and number theory anticipate new forms of quantum matter associated to hyperbolic lattices.

The purpose of this brief note is to report on some recent interactions between complex algebraic geometry, number theory, and the burgeoning field of quantum matter.[1] The propagation of waves in periodic media under a potential respecting the symmetry of the medium is governed by Bloch's theorem [1]. A direct result is that the admissible momenta of the wave are captured by a topological space called the *Brillouin zone*. When the underlying medium is $\mathbb{R}^2$ with the standard Euclidean inner product, and when the periodicity is given by a parallelogram or hexagon lattice Λ, the Brillouin zone is a torus. Multiple interpretations are available for this torus. Most immediately, it is the space of $U(1)$-representations of the translation group of the lattice. Equivalently, it is the space of $U(1)$-representations of the fundamental group of $\mathbb{R}^2/\Lambda$. Note here that $\mathbb{R}^2/\Lambda$ is itself a torus, but this is not the same as

---

[1] This report is the summary of an invited talk that I delivered on August 12, 2021 at the *Nankai Symposium on Mathematical Dialogues*, held at the Chern Institute of Mathematics in celebration of the 110th anniversary of Prof. S.-S. Chern and the 90th birthday of Sir R. Penrose. I express my gratitude to the organizers, Profs. Yang-Hui He, Cheng-Ming Bai, and Mo-Lin Ge, for the kind invitation and to Jiakang Bao, Ed Hirst, Hong-Qin Li, and Suvajit Majumder for all of their support in organizing this conference. For posterity, I would like to add that, due to the Covid-19 pandemic, the conference was held in a hybrid in-person/virtual format, and I was one of many virtual speakers and attendees. The conference was extremely lively and conversational, capturing exactly the spirit of an in-person conference. I commend the organizers for crafting what has been one of the finest virtual events I have had the privilege of attending.

S. Rayan (✉)
Centre for Quantum Topology and Its Applications (quanTA) and Department of Mathematics & Statistics, University of Saskatchewan, Saskatoon, SK, Canada
e-mail: rayan@math.usask.ca

Y.-H. He et al. (eds.), *Nankai Symposium on Mathematical Dialogues*,
https://doi.org/10.1007/978-981-19-2328-9_33

the Brillouin zone. The quotient is the configuration space, or position space, of the wave while the Brillouin zone is obtained from it as the so-called reciprocal space via a Fourier transform. By applying the Riemann-Hilbert correspondence to the Brillouin zone as a space of $U(1)$-representations, one can interpret it as a space of smooth data, namely flat unitary line bundles on the torus. From here, one can furthermore apply the Narasimhan-Seshadri correspondence [6] and interpret the Brillouin zone as a space of holomorphic data: holomorphic line bundles on the elliptic curve $E = \mathbb{R}^2/\Lambda$. In this latter interpretation, the Brillouin zone becomes the Jacobian of E, which is the moduli space of holomorphic line bundles on E.

While the considerations above apply to propagation in any periodic medium, it has been particularly fruitful to apply Bloch's theorem in condensed matter to electrical conductivity and resistivity. Here, the wave phenomenon is the motion of electrons in a highly crystalline material that, in two dimensions, can be modelled by the lattice Λ above. Here, the symmetric potential is incorporated into the Hamiltonian operator of Schrödinger's equation (with the correct periodic boundary conditions), which are the natural equations of motion for the electrons. From this point of view, the picture becomes one of *quantum* condensed matter. By studying the eigenvalues of the Hamiltonian as functions over the Brillouin zone, we initiate the celebrated electronic band theory: namely, the topology of the spectral surface of the Hamiltonian can be used to classify the material as an insulator or (semi)metal. This also anticipates the so-called *topological materials*, which have robust, topologically-protected conductivity properties. The 2016 Nobel Prize in Physics was awarded in recognition of both the theoretical prediction and the experimental realization of such materials.

For the purposes of most condensed matter and solid-state physics discourse, the interpretation of the Brillouin zone as a space of $U(1)$-representations or "phase factors" is entirely sufficient. The interpretations in our first paragraph above regarding flat bundles and holomorphic bundles, respectively, make no appearance in any quantum condensed matter theory literature—or more broadly any literature about periodic waves—that we are aware of. Thinking in these terms, however, allows us to generalize the theory of periodic quantum condensed matter to new geometries, thereby anticipating new forms of quantum matter. This is the foundation of what we term *hyperbolic band theory*, developed in [4].

Persisting in two dimensions, we replace the kinetic part of the Hamiltonian with the Laplace-Beltrami operator for the Poincaré metric. In other words, we replace the standard complex plane $\mathbb{C} \cong \mathbb{R}^2$ with the hyperbolic plane $\mathbb{H}$. The potential part is chosen to be a suitable function that is invariant with respect to the action of a (co-compact, strictly hyperbolic) Fuchsian group Γ, which defines a tessellation of $\mathbb{H}$—a periodic, crystalline geometry but with a generally noncommutative translation group now. We may also generalize the periodic boundary conditions appropriately. With this in hand, we show via direct construction [4] that there exist wave solutions ψ that satisfy a generalized equivariance property:

$$\psi \circ \gamma = \chi(\gamma)\psi,$$

where $\gamma \in \Gamma$ and $\chi : \Gamma \to U(1)$ is a representation. In other words, there exist wavefunctions that are weight-0 factors of automorphy. It is not so surprising that these staples of number theory should appear here as, on the one hand, they naturally generalize the quasi-periodicity of Euclidean Bloch waves and, on the other hand, a natural source of Γ-periodic potentials is given by generalizations of the types of infinite series (e.g. Eisenstein series) that typically occur in the theory of modular forms.

This program replaces the Brillouin zone with the space of representations of a higher-genus position space, namely $X = \mathbb{H}/\Gamma$. Specifically, when X has genus $g \geq 2$ (which can be achieved, for instance, when Γ is the translation group of a $4g$-gonal tessellation), the Brillouin zone is a $2g$-dimensional real, compact torus. The Fourier duality between position and momentum is now less obvious, especially considering the dimensional difference between the position surface and the momentum torus. By interpreting the Brillouin zone as the Jacobian of the Riemann surface X (which takes advantage of the natural complex structure on X coming from the quotient), we can appeal to the Abel-Jacobi map as a natural map from the symmetric product of X with itself g-many times to the Jacobian. Albeit non-canonical and only birational in a global sense, the map is almost everywhere a complex analytic isomorphism. In this way, complex algebraic geometry provides a wave-particle duality that completes the theory.

A natural question concerns our ability to calculate the band structure and make predictions in specific examples. One such example is explored at length in our article [4] for the Bolza surface, a genus-2 Riemann surface arising from a highly-symmetric octogonal tessellation, which we equip with both the zero potential (the "empty lattice") and with a generalized Eisenstein series. Here, numerical calculations verify expectations around degeneracies and branching of the spectrum. Another query concerns the classification of hyperbolic lattices in this context—in other words, their crystallography—for which a partial catalogue in two dimensions is attempted in [2]. A question concerning the extent to which Bloch's theorem holds for two-dimensional hyperbolic lattices, rather than simply the existence of *some* automorphic wavefunctions, is also natural. This is resolved in [5], which notably appeals to higher-rank representations of Γ and hence, by Narasimhan-Seshadri, to the moduli space of stable holomorphic bundles of arbitrary rank on X. Finally, one may ask a very compelling question about if and how hyperbolic matter might be experimentally realized. This actually predates our mathematical and theoretical investigations by a very short interval in time: in [3], success in constructing artificial photonic hyperbolic circuits is described. The authors lament the lack of a hyperbolic band theory, which [4] now provides.

We encourage interested readers to interact with the various articles listed above, which are by no means exhaustive as further articles emerge in this new subject of hyperbolic quantum matter. On the one hand, such matter may lead to novel models of topoelectric circuits and qubits for quantum computing, to name just one potential disruptive application. At a foundational level, the study of it is leading to new interactions between mathematics and physics by breathing algebraic geometry and number theory into condensed matter theory, subjects that have interacted only marginally if at all with this branch of physics.

References

1. F. Bloch, Über die Quantenmechanik der Elektronen in Kristallgittern, *Z. Phys.*52, 555–600 (1929).
2. I. Boettcher, A. Gorshkov, A. Kollár, J. Maciejko, S. Rayan, and R. Thomale, Crystallography of hyperbolic lattices, *Phys. Rev. B*105, no. 12, 125118 (2022).
3. A. Kollár, M. Fitzpatrick, and A. A. Houck, Hyperbolic lattices in circuit quantum electrodynamics, *Nature*571, 45–50 (2019).
4. J. Maciejko and S. Rayan, Hyperbolic band theory, *Sci. Adv.*7, no. 36, abe9170 (2021).
5. J. Maciejko and S. Rayan, Automorphic Bloch theorems for finite hyperbolic lattices, *Proc. Natl. Acad. Sci. U.S.A.*119, no. 9, e2116869119 (2022).
6. M. S. Narasimhan and C. S. Seshadri, Stable and Unitary Vector Bundles on a Compact Riemann Surface, *Ann. Math.*82, 540–567 (1965).

Excursions Inside Black Holes and Finite Temperature CFT's

D. Rodriguez-Gomez

Abstract Finite temperature allows to probe detailed aspects of a Conformal Field Theory (CFT), providing interesting new windows into its dynamics. In the case of CFT's with a gravity dual through the AdS/CFT correspondence, there is an interesting further twist as thermal correlation functions can contain information about the interior of the dual black hole encoded in a subtle way. In this talk we will study thermal correlation functions for heavy operators, focusing on 2- and 3-point functions. Allowing for higher curvature couplings, the 2-point function contains a contribution which can be understood in terms of geodesics reaching the singularity. As a consequence, the 2-point function can contain information about the travel time to the singularity.

1 Introduction and Summary

Among Quantum Field Theories (QFT), Conformal Field Theories in d dimensions (CFT_d) play a prominent role in modern Theoretical Physics. Since the most basic quantities in QFT are correlation functions of local operators, it is very interesting to study the structure of correlators in CFT_d and the restrictions which conformal invariance imposes on their form. This acquires further interest in view that CFT_d's are at the heart of holography, which is believed to be a key property of Quantum Gravity. From this point of view, it is natural to ask about the features that a generic holographic CFT_d–i.e. a CFT_d with a holographic dual in terms of gravity on an asymptotically AdS_{d+1} space–must have.

D. Rodriguez-Gomez (✉)
Department of Physics, Universidad de Oviedo, Oviedo, Spain
e-mail: d.rodriguez.gomez@uniovi.es

Instituto Universitario de Ciencias y Tecnologías Espaciales de Asturias (ICTEA), Oviedo, Spain

Y.-H. He et al. (eds.), *Nankai Symposium on Mathematical Dialogues*,
https://doi.org/10.1007/978-981-19-2328-9_34

In this contribution we will review recent progress [1–3] (to which we refer for all the technical details as well as for references) in the study of correlation functions for holographic CFT$_d$'s using the AdS/CFT correspondence (see the canonical reference [4] for a review), focusing on 2-point functions. On $\mathbb{R}^d$–we will mostly use Euclidean signature–, conformal invariance dictates the form of the 2-point function. The situation is more interesting on $S^1 \times \mathbb{R}^{d-1}$, since the background provides both a special direction–the coordinate τ along the circle–and a scale–the length β of the circle–, allowing for non-zero VEVs for primary operators. Note that this has the interpretation of a thermal field theory at temperature $T = \beta^{-1}$.

We will be interested on 2-point functions for scalar operators $\langle O_i(0) O_j(x)\rangle$ in holographic CFT$_d$'s, and we will concentrate on the regime $|x| \ll \beta$. This ensures that the operators are inserted within a ball which does not "wind around" the thermal circle, so that the Operator Product Expansion can be used to bootstrap correlation functions. For the particular case of the 2-point function of an operator with itself one finds [5]

$$\langle O(x) O(0)\rangle = \sum_{\mathcal{O}} \frac{a_{OO\mathcal{O}}}{\beta^{\Delta_{\mathcal{O}}}} \frac{1}{|x|^{2\Delta_O - \Delta_{\mathcal{O}}}} C^{(\nu)}_{J_{\mathcal{O}}}(\eta), \qquad a_{OO\mathcal{O}} = \frac{f_{OO\mathcal{O}}\, b_{\mathcal{O}}}{c_{\mathcal{O}}} \frac{J_{\mathcal{O}}!}{2^{J_{\mathcal{O}}}(\nu)_{J_{\mathcal{O}}}}, \tag{1}$$

where $\eta \equiv \tau/|x|$, $|x| = \sqrt{\tau^2 + \vec{x}^2}$ and $C^{(\nu)}_{J_{\mathcal{O}}}(\eta)$ are Gegenbauer polynomials ($\nu \equiv \frac{d-2}{2}$).

We will compute the 2-point functions through the gravitational dual–which is a planar black hole in AdS_{d+1}–in the supergravity limit.[1] When the dust settles, the structure in (1) is exactly recovered. Moreover, one can read off that, to leading order, the operators contributing to the sum in (1) are contractions of n energy-momentum tensors with spins $0, 2, \ldots, 2n$ which we denote by $\mathcal{T}^n$. Upon including subleading (in the supergravity leading sense) corrections, operators of the form $O\mathcal{T}^n$ contribute as well. These include the VEV of the scalar operator itself, which has been very recently argued in [6] to be sensible to the interior of the black hole in the gravity dual, as it carries information about the time of travel to the singularity. These subleading corrections are actually leading for 2-point functions of operators of different dimension,[2] which therefore stand as very interesting probes of Quantum Gravity, as they are, to begin with, dominated by the scalar VEV's sensible to the BH interior (see also [7]).

[1] In the paradigmatical example of $AdS_5 \times S_5 \leftrightarrow SU(N)\,\mathcal{N} = 4$ SYM example in Type IIB String Theory, this corresponds to large N and large 't Hooft coupling. More generically, it corresponds to the limit of large central charge and strong coupling.

[2] On $\mathbb{R}^d$, covariance under the full conformal group requires $\langle O_i(x_1) O_j(x_2)\rangle \sim \delta_{\Delta_{O_i}\Delta_{O_j}}$. In a more general background, such as $S^1 \times \mathbb{R}^{d-1}$, this need not be the case.

2 Holographic 2-Point Functions

The holographic dual to a CFT_d with a large number of degrees of freedom at finite temperature is in terms of a classical gravitational theory on the fixed background of an AdS_{d+1} black hole given by (we quote the Euclidean version related to the Lorentzian signature by $t = i\tau$)

$$ds^2 = \frac{R^2}{z^2}\left(f(z)\,d\tau^2 + \frac{dz^2}{f(z)} + d\vec{x}^2_{d-1}\right), \qquad f(z) = 1 - \frac{z^d}{z_0^d}, \qquad z_0 = \frac{d}{4\pi}\beta\,. \tag{2}$$

Focusing on the scalar sector of the theory, each operator O_i of dimension Δ_{O_i} in the boundary theory corresponds to a scalar field in the bulk ϕ_i with mass m_i^2 related to Δ_{O_i} in the standard way [4]. The leading terms in the effective action for the ϕ_i must be of the form

$$S \sim \int \sqrt{g}\left(\frac{1}{2}\partial\phi_i^2 + \frac{1}{2}m_i\phi_i^2 + \lambda_{ijk}\phi_i\phi_j\phi_k + \alpha_i\phi_i\,W^2\right), \qquad W^2 = \frac{d\,(d-2)\,(d-1)^2}{R^2}\frac{z^d}{z_0^d}\,; \tag{3}$$

where we included couplings to W^2–the Weyl tensor squared–which is non-zero in the AdS_{d+1} back hole background. Note that the magnitude of W^2 is biggest close to the singularity. Note that in a full-fledged String Theory construction, say in the paradigmatical $AdS_5 \times S_5 \leftrightarrow \mathcal{N} = 4$ SYM example in Type IIB String Theory, these would be higher curvature couplings and thus suppressed by extra powers of α' (and therefore, extra powers of the 't Hooft coupling/N).

The 2-point function $\langle O_i(x_1)O_j(x_2)\rangle$ in the boundary theory arises from an expansion on Witten diagrams as in Fig. 1 below. In these diagrams, blue lines correspond to the appropriate propagators–bulk-to-boundary or bulk-to-bulk–of the bulk theory. A dramatic simplification occurs in the large Δ limit, since the equation for the Green function can be solved through the geodesic approximation (essentially the WKB approximation) finding that $G(x_1, x_2) \sim e^{-\Delta L(x_1,x_2)}$, where $\Delta \sim mR$ and $L(x_1, x_2)$ is the length of a geodesic connecting x_1 and x_2.[3] Moreover, the large Δ limit allows to perform the integration over all the bulk of each interaction vertex through the saddle point approximation.

2.1 The Leading Term

The leading contribution to the 2-point function $\langle O_i(x_1)O_j(x_2)\rangle$ is given by diagram (1) in Fig. 1. As it is well-known, this contribution can be written as if it came from an

[3] Note that x stands for all the AdS_{d+1} coordinates, i.e. $x = (t, \vec{x}, z)$. Therefore this recipe includes both bulk-to-boundary as well as bulk-to-bulk propagators upon suitably choosing the starting/ending points.

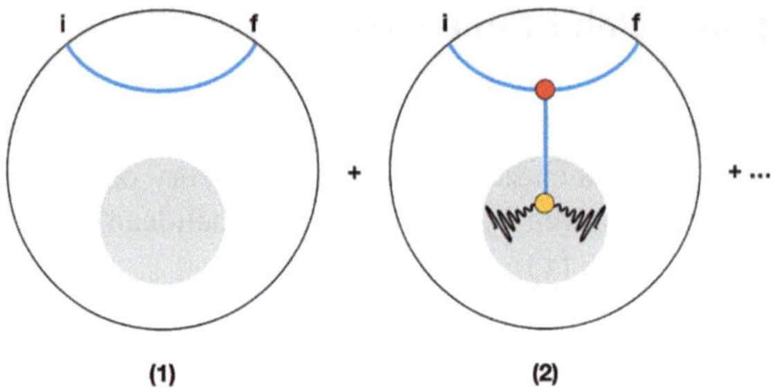

Fig. 1 Leading terms in the expansion on Witten diagrams of the 2-point function arising from (3). Blue lines correspond to propagators, red vertices to λ_{ijk}, yellow vertices to α_i and wiggly lines to W^2. The shaded blob is a cartoon of the black hole. With no loss of generality, we can take the insertion points of the operators at $x_i = (-\tau^\circ, -\vec{x}^\circ, z=0)$ and $x_f = (\tau^\circ, \vec{x}^\circ, z=0)$

interacting diagram by breaking the full geodesic in two geodesic arms meeting at a generic bulk point (one from the boundary point x_i to some bulk point $x = (\tau, \vec{x}, z)$, another from x to the other boundary point x_f), upon integrating x over the whole bulk. Thus

$$(1) = \int_{\text{Bulk}} dx\, e^{-\Delta S_{\text{eff}}}\,, \qquad S_{\text{eff}} = L(x_i, x) + L(x,\, x_f)\, \delta_{\Delta_{O_i}\Delta_{O_j}} \tag{4}$$

Note that the diagram is proportional to $\delta_{\Delta_{O_i}\Delta_{O_j}}$, that is, the leading contribution can only give non-zero correlators between operators of equal dimension $\Delta_{O_i} = \Delta_{O_j} = \Delta$. In the large Δ limit we can do the integral through the saddle point approximation. To begin with, symmetry fixes $x = (0, \vec{0}, z)$. Finally, the saddle point equation for z can be easily solved finding the corresponding $z_\star$, and the remaining task is simply to evaluate the integrand in (4) on the saddle point.

This is however subtle. The reason is that, while computing geodesics in black hole backgrounds is a much simpler problem than solving the Green function equation, it turns out that one can only find an implicit form of the geodesic. To be slightly more precise, parametrizing the geodesic by the z coordinate, one can easily find $\tau(z)$ and $\vec{x}(z)$. These functions depend on integration constants–appearing as conserved quantities which we will denote μ, $\vec{\nu}$–which are fixed by demanding the boundary conditions (in particular, for bulk-to-boundary propagators, that at the boundary the geodesic hits the point $x_{i,f}$ and that the other endpoint is at $(0, \vec{0}, z_\star)$). In principle, one may imagine inverting these equations to find $\mu = \mu(\tau^\circ, \vec{x}^\circ)$, $\vec{\nu} = \vec{\nu}(\tau^\circ, \vec{x}^\circ)$, and using this to explicitly write the $\tau(z)$, $\vec{x}(z)$ functions as well as the on-shell action after evaluation on the saddle point. However it turns out that, as soon as $T \neq 0$, one cannot perform such inversion, and hence one has to live with the geodesic implicitly defined as a function of $(\mu, \vec{\nu})$, in addition to the boundary conditions defining $(\mu, \vec{\nu})$ in terms of the boundary data of the geodesic (that is, the insertion points of the operators in the boundary of the AdS space).

Restricting for simplicity to correlators with $\vec{x}^\circ = 0$–which amounts to $\vec{\nu} = 0$–, one can see that the large μ region corresponds to small τ, in such a way that one can find an approximate expression μ in terms of inverse powers of $(\tau^\circ)^{-1}$. Thus, in that region one can find a power expansion for the correlation function. When the dust settles (we quote the $d = 4$ result. See also [8]).

$$(1) = \frac{1}{\tau^{2\Delta}}\left(1 + \frac{\Delta\,\pi^4\,T^4\,\tau^4}{40} + \frac{\Delta\,\pi^8\,T^8\tau^8}{28800}\,(9\,\Delta + 22) + \frac{\Delta\,\pi^{12}\,T^{12}\,\tau^{12}}{14976000}\,(39\,\Delta^2 + 286\,\Delta + 712) + \mathcal{O}\left(\tau^{16}\right)\right). \tag{5}$$

Here we have relabelled $2\tau^\circ = \tau$.

With a bit more effort, one can include the $\vec{x}$-dependence as well, finding (likewise $2\vec{x} = \vec{x}^\circ$)

$$(1)\rangle = \frac{1}{|x|^{2\Delta}}\left[1 + \frac{\Delta\,\pi^4\,T^4}{120}\,C_2^{(1)}(\eta)\,|x|^4 + \frac{\Delta^2\,\pi^8\,T^8}{28800}\left(C_4^{(1)}(\eta) + C_2^{(1)}(\eta) + C_0^{(1)}(\eta)\right)|x|^8 + \cdots\right]. \tag{6}$$

A similar structure emerges in d dimensions, with the relevant Gegenbauers in each case, recovering the expected form for the 2-point functions (*cf.* (1)). Moreover, one can see that the operators contributing to the 2-point function in d dimensions are dimension dn operators with spin $J = 0, 2, 4\ldots, 2n$ which are naturally identified with the n possible contractions of n energy-momentum tensors $\mathcal{T}^n$. The contribution of $n = 1$, the energy-momentum tensor itself, is particularly interesting. In generic d, one has

$$\langle O(\tau)\,O(0)\rangle = \frac{1}{\tau^{2\Delta}}\,\exp\left[\alpha_d\,\Delta\,T^d\,\tau^d\,(1 + \mathcal{O}((T\tau)^d))\right], \qquad \alpha_d \equiv \frac{\pi^{d+1}\,\Gamma(d)}{d^d\,\Gamma(\frac{d+3}{2})\,\Gamma(\frac{d-1}{2})}. \tag{7}$$

Comparing with (1) and using the superconformal Ward identity which sets $f_{OO\mathcal{T}} \sim \Delta_O$, one can find a prediction for the thermal free energy as a function of $c_\mathcal{T}$

$$F = -\frac{2^{2d-1}\,(d-1)\,\pi^{\frac{3d}{2}}\,\Gamma(\frac{d}{2})}{d^d\,\Gamma(d+2)}\,c_\mathcal{T}\,T^d\,. \tag{8}$$

One can check that all known examples fit in this formula. Moreover, in the $d = 4$ case, one can easily include the α' correction, finding perfect agreement with the literature.

2.2 The Subleading Term

The subleading term corresponds to diagram (2) in Fig. 1. In the geodesic approximation $\Delta_{O_i} \gg 1$, we may write it as

$$(2) = \frac{d\,(d-2)\,(d-1)^2}{R^2}\,\lambda_{ijk}\int_{\text{bulk}} du\, e^{-\Delta_{O_i}\,L(x_1,u)-\Delta_{O_j}\,L(x_2,u)}\int_{\text{bulk}} dv\, e^{-\Delta_{O_k}\,L(u,v)-d\,\log\frac{v_z}{z_0}}\,, \tag{9}$$

After some manipulations, one can show that the geodesic connecting the two bulk points of interaction is vertical–i.e. it runs along z–. Finally, we can write (9) as

$$(2) = \langle O_k\rangle\,\lambda_{ijk}\int_{\text{bulk}} du\, e^{-\Delta_{O_i}\,L(x_1,u)-\Delta_{O_j}\,L(x_2,u)+\Delta_{O_k}\,L_{\text{vert}}(0,u)}\,. \tag{10}$$

where

$$\langle O_k\rangle = \int_{\text{bulk}} dv\, e^{-\Delta_k\,\ell}\,W^2\,, \qquad L_{\text{vert}}(0,u) = \int_0^{u_z} dz\frac{1}{z\,\sqrt{f(z)}}\,; \tag{11}$$

being ℓ the length of the vertical–i.e. along z–geodesic from the boundary all the way to the singularity which captures the thermal VEV. Thus VEV's other than those of the powers of the energy-momentum tensor enter, in the supergravity regime of the holographic duality when $c_{\mathcal{T}}$ is large, through a subleading effect. Moreover, since the contribution in (10) does not generically vanish for $\Delta_{O_i} \neq \Delta_{O_j}$, the diagram (2) actually captures the leading contribution to off-diagonal correlators, and is controlled by the thermal VEV's which, in a precise sense, explore the interior of the black hole [6].

To get a flavor of this contribution, let us assume $\Delta = \Delta_{O_i} = \Delta_{O_j} \gg \Delta_{O_k}$. In the regime $T|x| \ll 1$ one can solve the saddle equations and the inversion of the conserved momenta in terms of the boundary insertion points as above in a series expansion, finding

$$(2) = \langle O_k\rangle\,\lambda_{iik}\,C_0^{(\nu)}(\eta)\,|x|^{\Delta_{O_k}-2\Delta} + \frac{\Delta_{O_k}\,c_d\,\langle O_k\rangle\,\lambda_{iik}}{\beta^d}\,C_2^{(\nu)}(\eta)\,|x|^{\Delta_{O_k}+d-2\Delta} + \cdots\,,$$
$$c_d = \frac{1}{2d}\left(\frac{2\pi}{d}\right)^d\left(1-d+\frac{(d-1)\,d\,\sqrt{\pi}\,\Gamma(1+\frac{d}{2})}{2\,\Gamma(\frac{3+d}{2})}\right). \tag{12}$$

Comparing with (1), we recognize here the contribution of dimension $\Delta_{O_k} + dn$ operators with spin $0, 2, \ldots 2n$ which are naturally identified with $O_k\mathcal{T}^n$.

Acknowledgements I would like to thank H. Krishna and J. Russo. I am specially grateful to J. Russo for so many interesting discussions and collaborations. This work is partially supported by the Spanish government grant MINECO-16-FPA2015-63667-P, and also by the Principado de Asturias through the grant FC-GRUPIN-IDI/2018/000174.

References

1. D. Rodriguez-Gomez and J. G. Russo, "Correlation functions in finite temperature CFT and black hole singularities," JHEP **06** (2021), 048 https://doi.org/10.1007/JHEP06(2021)048arXiv:2102.11891 [hep-th].
2. D. Rodriguez-Gomez and J. G. Russo, "Thermal correlation functions in CFT and factorization," JHEP **11** (2021), 049 https://doi.org/10.1007/JHEP11(2021)049arXiv:2105.13909 [hep-th].
3. H. Krishna and D. Rodriguez-Gomez, "Holographic thermal correlators revisited," arXiv:2108.00277 [hep-th].
4. O. Aharony, S. S. Gubser, J. M. Maldacena, H. Ooguri and Y. Oz, "Large N field theories, string theory and gravity," Phys. Rept. **323** (2000), 183–386 https://doi.org/10.1016/S0370-1573(99)00083-6arXiv:hep-th/9905111 [hep-th].
5. L. Iliesiu, M. Koloğlu, R. Mahajan, E. Perlmutter and D. Simmons-Duffin, "The Conformal Bootstrap at Finite Temperature," JHEP **10** (2018), 070 arXiv:1802.10266 [hep-th].
6. M. Grinberg and J. Maldacena, "Proper time to the black hole singularity from thermal one-point functions," JHEP **03** (2021), 131 https://doi.org/10.1007/JHEP03(2021)131arXiv:2011.01004 [hep-th].
7. J. Engelsöy and B. Sundborg, "Tidal excitation as mixing in thermal CFT," https://doi.org/10.1007/JHEP08(2021)085arXiv:2106.06520 [hep-th].
8. A. L. Fitzpatrick and K. W. Huang, "Universal Lowest-Twist in CFTs from Holography," JHEP **08** (2019), 138 arXiv:1903.05306 [hep-th].

Symmetry: From Physics and Calabi-Yau Threefolds to Algebra and Gorenstein Rings

Hal Schenck

Abstract We give an overview of a pair of constructions: on the geometric side, we describe Calabi-Yau manifolds, and on the algebraic side, we discuss Gorenstein rings. With certain hypotheses, Gorenstein rings give rise to Calabi-Yau manifolds [24]. Calabi-Yau manifolds are of interest for many reasons, one of which is the central role they play in physics of string theory. The first half of this note gives a brisk review of the geometry necessary to define Calabi-Yau manifolds, and the second half describes the construction of a special type of Gorenstein ring. The objects resulting from both constructions possess an internal symmetry; an open question is to find the mirror of a Calabi-Yau threefold constructed from a Gorenstein ring.

Keywords Calabi-Yau Manifold · Gorenstein ring · Inverse system · Free resolution

2000 Mathematics Subject Classification Primary 14J32 · Secondary 13D02

1 Introduction

Candelas-Horowitz-Strominger-Witten show in [6] that Calabi-Yau threefolds (CY_3) play a central role in string theory. This was further developed in works by Candelas-Lynker-Schimmrigk [8] and Candelas-de la Ossa-Green-Parkes [7]. For additional background, see the book of Cox-Katz [10].

In Sect. 2 we review some basic definitions (manifolds, tangent bundle, exterior powers, canonical bundle) needed to discuss the geometry of Calabi-Yau threefolds. We sketch several classical constructions, including the work of Batyrev [1] on CY_3's as hypersurfaces in toric varieties.

Schenck supported by NSF 2048906.

H. Schenck (✉)
Mathematics Department, Auburn University, Auburn, AL, USA
e-mail: hks0015@auburn.edu

Y.-H. He et al. (eds.), *Nankai Symposium on Mathematical Dialogues*,
https://doi.org/10.1007/978-981-19-2328-9_35

$$
\begin{array}{ccccccc}
 & & & 1 & & & \\
 & & 0 & & 0 & & \\
 & 0 & & 1 & & 0 & \\
1 & & 52 & & 52 & & 1 \\
 & 0 & & 1 & & 0 & \\
 & & 0 & & 0 & & \\
 & & & 1 & & &
\end{array}
$$

total	1	7	22	22	7	1
0	1	–	–	–	–	–
1	–	6	1	–	–	–
2	–	1	21	21	1	–
3	–	–	–	1	6	–
4	–	–	–	–	–	1

Fig. 1 The Hodge diamond (left) and Betti table (right) for a certain CY_3 in $\mathbb{P}^8$

In Sect. 3 we focus on the construction of CY_3's coming from arithmetically Gorenstein (AG) rings, and outline Macaulay's 1916 construction [21] of Artinian Gorenstein rings via inverse systems. The example from [24] below illustrates the symmetry in the constructions.

Example 1.1 Figure 1 above depicts the Hodge diamond (see Sect. 2.2) and Betti table (see Sect. 3.1) for a Calabi-Yau threefold $V(I) \subseteq \mathbb{P}^8$ of degree 21 corresponding to an arithmetically Gorenstein (AG) ideal $I \subseteq \mathbb{K}[x_0, \ldots, x_8]$ having one cubic and six quadratic generators.

The Hodge numbers of this example do not arise from a CY_3 constructed using the methods outlined in Sect. 2.1. The resulting CY_3 corresponds to a point on the line connecting the Hodge numbers of the degree 13 and 14 examples appearing in Fig. 1 of the paper [25] of Tonoli.

2 Geometry

In order to discuss Calabi-Yau manifolds and the example above, we begin with a quick, intuitive, and completely non-rigorous reminder of the geometry needed to define Calabi-Yau varieties, followed by a description of the beautiful construction of Batyrev [1].

2.1 Manifolds, Tangent Bundle, Exterior Power and Canonical Bundle

This section is intended to give intuition for the non-expert, so we start with the notion of a real (or complex) compact n-dimensional manifold X, which is a topological space such that every point $p \in X$ has a neighborhood $N_\delta(p)$ diffeomorphic (or holomorphic) to an open neighborhood in $\mathbb{R}^n$ or $\mathbb{C}^n$. In particular, the tangent space $T_p(X)$ at every point $p \in X$ is an $n-$dimensional vector space (Fig. 2).

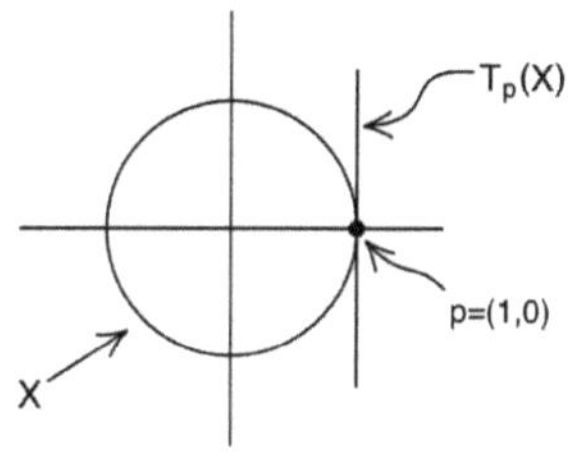

Fig. 2 Tangent space $T_p(X)$ when $p = (1, 0)$ and X is the unit circle

Example 2.1 The unit circle $X = V(x^2 + y^2 - 1) \subseteq \mathbb{R}^2$ is a one dimensional compact real manifold, and we learned in calculus how to define the tangent line to a point $p \in X$.

We gather all the tangent spaces together into a *bundle* to form the *tangent bundle* $\mathcal{T}_X$ of X. The next concept we need is that of an exterior power, which we describe in the vector space setting:

Definition 2.2 For a vector space V with basis $\{e_1, \ldots, e_n\}$, the ith exterior power of V is a vector space $\Lambda^i(V)$, with basis $\{e_{j_1} \wedge \cdots \wedge e_{j_i} \mid j_1 < j_2 < \cdots < j_i\}$. Therefore $\dim(\Lambda^i(V)) = \binom{n}{i}$.

We can carry out this construction when V is the tangent space to X to obtain exterior powers of the tangent space (or, more generally, exterior powers of the tangent bundle). The dual space to the tangent space $T_p(X)$ is the *cotangent space* to X at p, written as $\Omega^1_p(X)$. Just as we generalized from tangent space to tangent bundle, we can do the same to obtain the cotangent bundle $\Omega^1(X)$, and we write $\Omega^i(X)$ for the ith exterior power of the cotangent bundle. When X is of dimension n, then at any point $p \in X$ we have $\dim T_p(X) = n = \dim \Omega^1_p(X)$, so from Definition 2.2 $\dim \Lambda^n(\Omega^1_p(X)) = 1$. The corresponding bundle is therefore one-dimensional, and is known as the *canonical bundle*, denoted K_X.

Remark 2.3 Under certain conditions, the concepts described above extend to singular varieties. This is relevant, because if X is a CY_3, then there is a mirror CY_3 X^o. Mirroring interchanges the Hodge numbers: $h^{11}(X) = h^{21}(X^o)$, and $h^{21}(X) = h^{11}(X^o)$. However, even if X is smooth, X^o may not be.

2.2 *Calabi-Yau Manifolds, Hodge Theory and Hodge Diamond*

Definition 2.4 For X a smooth complex n-dimensional projective variety, the Hodge numbers of X are the dimensions $h^{p,q}(X)$ of the spaces $H^q(\Omega^p(X))$. The Hodge numbers have [26] two symmetries

$$h^{n-p,n-q}(X) = h^{p,q}(X) = h^{q,p}(X)$$

$$
\begin{array}{ccccccc}
 & & & 1 & & & \\
 & & 0 & & 0 & & \\
 & 0 & & h^{11} & & 0 & \\
1 & & h^{12} & & h^{12} & & 1 \\
 & 0 & & h^{11} & & 0 & \\
 & & 0 & & 0 & & \\
 & & & 1 & & &
\end{array}
$$

Fig. 3 The Hodge diamond of a CY threefold

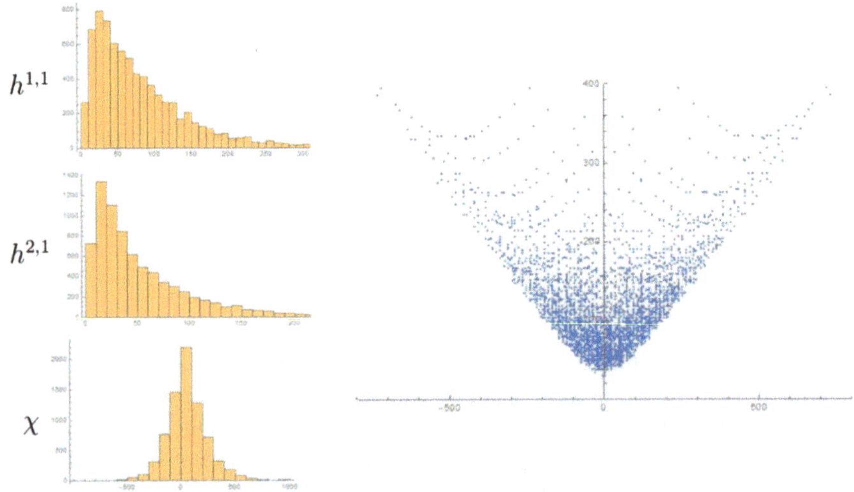

Fig. 4 Scatterplot of CY's with $h^{11} + h^{12}$ on y-axis, and $\chi(X) = 2(h^{11} - h^{21})$ on x-axis

A smooth variety X of dimension n is Calabi-Yau if $K_X \simeq \mathcal{O}_X$ and

$$H^1(\mathcal{O}_X), \ldots, H^{n-1}(\mathcal{O}_X) = 0.$$

The Hodge numbers are typically displayed as a Hodge diamond. The symmetries noted above, combined with the vanishing of $h^{10}(X)$ and $h^{20}(X)$ for a CY_3 means that there are only two interesting Hodge numbers for a CY_3: h^{11} and h^{12}, so the Hodge diamond has the form above (Fig. 3):

Different Calabi-Yau threefolds result in different physical models, so it is of interest to find Calabi-Yau threefolds which have numerical invariants which differ from those already constructed. Figure 4 above respresents the current state of the art, and appears in [17].

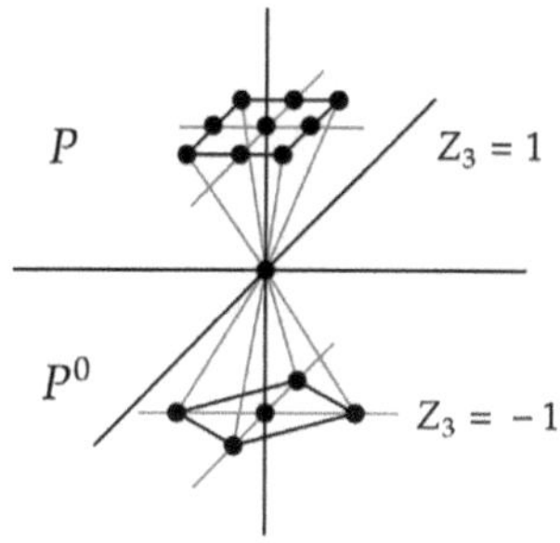

Fig. 5 The reflexive polytope P and the polar dual P^o

2.3 Batyrev's Construction

In [15] Green-Hübsch-Lütken determine all CICY's: complete intersection CY_3's in $\prod_{i=1}^{m} \mathbb{P}^{n_i}$. As $\prod_{i=1}^{m} \mathbb{P}^{n_i}$ is a toric variety [11], this suggests a possible generalization. In [1], Batyrev obtains such a generalization, as well as a beautiful construction for mirror pairs.

Definition 2.5 Let $X \subseteq \mathbb{R}^n$ be a finite set of points. A convex polytope P is the set of all convex combinations $C(X)$ of the points of X. P is a lattice polytope if the points of X are in $\mathbb{Z}^n$, and reflexive if there is exactly one interior lattice point (after translation we may assume the interior point is the origin). The polar polytope P^o is defined as $\{y \in \mathbb{R}^n \mid \langle x, y\rangle \leq -1 \text{ for all } x \in P\}$.

Example 2.6 If $P = C\{(-1,-1), (-1,1), (1,-1), (1,1)\}$, then $P^o = C\{(-1,0), (1,0), (0,-1), (0,1)\}$. This can be visualized in general as follows: embed P in the hyperplane $z_{n+1} = 1 \subseteq \mathbb{R}^{n+1}$, and let P' be the cone with the origin. Then the convex hull of the points where the facet normals to P' meet the hyperplane $z_{n+1} = -1 \subseteq \mathbb{R}^{n+1}$ define P^o. We depict this in Fig. 5 above.

Every toric surface obtained from a lattice polygon is projective, with hyperplane section a curve of genus equal to the number of interior lattice points [11]. So the examples above both have hyperplane sections which are curves of genus one, hence Calabi-Yau.

Theorem 2.7 (Sect. 4 of [1], paraphrased) *A 4-dimensional reflexive lattice polytope P defines a 4-dimensional toric Fano variety X_P, as does the polar polytope P^o. Both X_P and X_{P^o} have mild singularities, and appropriate hypersurfaces in X_P and X_{P^o} yield mirror CY_3's.*

3 Algebra

We now shift gears, and study algebraic objects. The main object of investigation is a homogeneous prime ideal $I \subseteq S = \mathbb{K}[x_0, \ldots, x_n]$ with $\mathbb{K}$ algebraically closed. The corresponding zero locus (solutions to the system of equations defined by I) is a irreducible variety $X = V(I) \subseteq \mathbb{P}^n$. Much of the geometric information about $V(I)$ may be extracted from a *free resolution*.

3.1 Free Resolution, Betti Table, Complete Intersection

For an ideal I as above, the Hilbert syzygy theorem [13] guarantees that there is a *minimal free resolution* of length at most $n+1$.

Definition 3.1 A minimal free resolution for S/I is an exact sequence of graded free modules

$$0 \longleftarrow S/I \longleftarrow F_0 \xleftarrow{d_1} F_1 \xleftarrow{d_2} \cdots \longleftarrow F_{n+1} \longleftarrow 0, \text{ where } F_i = \bigoplus_{j \in \mathbb{Z}} S(-j)^{b_{ij}} \text{ and } b_{ij} \in \mathbb{Z}_+.$$

The differentials d_i are matrices of homogeneous polynomials of positive degree (or zero). The degrees b_{ij} are the *Betti numbers* and are encoded in a *Betti table*, [13]. The top row of the Betti table lists the ranks of the F_i, and the entry of the table in position (i, j)–counting over and down–is $b_{i,i+j}$.

Example 3.2 For the ideal I_C of the twisted cubic in $\mathbb{P}^3$, we have

$$0 \longleftarrow S/I_C \longleftarrow S \xleftarrow{\begin{bmatrix} x_1^2 - x_0x_2 & x_1x_2 - x_0x_3 & x_2^2 - x_1x_3 \end{bmatrix}} S(-2)^3 \xleftarrow{\begin{bmatrix} -x_2 & x_3 \\ x_1 & -x_2 \\ -x_0 & x_1 \end{bmatrix}} S(-3)^2 \longleftarrow 0$$

We see $b_{00} = 1$, $b_{12} = 3$, and $b_{23} = 2$, so the Betti table has the form:

total	1	3	2
0	1	–	–
1	-	3	2

Example 3.3 Instead of the specific quadrics which appear in Example 3.2, consider 3 general quadrics $\{f_1, f_2, f_3\}$. By Bezout's theorem [13] they cut out a zero scheme of degree 8, rather than a curve as in Example 3.2. So they define a complete intersection (CI): the codimension of $V(I)$ is equal to the number of generators of I. For a CI, the differentials come from the "obvious" (or Koszul) relations:

$$0 \longleftarrow S/I_C \longleftarrow S \xleftarrow{\begin{bmatrix} f_1 & f_2 & f_3 \end{bmatrix}} S(-2)^3 \xleftarrow{\begin{bmatrix} f_2 & -f_3 & 0 \\ -f_1 & 0 & f_3 \\ 0 & f_1 & -f_2 \end{bmatrix}} S(-4)^3 \xleftarrow{\begin{bmatrix} f_3 \\ f_2 \\ f_1 \end{bmatrix}} S(-6) \longleftarrow 0$$

The Betti table for this resolution is:

total	1	3	3	1
0	1	–	–	–
1	–	3	–	–
2	–	–	3	–
3	–	–	–	1

The odd shifting in the labelling of the Betti table makes it easy to read off the *Castelnuovo-Mumford* regularity [12] of S/I, which is a key numerical invariant. The connection to Calabi-Yau threefolds comes from the fact that an AG ring of regularity four defines a CY_3, see for example [24].

3.2 Gorenstein Rings, Macaulay Construction, Computation

Complete intersections are examples of *arithmetically Gorenstein* (AG) rings. A ring S/I is AG if $Ext_S^{\mathrm{codim}(I}(S/I, S) \simeq S/I(a)$ for some a, see [12]. A more concrete way to express this is that the last differential in the minimal free resolution is (after a change of basis) equal to the first differential, as in Example 3.3 above.

Every codimension two Gorenstein ideal is a CI, and in [5] Buchsbaum-Eisenbud prove that a codimension three Gorenstein ideal is generated by the Pfaffians of a skew-symmetric matrix. This class of examples is investigated from the CY_3 perspective in [19, 23, 25] and subsequent papers. There has also been much work on the codimension four case: see [2–4, 9, 20, 22, 24].

If S/I is AG, then we may quotient by a regular sequence of linear forms, reducing to an Artinian (finite dimension as a $\mathbb{K}$-vector space) Gorenstein ring with the same Betti table. In particular, to study the behavior of Betti numbers and free resolutions, we may restrict to the Artinian case. It is important to note (see [24] for example) that not every Artinian Gorenstein ring can be lifted non-trivially to the ring of a smooth (or even irreducible) CY_3.

In light of the last remark, it might seem wasteful to study Artinian Gorenstein rings–after all, there are, in a sense, too many such objects, because not every Artinian Gorenstein ring lifts. But there is a compelling reason to study this case, thanks to a 1916 result of Macaulay:

Theorem 3.4 ([21]) *Any graded Artinian Gorenstein ring arises from an inverse system.*

What is this inverse system construction? Let F be a homogeneous polynomial in S of degree d over a field $\mathbb{K}$ of characteristic zero. The set of polynomial differential operators $P(\frac{\partial}{\partial x_0}, \dots, \frac{\partial}{\partial x_n})$ which annihilate F generate an ideal I_F in $T = \mathbb{K}[\frac{\partial}{\partial x_0}, \dots, \frac{\partial}{\partial x_n}]$, which is called the inverse system $I_F \subseteq T$. The corresponding ring T/I_F is an Artinian Gorenstein ring of regularity d (Fig. 6).

Example 3.5 The polynomial $F = x_0x_1x_2$ is obviously annihilated by $\{\frac{\partial}{\partial x_0^2}, \frac{\partial}{\partial x_1^2}, \frac{\partial}{\partial x_2^2}\}$, hence we see that $I_F = \langle x_0^2, x_1^2, x_2^2\rangle$, which is a complete intersection, so also arithmetically Gorenstein.

Example 3.6 A Gorenstein ring which could arise as an Artinian quotient of an AG ring representing a CY_3 must have (see remarks above) regularity four, so will come from the inverse system of a quartic. In [20, 24] we determine all potential

Fig. 6 Macaulay2 code to compute the inverse system of a general quartic in four variables

```
i1 : R=ZZ/101[x_0..x_3];

i2 : F=random(R^{4},R^1); --get a random degree 4 form

              1       1
o2 : Matrix R  <--- R

i3 : minimalBetti inverseSystem F

            0  1  2  3 4
o3 = total: 1 16 30 16 1
         0: 1  .  .  . .
         1: .  .  .  . .
         2: . 16 30 16 .
         3: .  .  .  . .
         4: .  .  .  . 1
```

Artinian reductions of CY_3's in $\mathbb{P}^7$. We illustrate how to carry out the computation in Macaulay2 [14] for a generic quartic F in four variables in Fig. 6.

It is typically not easy to (nontrivially) lift ideals; but in this case the ideal is given by the 3×3 minors of a generic 4×4 matrix. This was observed by Bertin in [2], making it easy to obtain a CY_3 with this resolution (determined in [16]). But computing the Hodge diamond is computationally intensive.

Example 3.7 A Huneke-Ulrich ideal I is obtained from the Pfaffian of a $2n \times 2n$ skew-symmetric matrix M whose entries are variables, together $M \cdot Y$, where Y is $2n \times 1$ vector of variables disjoint from those in M. In [18], Huneke-Ulrich show that such ideals are Gorenstein of deviation two, which means that I has 2 more generators than $\mathrm{codim}(I)$. If we carry this out for $n = 3$ we obtain an ideal with Betti table as in Example 1.1, with $V(I) \subseteq \mathbb{P}^{20}$ of codimension five. Quotienting by twelve generic linear forms yields a CY_3 in $\mathbb{P}^8$, with $V(I)$ smooth. While the standard packages to compute cohomology tend, as in the previous example, to be computationally intensive, in [24] code is developed which allows us to compute the Hodge diamond pictured in Example 1.1.

Acknowledgements I thank the organizers of the Nankai symposium and Chern institute for an interesting and informative conference. This report is an abridged version of [24], written with M. Stillman and B. Yuan.

References

1. V. Batyrev, Dual polyhedra and mirror symmetry for Calabi-Yau hypersurfaces in toric varieties, J. Algebraic Geom. 3 (1994), 493–535.
2. M. Bertin, Examples of Calabi–Yau 3-folds in $\mathbb{P}^7$ with $\rho = 1$, Canad. J. Math. 61, (2009), 1050–1072.
3. G. Brown, K. Georgiadis, Polarized Calabi-Yau 3-folds in codimension 4, Math. Nachr. 290 (2017), 710–725.

4. G. Brown, A. Kasprzyk, L. Zhu, Gorenstein formats, Canonical and Calabi-Yau threefolds, Experimental Mathematics (2019) 46–74.
5. D. Buchsbaum, D. Eisenbud, Algebra structures for finite free resolutions, and some structure theorems for ideals of codimension three, Amer. J. Math. 99 (1977), 447–485.
6. P. Candelas, G. Horowitz, A. Strominger, E. Witten, Vacuum configurations for superstrings, Nuclear Physics B. 258 (1985) 46–74.
7. P. Candelas, X. de la Ossa, P. Green, L. Parkes, A pair of Calabi–Yau manifolds as an exactly soluble superconformal field theory, Nuclear Physics B. 359 (1991): 21–74.
8. P. Candelas, M. Lynker, R. Schimmrigk, Calabi–Yau manifolds in weighted $\mathbb{P}^4$, Nuclear Physics B. 341 (1990), 383–402.
9. S. Coughlan, L. Golebiowski, G. Kapustka, M. Kapustka, Arithmetically Gorenstein Calabi-Yau threefolds in $\mathbb{P}^7$, Electron. Res. Announc. Math. Sci. 23 (2016), 52–68.
10. D. Cox, S. Katz, Mirror Symmetry and Algebraic Geometry, AMS Math Surveys and Monographs, (1999).
11. D. Cox, J. Little, H. Schenck, Toric Varieties, AMS Graduate Studies in Mathematics, (2010).
12. D. Eisenbud, Commutative Algebra with a view towards Algebraic Geometry, Graduate Texts in Mathematics, vol. 150, Springer, Berlin-Heidelberg-New York, 1995.
13. D. Eisenbud, The geometry of syzygies, Graduate Texts in Mathematics, vol. 229, Springer, Berlin-Heidelberg-New York, 2005.
14. D. Grayson, M. Stillman, Macaulay 2: a software system for algebraic geometry and commutative algebra, http://www.math.uiuc.edu/Macaulay2
15. P. Green, T. Hübsch, C. Lütken, All the Hodge numbers for all Calabi-Yau complete intersections, Class. Quantum Grav. 6 (1989), 105–124.
16. T. Gulliksen, O. Negård, Un complexe résolvant pour certains idéaux déterminantiels, C. R. Acad. Sci. Paris 274, (1972), 16–18.
17. Y. H. He, The Calabi-Yau landscape: from geometry to physics to machine learning, Springer, to appear.
18. C. Huneke, B. Ulrich, Divisor class groups and deformations, Amer. J. Math. 107 (1985), 1265–1303.
19. G. Kapustka, M. Kapustka, A cascade of determinantal Calabi-Yau threefolds, Math. Nachr. 283 (2010), 1795–1809.
20. G. Kapustka, M. Kapustka, K. Ranestad, H. Schenck, M. Stillman, B. Yuan Quaternary quartics and Gorenstein rings, memoirs of the AMS, 2026.
21. F.S. Macaulay, Algebraic Theory of Modular Systems, Cambridge University Press, 1916.
22. M. Reid, Gorenstein in codimension 4: the general structure theory Adv. Stud. Pure Math, 65 (2015), 201–227.
23. E. Rødland, The Pfaffian Calabi-Yau, its mirror, and their link to the Grassmannian G(2,7), Compositio Math. 122 (2000), 135–149.
24. H. Schenck, M. Stillman, B. Yuan, Calabi-Yau threefolds in $\mathbb{P}^n$ and Gorenstein rings, Advances in Theoretical and Mathematical Physics, 26 (2022), 764–792.
25. F. Tonoli, Construction of Calabi-Yau 3-folds in $\mathbb{P}^6$, J. Algebraic Geom. 13 (2004), 209–232.
26. C. Voisin, Hodge Theory and Complex Algebraic Geometry I, II, Cambridge Univ. Press, Cambridge, (2003).

The Global Effects of Local Graph Structure

Alex Scott

A fundamental question about mathematical objects is how local structure and global structure interact. What are the large-scale effects of local structure? How are global properties reflected in local structure? The relationship between local and global structure is subtle, and is an important theme in many fields (combinatorics, statistical physics, complexity, geometric group theory, model theory, and across much of applied mathematics and computer science).

Our focus here will be on combinatorics, and in particular on the local/global structure of graphs. Perhaps the first question is: does a graph need to have any structure at all? In other words, can a graph be totally disordered or random? The simplest (or most highly ordered) graphs are complete graphs and graphs with no edges. A *complete subgraph* of a graph G is a set of vertices, every pair of which is joined by an edge; an *independent set* is a set of vertices, no pair of which is adjacent. It is then natural to ask: if G is a large graph, must it contain a large complete subgraph or independent set?[1]

This question was answered by Ramsey [15], who showed in 1930 that every infinite graph contains an infinite complete subgraph or independent set. The finite version of this result (which is equivalent by compactness) is the following:

Theorem 1 (Ramsey's Theorem) *For every $k \geq 1$ there is an integer $R(k)$ such that every graph with at least $R(k)$ vertices contains a complete subgraph or independent set of size k.*

[1] Note that we have to include both complete subgraphs and independent sets as possibilities, as G might itself be a complete graph, or it might have no edges.

Nankai Symposium, August 2021.

A. Scott (✉)
University of Oxford, Oxford, England
e-mail: scott@maths.ox.ac.uk

Y.-H. He et al. (eds.), *Nankai Symposium on Mathematical Dialogues*,
https://doi.org/10.1007/978-981-19-2328-9_36

This elementary but foundational result shows that 'large' graphs contain 'large' homogeneous structures. But how large is large? A quantitative version of Ramsey's Theorem was proved in 1933 by Erdős and Szekeres [10], who showed the 'off-diagonal' result that every graph with at least $\binom{r+s-2}{r-1}$ vertices contains a complete subgraph with r vertices or an independent set with s vertices. A straightforward calculation gives the following corollary.

Theorem 2 *Every graph on n vertices contains a complete subgraph or independent set of size at least* $c_1 \log n$.

Can this be improved? It turns out that this bound is within a constant factor of being best possible. In fact, by considering random graphs, it can be shown that most graphs do not have cliques or independent sets of more than logarithmic size. This was shown by Erdős [8], in one of the earliest applications of the probabilistic method.

Theorem 3 *For some* $c_2 > 0$ *and every* $n \geq 2$*, there are graphs on n vertices do not contain a complete subgraph or independent set of size more than* $c_2 \log n$.

In fact, this bound is quite simple to prove: consider a random graph on n vertices, in which each possible edge is present independently with probability $1/2$. The expected number of independent sets or complete subgraphs on k vertices is $2\binom{n}{k}2^{-\binom{k}{2}} < (n2^{-(k-1)/2})^k$, which tends to 0 if $k > c_2 \log n$ (where c_2 is an appropriately chosen constant). By Markov's Inequality, the probability that there is an independent set or complete subgraph of size more than $c_2 \log n$ also tends to 0.

Thus most graphs have complete subgraphs and independent sets of only logarithmic size. But what if we are given some information about the graph? In particular, how does the picture change if there is some restriction on the local structure of a graph? It turns out that the picture is then very different.

In order to discuss this, we must first fix what we mean by 'local structure'. There are two natural notions of substructure for graphs:

- A *subgraph* of G is any graph obtained from G by deleting vertices and edges.
- An *induced subgraph* of G is any graph obtained by deleting vertices (and only edges that are incident with deleted vertices).

For example, every graph is a subgraph of some complete graph; but every induced subgraph of a complete graph is complete. Our interest here will be in *induced subgraphs*.

So, what happens if we constrain the local structure, by looking at graphs that do not contain some specific graph H as an induced subgraph? Let us say that a graph is H*-free* if it does not contain an induced copy of H. Erdős and Hajnal [9] conjectured in 1989 that H-free graphs exhibit a radically different behaviour: the size of independent sets or complete subgraphs jumps from *logarithmic* to *polynomial*.

Conjecture 4 (Erdős and Hajnal [9]) *For every graph H, there is a constant* $c = c(H) > 0$ *such that the following holds: for* $n \geq 2$*, every H-free graph with n vertices has a complete subgraph or independent set with at least* n^c *vertices.*

The Erdős-Hajnal Conjecture has become one of the central conjectures in graph theory, and has inspired a substantial body of research. Erdős and Hajnal proved that the conjecture holds with the weaker bound $e^{c\sqrt{\log n}}$ in place of n^c, but this remains the best result for general graphs H.

For what graphs H has the the Erdős-Hajnal Conjecture been proved? Embarrassingly few....

- The conjecture holds when H is a complete graph (this follows easily from the quantitative version of Ramsey's theorem).
- The conjecture holds when H is three-edge path (this gives a well understood class of graphs known as *cographs*). This follows from a simple structural argument. A more complex structural argument shows the conjecture for the five-vertex graph known as the *bull* (Chudnovsky and Safra [4]).
- If it's true for H then it is also true for $\overline{H}$ (this follows easily from the statement of the conjecture).

An influential paper of Alon et al. [1] considered the effects of *substitution*. The operation of *substituting* a graph H_1 into a vertex v of a graph H_2 replaces v by a copy of H_1, whose vertices are adjacent to all the neighbours of v in G_2. This was a natural operation to investigate, as it had previously played an important role in the investigation of perfect graphs: Lovász [13, 14] showed that the class of perfect graphs is closed under substitution en route to proving the Weak Perfect Graph Conjecture. Using very different methods, Alon et al. [1] showed the following.

Theorem 5 (Alon et al. [1]) *The class of graphs H for which Erdős-Hajnal holds is closed under substitution.*

Until recently, the only graphs H for which the Erdős-Hajnal Conjecture was known were the graphs listed above, together with graphs that can be obtained from them by repeated applications of complementation and substitution. A particularly frustrating open case was the five-vertex cycle C_5. This has been the focus of significant attention over the last thirty years. Indeed, it was specifically highlighted by Erdős and Hajnal [9] and also by Gyárfás [11].

Recently, there was some progress.

Theorem 6 (Chudnovsky et al. [7]) *The Erdős-Hajnal Conjecture holds when H is a cycle of length 5.*

The next target is clear: up to complementation, the five-vertex path P_5 is the only graph on five vertices for which the Erdős-Hajnal Conjecture is not known.

Another line of development that has been very fruitful recently has been to consider the effects of forbidding more than one induced subgraph. A *hereditary class* of graphs is a class of graphs that is closed under taking induced subgraphs. Two important properties of hereditary classes are the following:

- A hereditary class $\mathcal{G}$ of graphs has the *Erdős-Hajnal property* if there is $c > 0$ such that every $G \in \mathcal{G}$ has an independent set or complete subgraph with at least $|G|^c$ vertices.

- A hereditary class $\mathcal{G}$ of graphs has the *strong Erdős-Hajnal property* if there is $\delta > 0$ such that every $G \in \mathcal{G}$ with at least two vertices has disjoint sets A, B of at least δn vertices such that the pair (A, B) is either complete or anticomplete (either all edges between A and B are present or none are).

The Erdős-Hajnal Conjecture says that the class of H-free graphs has the Erdős-Hajnal property. It is not hard to show that the strong Erdős-Hajnal property imples the Erdős-Hajnal property.

So when does the strong Erdős-Hajnal property hold for the class of H-free graphs? By considering random graphs, it is easily shown that H must be a forest (take sparse random graph, with $p \sim \log n/n$). Unfortunately, the same argument (using very dense random graphs) shows that the complement $\overline{H}$ must also be a forest. But then H has at most four vertices, and the conjecture is already known for these cases.

However, the situation is more interesting if we forbid more than one induced subgraph. The starting point here is a result of Bousquet et al. [2], which states that for every positive integer t, the class of graphs G such that neither G nor its complement contains a t-vertex path as an induced subgraph satisfies the strong Erdős-Hajnal property. It follows that the Erdős-Hajnal property holds if we exclude two graphs: *both* a path on t vertices *and* its complement.

This was extended by Choromanski et al. [3] and then further by Liebenau et al. [12] who conjectured that the path could be replaced by any tree. This conjecture has now been proved.

Theorem 7 (Chudnovsky et al. [5]) *Let T be a tree. Then the class of graphs G such that neither G nor $\overline{G}$ contains an induced copy of T satisfies the strong Erdős-Hajnal property.*

It follows that, if we forbid both a tree and the complement of a tree as induced subgraphs, then we get complete subgraphs or independent sets of polynomial size (the Erdős-Hajnal Conjecture says that it would be enough to forbid just one of these two induced subgraphs). Furthermore, since we *must* exclude both a tree and the complement of a tree to get the strong Erdős-Hajnal property, this result characterizes hereditary classes satisfying the strong Erdős-Hajnal property.

For more general H, there is an analogue of Theorem 7 where we exclude subdivisions of H. A *subdivision* of a graph H is any graph obtained by added new vertices along the edges of H (or equivalently replacing some of the edges by paths).

Theorem 8 (Chudnovsky et al. [6]) *For any graph H, the class of graphs G such that neither G nor $\overline{G}$ contains an induced subdivision of H satisfies the strong Erdős-Hajnal property*

There isn't space here for any detailed discussion of the proofs, but it seems likely that any successful strategy for attacking the conjecture will combine structural understanding with (pseudo)random techniques (indeed, many proofs already begin by applying Szemerédi's Regularity Lemma [17], often by means of a valuable result of Rödl [16]).

In any case, it is clear that the relationship between local and global structure is still somewhat mysterious. The Erdős-Hajnal Conjecture remains a central problem in graph theory, and an intriguing challenge.

References

1. N. Alon, J. Pach and J. Solymosi, Ramsey-type theorems with forbidden subgraphs, *Combinatorica* **21** (2001), 155–170.
2. N. Bousquet, A. Lagoutte, and S. Thomassé, The Erdős-Hajnal conjecture for paths and antipaths, *J. Combinatorial Theory, Ser. B*, **113** (2015), 261–264.
3. K. Choromanski, D. Falik, A. Liebenau, V. Patel, and M. Pilipczuk, Excluding hooks and their complements, *Electronic J. Combinatorics* **25** (2018) P3.27.
4. M. Chudnovsky and M. Safra, The Erdős-Hajnal conjecture for bull-free graphs, *J. Combinatorial Theory, Ser. B*, **98** (2008), 1301–1310.
5. M. Chudnovsky, A. Scott, P. Seymour and S. Spirkl, Pure pairs. I. Trees and linear anticomplete pairs, *Advances in Mathematics* **375** (2 December 2020), 107396.
6. M. Chudnovsky, A. Scott, P. Seymour and S. Spirkl, Pure pairs. II. Excluding all subdivisions of a graph, *Combinatorica* **41** (2021), 379–405.
7. M. Chudnovsky, A. Scott, P. Seymour and S. Spirkl, Erdős-Hajnal for graphs with no 5-hole, *arXiv:2102.04994*
8. P. Erdős, Some remarks on the theory of graphs, *Bulletin of the American Mathematical Society* **53** (1947), 292–294.
9. P. Erdős and A. Hajnal, Ramsey-type theorems, *Discrete Applied Math.* **25** (1989), 37–52.
10. P. Erdős and G. Szekeres, A combinatorial problem in geometry, *Compositio Mathematica* **2** (1935), 463–470.
11. A. Gyárfás, Reflections on a problem of Erdős and Hajnal, *in: The Mathematics of Paul Erdős*, (R. L. Graham and J. Nešetřil, eds.), Algorithms and Combinatorics **14**, Volume II, Springer-Verlag, Heidelberg, 1997, 93–98.
12. A. Liebenau, M. Pilipczuk, P. Seymour and S. Spirkl, Caterpillars in Erd"os-Hajnal, *J. Combinatorial Theory, Ser. B* **136** (2019), 33–43.
13. L. Lovász, Normal hypergraphs and the perfect graph conjecture, *Discrete Mathematics* **2** (1972), 253–267.
14. L. Lovász, Perfect Graphs, *in: Selected Topics in GraphTheory, Volume 2* (L. W. Beineke, R. J. Wilson, eds.), Academic Prcess, London-New York, 1983, 55–87.
15. F.P. Ramsey, On a Problem of Formal Logic, *Proceedings of the London Mathematical Society* (1930), 264–286.
16. V. Rödl, On universality of graphs with uniformly distributed edges, *Discrete Math.* **59** (1986), 125–134.
17. E. Szemerédi, Regular partitions of graphs, *Problèmes combinatoires et théorie des graphes* (Colloq. Internat. CNRS, Univ. Orsay, Orsay, 1976), Colloq. Internat. CNRS, vol. 260, Paris: CNRS, pp. 399–401.

Decomposition of Quantum Field Theories

Eric Sharpe

Briefly, decomposition [1] is the observation that some quantum field theories are equivalent to disjoint unions of other quantum field theories, known as 'universes.' When this happens, we say that the quantum field theory decomposes, and decomposition can be applied to give insight into its properties. There has been a great deal of work done on decomposition and its applications since the paper [1] appeared, too much to summarize here, so we shall only give a short overview of a handful of highlights.

In principle, for one quantum field theory to be a disjoint union of others means that the theory contains topological operators Π_i such that

$$\Pi_i \Pi_j = \delta_{ij} \Pi_j, \quad \sum_i \Pi_i = 1. \tag{1}$$

Given such a set of topological operators, it is easy to see that correlation functions decompose:

$$\langle \mathcal{O}_1 \cdots \mathcal{O}_m \rangle = \sum_i \langle \Pi_i \mathcal{O}_1 \cdots \mathcal{O}_m \rangle = \sum_i \langle (\Pi_i \mathcal{O}_1) \cdots (\Pi_i \mathcal{O}_m) \rangle = \sum_i \langle \tilde{\mathcal{O}}_1 \cdots \tilde{\mathcal{O}}_m \rangle_i. \tag{2}$$

One efficient way to check for decomposition is via partition functions. If a theory decomposes, then on a connected spacetime, the partition function is a sum of the partition functions of the constituent universes:

This note is a contribution to the proceedings of the Nankai symposium, August 2021.

E. Sharpe (✉)
Department of Physics, Virginia Tech, Blacksburg, VA, USA
e-mail: ersharpe@vt.edu

Y.-H. He et al. (eds.), *Nankai Symposium on Mathematical Dialogues*,
https://doi.org/10.1007/978-981-19-2328-9_37

$$Z = \sum_i Z_i. \tag{3}$$

In d spacetime dimensions, a theory decomposes if it has a (possibly noninvertible) global $(d-1)$-form symmetry [2, 3]. In this short review, I will focus on two-dimensional theories with linear one-form symmetries (though decomposition is understood more generally). For this review, intuitively, a one-form symmetry is a 'group' (technically, a 2-group) that exchanges nonperturbative sectors. Consider for example a G gauge theory in which the massless matter is invariant under a finite abelian subgroup $K \subseteq G$, which for simplicity here we take to lie in the center of G. Then, nonperturbative sectors (principal G bundles) are invariant under

$$(G \text{ bundle}) \mapsto (G \text{ bundle}) \otimes (K \text{ bundle}), \tag{4}$$

or intuitively, translations of a gauge field as

$$A \mapsto A + A'. \tag{5}$$

This is the one-form symmetry, involving an action of a 'group' of principal K bundles (rather than elements of K). This (analogue of a) group is denoted BK (in the math literature) and $K^{(1)}$ (in the physics literature).

Two-dimensional theories with global one-form symmetries can be described in several ways, such as [1, 4–6]

- gauge theories with trivially-acting subgroups,
- theories with restrictions on instantons,
- theories with multiple topological (dimension zero) local operators,
- sigma models on gerbes. (Gerbes are fiber bundles with fibers equal to 'groups' of one-form symmetries BG.)

Decomposition (into 'universes') relates these pictures. It also solves technical issues with such theories. For example, one ordinarily does not restrict instantons, as this violates the cluster decomposition of quantum field theory. The fact that such theories are equivalent to disjoint unions, effectively solves that problem. (A disjoint union also violates cluster decomposition, but in a mild and controllable fashion.) Decomposition has been checked in numerous ways, described extensively in references elsewhere; here, we will merely observe recent numerical verification in lattice models [7].

To make this more concrete, consider a G gauge theory in two dimensions, with finite $K \subset G$ acting trivially, where for simplicity we assume K is in the center for the moment.

For reasons discussed above, this quantum field theory has a BK one-form symmetry. It also decomposes. Formally, the projection operators are constructed from twist fields/Gukov-Witten operators that correspond to elements of the center of the group algebra $\mathbb{C}[K]$. Existence of projectors, forming a basis for the center, is ultimately a consequence of Wedderburn's theorem in mathematics. Universes are in one-to-one correspondence with irreducible representations of K.

In this case, decomposition is the statement

$$\text{QFT}\,(G \text{ gauge theory}) \;=\; \coprod_{\hat{K}} \text{QFT}\,(G/K \text{ gauge theory with discrete theta angles/torsion})\,.$$

Formally, the partition function of the disjoint union can be written

$$Z = \sum_{\theta \in \hat{K}} \int [DA] \exp(-S) \exp\left(\theta \int w_2(A)\right), \tag{6}$$

$$= \int [DA] \exp(-S) \left(\sum_{\theta \in \hat{K}} \exp\left(\theta \int w_2(A)\right)\right), \tag{7}$$

but the factor

$$\sum_{\theta \in \hat{K}} \exp\left(\theta \int w_2(A)\right) \tag{8}$$

is proportional to a projection operator, which cancels out some of the nonperturbative sectors. This formal argument reveals that, for the correct weightings, a disjoint union of several QFTs is equivalent to a single QFT with a restriction on nonperturbative sectors. In this fashion, we see that restrictions on nonperturbative sectors arise as a "multiverse interference effect" in decomposition.

Let us briefly outline how this works in pure Yang-Mills theory in two dimensions for gauge groups with nontrivial centers, see [8] for the details. For example, due to the fact that $SU(2)$ has a central $\mathbb{Z}_2$, pure $SU(2)$ gauge theory in two dimensions is the same as a disjoint union of pure $SO(3)_+$ and $SO(3)_-$ gauge theories, where the $\pm$ denote discrete theta angles. Perturbatively, the $SU(2)$ and $SO(3)_\pm$ theories are identical; differences lie in nonperturbative sectors. Specifically, the $SU(2)$ instantons (bundles) are a subset of $SO(3)$ instantons (bundles), and the discrete theta angles weight the non-$SU(2)$ $SO(3)$ bundles by signs, so that they cancel out of the sum. As a result, summing over the $SO(3)$ theories projects out the non-$SU(2)$ bundles, giving the $SU(2)$ theory. In fact, this decomposition by the center one-form symmetry is a special case of a more radical decomposition of pure two-dimensional Yang-Mills theory into a disjoint union of invertible field theories indexed by irreducible representations, discussed in [9]. (The details are unfortunately beyond the scope of this review.)

Let us next walk through an example of an orbifold. Consider $[X/D_4]$, where D_4 is the eight-element dihedral group with center $\mathbb{Z}_2$, and assume that center acts trivially. Using the fact that $D_4/\mathbb{Z}_2 = \mathbb{Z}_2 \times \mathbb{Z}_2$, decomposition predicts [1]

$$\text{QFT}\,([X/D_4]) \;=\; \text{QFT}\left([X/\mathbb{Z}_2 \times \mathbb{Z}_2] \coprod [X/\mathbb{Z}_2 \times \mathbb{Z}_2]_{\text{d.t.}}\right), \tag{9}$$

where the second universe has discrete torsion.

In terms of partition functions, only some of the $\mathbb{Z}_2 \times \mathbb{Z}_2$ sectors are present, others (forming a modular orbit) are omitted, as discussed in [1]—another example of a restriction on instantons. Here, the omitted sectors are multiplied by signs when one turns on discrete torsion in the $\mathbb{Z}_2 \times \mathbb{Z}_2$ orbifold, so adding together the two orbifolds cancels out the appropriate sectors—an example of a multiverse interference effect.

So far we have discussed examples in which the trivially-acting subgroup lies in the center, but decomposition applies more generally, as discussed in [1]. For example, consider the orbifold $]X/\mathbb{H}]$, where $\mathbb{H}$ is the eight-element group of unit quaternions, and the subgroup that acts trivially is $\langle i \rangle = \mathbb{Z}_4$, which includes the $\mathbb{Z}_2$ center of $\mathbb{H}$ as a proper subset. In this case, it can be shown

$$\mathrm{QFT}\,([X/\mathbb{H}]) \;=\; \mathrm{QFT}\left(X \coprod [X/\mathbb{Z}_2] \coprod [X/\mathbb{Z}_2]\right). \tag{10}$$

In this case, one has three universes, which are not all the same.

We have discussed decomposition in two dimensions; however, analogues also exist in other dimensions. Specifically, a d-dimensional theory with a $(d-1)$-form symmetry is expected to decompose, as discussed in [2, 3].

Having described the basics of decomposition, let us now outline a few applications.

One of the first applications was to understand string compactifications on generalizations of spaces known as stacks and gerbes [5]. A gerbe is a fiber bundle whose fibers are higher groups BG, and so a well-defined sigma model on a gerbe has a global BG symmetry, corresponding to translations along the fibers. Applying decomposition, we see that a two-dimensional sigma model on a gerbe is equivalent to a disjoint union of sigma models on ordinary spaces. This had the salutary effect of resolving technical issues with cluster decomposition in such models, one of a number of technical challenges in that context.

Now, there exists a notion of Gromov-Witten theory for stacks and gerbes, and so decomposition immediately makes a prediction for Gromov-Witten theory, namely that Gromov-Witten invariants of gerbes are the same as sums of Gromov-Witten invariants of spaces. This prediction was checked rigorously in a series of papers that began in [10].

Another set of applications was to understand phases of gauged linear sigma models (GLSMs), where decomposition has been used to describe geometries nonperturbatively, resolving puzzles in certain models (see e.g. [11]). Another application is to elliptic genera, whose decomposition can be used as a computational tool, see e.g. [12]. Another more recent application is to anomaly resolution in orbifolds, where decomposition can be applied to a generalization of quantum symmetries to understand how anomaly-resolution procedures operate, see e.g. [13]. These ideas also play a role in understanding SCFT moduli spaces, where one finds for example that (as a result of self-dualities) the moduli spaces admit gerbe structures, see e.g. [14–16].

References

1. S. Hellerman, A. Henriques, T. Pantev, E. Sharpe and M. Ando, "Cluster decomposition, T-duality, and gerby CFT's," Adv. Theor. Math. Phys. **11** (2007) 751–818, arXiv:hep-th/0606034.
2. Y. Tanizaki and M. Ünsal, "Modified instanton sum in QCD and higher-groups," JHEP **03** (2020) 123, arXiv:1912.01033.
3. A. Cherman and T. Jacobson, "Lifetimes of near eternal false vacua," Phys. Rev. D **103** (2021) 105012, arXiv:2012.10555.
4. T. Pantev and E. Sharpe, "Notes on gauging noneffective group actions," arXiv:hep-th/0502027.
5. T. Pantev and E. Sharpe, "String compactifications on Calabi-Yau stacks," Nucl. Phys. B **733** (2006) 233-296, arXiv:hep-th/0502044.
6. T. Pantev and E. Sharpe, "GLSM's for gerbes (and other toric stacks)," Adv. Theor. Math. Phys. **10** (2006) 77–121, arXiv:hep-th/0502053.
7. M. Honda, E. Itou, Y. Kikuchi and Y. Tanizaki, "Negative string tension of higher-charge Schwinger model via digital quantum simulation," arXiv:2110.14105.
8. E. Sharpe, "Decomposition in diverse dimensions," Phys. Rev. D **90** (2014) 025030, arXiv:1404.3986.
9. M. Nguyen, Y. Tanizaki and M. Ünsal, "Noninvertible 1-form symmetry and Casimir scaling in 2D Yang-Mills theory," Phys. Rev. D **104** (2021) 065003, arXiv:2104.01824.
10. E. Andreini, Y. Jiang, H.-H. Tseng, "On Gromov-Witten theory of root gerbes," arXiv:0812.4477.
11. A. Caldararu, J. Distler, S. Hellerman, T. Pantev and E. Sharpe, "Non-birational twisted derived equivalences in abelian GLSMs," Commun. Math. Phys. **294** (2010) 605–645, arXiv:0709.3855.
12. R. Eager and E. Sharpe, "Elliptic genera of pure gauge theories in two dimensions with semisimple non-simply-connected gauge groups," Commun. Math. Phys. **387** (2021) 267–297, arXiv:2009.03907.
13. D. G. Robbins, E. Sharpe and T. Vandermeulen, "Anomaly resolution via decomposition," Int. J. Mod. Phys. A **36** (2021) 2150220, arXiv:2107.13552.
14. W. Gu and E. Sharpe, "Bagger–Witten line bundles on moduli spaces of elliptic curves," Int. J. Mod. Phys. A **31** (2016) 1650188, arXiv:1606.07078.
15. R. Donagi, M. Macerato and E. Sharpe, "On the global moduli of Calabi-Yau threefolds," arXiv:1707.05322.
16. R. Donagi, M. Macerato and E. Sharpe, "Global aspects of moduli spaces of 2d SCFTs," arXiv:1906.11254.

Accessible Information: From Coins to Quantum Gravity

Joan Simon

Abstract We revisit the universal role played by the mathematics of information in physics, focusing on entangled states in quantum mechanics, the Unruh effect in quantum field theory and, time permitting, the connection between these two in gauge/gravity duality.

1 What is this Talk About?

Given the multi-disciplinary nature of this conference, it seems appropriate to stress the universality of *information theory*, not only for the natural sciences, but also for the social ones.

Consider different tasks such as flipping a coin, performing a measurement in quantum mechanics, observing the exterior of a black hole or guessing an english word when only a subset of its letters are known. Their specific details are manifestly different. However, they all share a common feature: the observer performing the task faces some intrinsic uncertainty regarding the outcome because of her ignorance on the precise *state* of the given system.

We refer to such uncertainty as *information*. The purpose of this talk is to review how to quantify this notion in a classical and a quantum set-up, stressing the common underlying mathematical structure and the overall shared logic.

2 How to Quantify Information

When we flip a coin, we know there are only two outcomes, but there exists an intrinsic uncertainty since each occurs with some probability. How much *information* do we gain *after* the flip ? Is it the same for a fair and a biased coin ?

J. Simon (✉)
School of Mathematics and Maxwell Institute, University of Edinburgh, Peter Guthrie Tait Road, Edinburgh EH9 3FD, UK
e-mail: J.Simon@ed.ac.uk

Y.-H. He et al. (eds.), *Nankai Symposium on Mathematical Dialogues*,
https://doi.org/10.1007/978-981-19-2328-9_38

If we view a coin as a particular example of a discrete random variable X with outcomes x_i occurring with probabilities p_i, where $i = 1, \ldots n$ labels the number of outcomes, quantifying our uncertainty is equivalent to finding a function $h(p_i)$ satisfying some properties capturing our intuitive notion of uncertainty.

The *entropy* $h(x_i)$ *of an event* x_i can be shown to be uniquely fixed by

$$h(x_i) = -\log_2 p_i \,, \tag{1}$$

under some continuity assumption, requiring the entropy to be additive for independent events and choosing some specific normalisation.

When a rare event occurs, (1) tells us we gain more information than when a typical event does. However, this fact does not take into account its occurrence is rare. This notion is captured by the expectation value of $h(x_i)$ over the full random variable. This defines Shannon's entropy [1]

$$\boxed{H(X) \equiv \langle h(X) \rangle = -\sum_{x_i \in X} p_i \log_2 p_i} \tag{2}$$

Returning to the coin example, (2) confirms fair coins carry more information, than biased ones. Indeed, if you are told a coin is biased and you are asked to make a guess, your strategy should be clear : you will choose the most probable outcome, i.e. you gain less information when you flip a biased coin.

A game with a deep lesson. To stress the universality of Shannon's theory and its link to the natural sciences, let us consider the following toy example. Consider a set of 4 cells. Each cell can be white or black, equivalently, 0 or 1, respectively. If order, or location, matters, there are 2^4 inequivalent cell configurations.

Consider an observer who knows there are 4 cells but can *only* measure the number of black cells (E). Whenever $E = 0$ or $E = 4$, there is a unique configuration compatible with that data. However, whenever $E = n$ with $n = 1, 2, 3$, there are

$$d_E = \binom{4}{E} \tag{3}$$

compatible configurations with our data. Assigning equal probability $p_i = \frac{1}{d_E}$ to each of these, Shannon's entropy (2) equals

$$S(E) = -\sum_{i=1}^{d_E} p_i \log_2 p_i = \log_2 d_E \,. \tag{4}$$

Readers familiar with statistical mechanics will recognise this discussion as a particular example of the microcanonical ensemble. However, Shannon's framework provides a broader and more universal perspective.

What did we learn from this example? *Coarse-graining* our description, i.e. having *no access to the small cells*, generated some uncertainty quantified by Shannon's entropy (2). In general, entropy is *relative to an observer's knowledge.*

3 Quantum Information

In quantum mechanics, a physical system is assigned to a Hilbert space $\mathcal{H}$. Its state is described by a *density operator* ρ_0, i.e. a positive definite hermitian operator with tr(ρ_0)=1, and its time evolution $\rho(t)$ is determined by the action of a unitary operator $U(t)$ so that $\rho(t) = U(t)\rho_0 U^\dagger(t)$.

Measurements are characterised by a collection of operators $\{M_m\}$ satisfying $\sum_m M_m^\dagger M_m = \mathbb{I}$. The outcome of the measurement is labelled by m and it occurs with probability

$$p(m) = \text{tr}\left(\rho(t)\, M_m^\dagger M_m\right). \tag{5}$$

Consider a single qubit, i.e. $\mathcal{H} = \mathbb{C}^2$. Pick a normalised vector $|\psi\rangle = \alpha|0\rangle + \beta|1\rangle$, i.e. $|\alpha|^2 + |\beta|^2 = 1$, so that $\rho_0 = |\psi\rangle\langle\psi|$. Since the state of the system depends on arbitrary continuous parameters, one may wonder whether there is an infinite amount of information to be gained. However, this expectation is *not* correct. For example, there exists a measurement $\{P_\psi = |\psi\rangle\langle\psi|,\, \mathbb{I} - P_\psi\}$ identifying ρ_0 with *certainty*. From this perspective, there is nothing to be gained by such measurement and according to our earlier discussion, there would be no entropy to be assigned to this situation.

Should we conclude this is always the case? In our classical discussion, coarse-graining generated uncertainty. How can we coarse-grain in quantum mechanics? Let us consider $\mathcal{H} = \mathbb{C}^2_{\text{A}} \otimes \mathbb{C}^2_{\text{B}}$ and an observer that can only measure operators of the form $M_{\text{A}} \otimes \mathbb{I}_{\text{B}}$. Given a vector

$$|\psi\rangle_{\text{AB}} = a\,|0\rangle_{\text{A}} \otimes |0\rangle_{\text{B}} + b\,|1\rangle_{\text{A}} \otimes |1\rangle_{\text{B}} \in \mathcal{H}_{\text{A}} \otimes \mathcal{H}_{\text{B}}, \tag{6}$$

applying the axioms of quantum mechanics, the expectation value of any such measurement equals

$$\begin{aligned} \langle\psi|M_{\text{A}} \otimes \mathbb{I}_{\text{B}}|\psi\rangle &= |a|^2\langle 0|M_{\text{A}}|0\rangle + |b|^2\langle 1|M_{\text{A}}|1\rangle = \text{tr}_{\mathcal{H}_{\text{A}}}\left(\rho_{\text{A}}\, M_{\text{A}}\right) \\ \rho_{\text{A}} \equiv \text{tr}_{\text{B}}|\psi\rangle_{\text{AB}}\langle\psi| &= \text{tr}_{\text{B}}\rho_{\text{AB}} = |a|^2|0\rangle\langle 0| + |b|^2|1\rangle\langle 1| \end{aligned} \tag{7}$$

What we learn is that any such *local* measurement, i.e. measurement performed in $\mathbb{C}^2_{\text{A}}$, is compatible, and in fact can *not* be distinguished, from the *ensemble* described by the *reduced density operator* ρ_{A} characterised by the classical probabilities $|a|^2$ and $|b|^2$.

By analogy with Shannon's entropy, given any density operator, we can define the von Neumann entropy [2]

$$\boxed{S(\rho) = -\mathrm{tr}\rho \log \rho} \tag{8}$$

Whenever $\rho = |\psi\rangle\langle\psi|$, it follows $S(\rho) = 0$ because it has a single eigenvalue. However, when $\rho = \sum_i p_i \rho_i$, the entropy is generically non-vanishing. In our example above,

$$S(\rho_A) = -|a|^2 \log_2 |a|^2 - |b|^2 \log_2 |b|^2 = -|a|^2 \log_2 |a|^2 - (1 - |a|^2) \log_2 (1 - |a|^2) \,. \tag{9}$$

The lesson we extract from this quantum discussion is similar to the one we reached in the classical discussion: not having access to the subsystem $\mathbb{C}^2_B$, generated some uncertainty captured by the reduced density matrix ρ_A and quantified by (8).

However, our quantum discussion uncovers a fundamentally new feature that is *non-classical*. Consider the basis of Bell states in $\mathbb{C}^2 \otimes \mathbb{C}^2$

$$|\phi^\pm\rangle = \frac{1}{\sqrt{2}} (|00\rangle \pm |11\rangle) \,, \qquad |\psi^\pm\rangle = \frac{1}{\sqrt{2}} (|01\rangle \pm |10\rangle) \tag{10}$$

where $|nm\rangle$ is short notation for $|n\rangle \otimes |m\rangle$. Pick any of these vectors, call it $|\psi_{\text{Bell}}\rangle$ and construct $\rho_{\text{Bell}} = |\psi_{\text{Bell}}\rangle\langle\psi_{\text{Bell}}|$. There exists a projective measurement

$$\begin{aligned} P_{\phi^\pm} &= |\phi^\pm\rangle\langle\phi^\pm| \,, \qquad P_{\psi^\pm} = |\psi^\pm\rangle\langle\psi^\pm| \\ \mathbb{I}_4 &= P_{\phi^+} + P_{\phi^-} + P_{\psi^+} + P_{\psi^-} \end{aligned} \tag{11}$$

that is capable of identifying ρ_{Bell} with *certainty*. Consequently, $S(\rho_{\text{Bell}}) = 0$. Any *non-local* observer with *access* to both qubits would reach this conclusion. However, observers having only access to a single qubit are only sensitive to the reduced density matrix

$$\rho_A = \mathrm{tr}_B \rho_{\text{Bell}} = \frac{1}{2}\mathbb{I}_A \quad \Rightarrow \quad S(\rho_A) = 1 \,. \tag{12}$$

Any quantum state having vanishing von Neumann entropy from the full quantum system perspective, but having non-zero entropy from the open (coarse-grained) subsystem is called *entangled*. Equivalently, *entanglement* is the quantum property allowing a non-local observer to know everything about the full system, while a local observer may be maximally ignorant about its subsystem.[1]

The above discussion may appear abstract, though it has systematically been checked experimentally in quantum laboratories. Let us connect it with *thermal* physics next. In quantum statistical mechanics, it is well known the state ρ_β describing the equilibrium of the system is the one maximising (8) subject to the constraint

[1] The key difference is that even if we allow both local observers to exchange classical information, the quantum correlations hidden in the original vector are *not* accessible to them separately. This is the feature not occurring classically, i.e. if we break a code into two subcodes and we bring them back together, we surely can reproduce the original code in this way. The same is *not* true in quantum mechanics for local observers.

of having some fixed energy, i.e. ρ_β extremises $(\delta I(E)/\delta\rho = \delta I(e)/\delta\lambda = 0)$ the functional

$$I[E] \equiv -\text{tr}\,(\rho\log\rho) + \lambda(\text{tr}\,(\rho\,H) - E) \quad\Rightarrow\quad \rho_\beta = \frac{e^{-\beta H}}{Z(\beta)} = \frac{e^{-\beta H}}{\text{tr}(e^{-\beta H})} \tag{13}$$

where $\beta = (k_B T)^{-1}$, with T the physical temperature and k_B Boltzmann's constant, is implicitly determined by the constraint $\text{tr}\,(\rho\,H) = E$. Using the energy eigenbasis diagonalising the Hamiltonian H, i.e. $\rho_\beta = \sum_i \frac{e^{-\beta E_i}}{Z(\beta)}|E_i\rangle\langle E_i|$, we can evaluate (8) for this equilibrium density operator

$$S(\rho_\beta) = \beta\,E + \log Z(\beta) = S_{\text{therm}}(\beta). \tag{14}$$

Thus, the von Neumann entropy reproduces the *thermodynamic* entropy. Moreover, the canonical ensemble ρ_β can be understood as the reduced density matrix of the pure state defined in a larger Hilbert space consisting of two copies $(\mathcal{H}\otimes\mathcal{H})$ of the original one $(\mathcal{H})$

$$|\Psi\rangle = \sum_i \frac{e^{-\beta E_i/2}}{\sqrt{Z(\beta)}}|E_i\rangle\otimes|E_i\rangle \qquad \text{so that} \qquad \rho_\beta = \text{Tr}_{\mathcal{H}}|\Psi\rangle\langle\Psi|. \tag{15}$$

Thus, thermodynamic entropy can be reinterpreted as entanglement entropy using this purification protocol. We shall return to the gravitational physics reincarnation of this mathematical result shortly.

4 Quantum Field Theory

The previous quantum mechanical discussion did not depend on the location of the quantum (sub)systems. However, *locality* is fundamental in any formulation of relativistic quantum field theory in order to capture the relativistic invariance of the relevant physical degrees of freedom.

In the following, we will ignore the mathematical subtleties concerned with the proper mathematical definition of the von Neumann entropy for a theory defined in the continuum. We will introduce an ultraviolet cut-off δ and describe the corresponding system as a spin lattice system.

Consider a time independent operator $\hat{\phi}(\mathbf{x})$ with $\mathbf{x}\in\mathbb{R}^3$. The relevant Hilbert space is $\otimes_{\mathbf{x}}\mathcal{H}_{\mathbf{x}}$. Introducing an orthonormal basis of eigenstates

$$\hat{\phi}(\mathbf{x})|\phi\rangle = \phi(\mathbf{x})\,|\phi\rangle \quad \text{where} \quad |\phi\rangle = \prod_{\mathbf{x}}|\phi(\mathbf{x})\rangle, \tag{16}$$

the natural extension of the qubit entanglement discussion here is to consider local observers restricted to a *region of space A*. It turns out, the von Neumann entropy of the reduced density matrix in region A in the vacuum $|0\rangle$, i.e. $\rho_A = \mathrm{Tr}_{\bar{A}}|0\rangle\langle 0|$ where $\bar{A}$ stands for the complementary region to the one where we have access to, equals

$$S(\rho_A) \propto |\partial A| + \dots \tag{17}$$

Notice this scales with the *area of the boundary of the region* ∂A, where dots stand for subleading contributions.

This result is universal. Notice this scaling is different from the volume scaling controlling the thermodynamic entropy in a relativistic quantum field theory at finite temperature. For the logic of this talk, what is relevant to stress here is that having no access to the complementary region to A, generates uncertainty, i.e. the vacuum of a relativistic quantum field theory is entangled. This is the same phenomenon we encountered with flipping coins or pairs of qubits.

4.1 Unruh's Effect

A particularly striking result in relativistic quantum field theory helping us to bridge the discussion to gravitational physics, and black holes in particular, is *Unruh's effect*. Consider observers in special relativity with a constant proper acceleration a, i.e. Rindler observers, in the x-direction. Their trajectories describe hyperbolas that asymptote to the light-cones $x = \pm t$, where t is the inertial time.

Rindler observers provide examples of observers confined to specific local sub-regions of Minkowski spacetime. This is a reflection of their non-inertial hyperbolic motion and because of that the light-cones $x = \pm t$ are actual causal boundaries for them. In particular, given a constant inertial $t = 0$ Cauchy surface Σ, it naturally splits into $\Sigma_L \cup \Sigma_R$, where Σ_L and Σ_R are Rindler observer Cauchy surfaces for their corresponding open subsystems, which are referred to as left and right Rindler wedges, respectively.

Given some local degrees of freedom $\phi(\mathbf{x})$, these naturally split into $\phi_R(x > 0)$ and $\phi_L(x < 0)$. This set-up allows us to explore the general quantum field theory framework described earlier. Indeed, given the vacuum $|0\rangle$ of the inertial Minkowski observer, with density matrix $\rho = |0\rangle\langle 0|$, one can ask what the reduced density matrix is for each of the Rindler wedges. Famously, the latter is given by

$$\rho_R = \frac{1}{Z} e^{-2\pi K_R}, \tag{18}$$

where K_R stands for the restriction of the boost generator to the right Rindler wedge. Physically, this describes a *thermal* density matrix with physical temperature, the *Unruh* temperature [3]

$$\boxed{T_{\text{Unruh}} = \frac{\hbar a}{2\pi c\, k_B}} \tag{19}$$

Let us stress how this result fits into the logic of the talk : non-inertial Rindler observers are confined to subregions of the Minkowski spacetime and they perceive the Minkowski vacuum as a thermal state at temperature (19). Thus, once more, not having access to one of the Rindler wedges generates uncertainty. The latter is encoded in perceiving a mixed state at finite temperature (18), pretty much as when we coarse-grain the microscopic states of the molecules in a gas, leading to a thermodynamic description of the equilibrium system as a thermal state.

5 Role of Information in Quantum Gravity

The equivalence principle tells us a non-inertial frame can locally not be distinguished from a gravitational field. It is natural to wonder whether we can extend Unruh's conclusion to gravitational physics. In particular, since the near horizon geometry of any finite temperature black hole is locally like the spacetime experienced by a Rindler observer, do black holes show similar behaviour to the one discussed for constant proper acceleration observers ?

Famously, it turns out the answer is yes. Hawking [4] discovered black holes radiate at finite temperature when quantum effects ($\hbar \neq 0$) are turned on. His calculation confirmed the earlier mathematical theorems [5] and physical arguments [6] assigning an entropy to the black hole

$$S_{\text{BH}} = \frac{A_{\text{h}}}{4\hbar G_{\text{N}}}, \tag{20}$$

where A_{h} stands for the area of the black hole event horizon and G_{N} is Newton's constant.

These physical properties are measured by an observer remaining *outside* of the black hole. The event horizon plays the same role as the $x = \pm t$ light-cone boundary for the Rindler observer's in the previous discussion. Thus, outside observers have no access to the interior of the black hole. Classically, black holes behave like thermal equilibrium systems for these asymptotic observers, extending our earlier ideas to gravitational physics.

This result has profound consequences. Even if we do not know a non-perturbative definition of quantum gravity, the semi-classical study of black holes allowed us to derive (20). The scaling of the thermodynamic entropy with the area is different from the volume scaling one finds in relativistic quantum field theories at finite temperature. This led to conjecture that quantum gravity should satisfy the *holographic principle* [7, 8] : a region with boundary of area A is fully described by no more than $\frac{A}{4\hbar G_{\text{N}}}$ degrees of freedom.

These ideas have been explicitly realised in the AdS/CFT correspondence [9]. Furthermore, one may wonder whether the black hole entropy (20) can be interpreted as entanglement entropy (17), since both scale with the area. In [10], Ryu and Takayanagi (RT) conjectured the entanglement entropy in a subregion of the boundary A in the holographic principle is captured by the area of an extremal surface in the volume enclosed by A. This claim generalises the black hole-entanglement entropy relation to arbitrary subregions of space and we know it to be true within the AdS/CFT set-up.

The AdS/CFT realisation of the holographic principle and the RT formula provide an explicit link between ideas in quantum information and gravitational physics. They strongly suggest the structure responsible for gluing spacetime, i.e. for keeping it together, is quantum entanglement [11, 12].

I hope this modest presentation illustrates the great value of multi-disciplinary research and the weight that good conceptual ideas can carry: the uncertainty behind the flipping of coins may have some common features with the mathematical description of what keeps spacetime together in General Relativity. This can be conceptually satisfactory since, after all, *information is physical* [13].

References

1. C. E. Shannon, "A mathematical theory of communication." The Bell system technical journal 27.3 (1948): 379–423.
2. J. von Neumann, "Thermodynamik quantummechanischer Gesamheiten", Gött. Nach. 1(1927), 273–291.
3. W. G. Unruh, "Notes on black hole evaporation," Phys. Rev. D **14** (1976), 870.
4. S. W. Hawking, "Particle Creation by Black Holes," Commun. Math. Phys. **43** (1975), 199–220 [erratum: Commun. Math. Phys. **46** (1976), 206].
5. J. M. Bardeen, B. Carter and S. W. Hawking, "The Four laws of black hole mechanics," Commun. Math. Phys. **31** (1973), 161–170
6. J. D. Bekenstein, "Black holes and entropy," Phys. Rev. D **7** (1973), 2333–2346
7. G. 't Hooft, "Dimensional reduction in quantum gravity," Conf. Proc. C **930308** (1993), 284-296 [arXiv:gr-qc/9310026 [gr-qc]].
8. L. Susskind, "The World as a hologram," J. Math. Phys. **36** (1995), 6377–6396 [arXiv:hep-th/9409089 [hep-th]].
9. J. M. Maldacena, "The Large N limit of superconformal field theories and supergravity," Adv. Theor. Math. Phys. **2** (1998), 231–252 [arXiv:hep-th/9711200 [hep-th]].
10. S. Ryu and T. Takayanagi, "Holographic derivation of entanglement entropy from AdS/CFT," Phys. Rev. Lett. **96** (2006), 181602 [arXiv:hep-th/0603001 [hep-th]].
11. M. Van Raamsdonk, "Building up spacetime with quantum entanglement," Gen. Rel. Grav. **42** (2010) 2323 [Int. J. Mod. Phys. D **19** (2010) 2429] [arXiv:1005.3035 [hep-th]].
12. J. Maldacena and L. Susskind, "Cool horizons for entangled black holes," Fortsch. Phys. **61** (2013) 781 [arXiv:1306.0533 [hep-th]].
13. R. Landauer, "Information is physical," Physics Today **44** 5,23 (1991).

The Structure of K-Theoretic Blowup Equations

Kaiwen Sun

Abstract We study the structure of K-theoretic blowup equations–both unity and vanishing. We find the full set of K-theoretic blowup equations for all rank one and many higher rank 5d $\mathcal{N} = 1$ gauge theories. Our strategy is starting from 5d Kaluza-Klein (KK) theories compactified from 6d SCFTs and translating all elliptic blowup equations to K-theoretic ones. Then we reduce the blowup equations by decoupling one at a time the hypermultiplets in various representations. We find the 5d dualities are highly non-trivially reflected in the map among blowup equations. We also study the blowup equations near the 5d SCFT point and find the enhanced global symmetry appears naturally. In particular, all r fields form Weyl orbits of the enhanced flavor group in the lattice of invariant Coulomb parameters. This is a short summary on a joint paper with Lee and Sun [1] for the Proceedings of the Nankai Symposium on Mathematical Dialogues in 2021 August celebrating 110th anniversary of the birth of Prof. S.-S. Chern.

Keywords Blowup equations · 5d supersymmetric gauge theories · Flavor enhancement · 5d dualities

1 Introduction

1.1 Blowup Equations

Recently there has been a resurgence of interest in blowup equations. In a modern sense, blowup equations are the functional equations of Nekrasov partition functions first proposed by Nakajima and Yoshioka for 4d $\mathcal{N} = 2$ $SU(N)$ gauge theories [2]. They are latter generalized to various 5d $\mathcal{N} = 1$ gauge theories in [1, 3–7], 6d $(1, 0)$ gauge theories [8–11] and their untwisted/twisted circle compactifications [12], 6d $(2, 0)$ and 5d $\mathcal{N} = 1^*$ theories [10, 13], and in general the refined topological

K. Sun (✉)
Korea Institute for Advanced Study, Dongdaemun-gu, Seoul, South Korea
e-mail: ksun@kias.re.kr

Y.-H. He et al. (eds.), *Nankai Symposium on Mathematical Dialogues*,
https://doi.org/10.1007/978-981-19-2328-9_39

string theory on local Calabi-Yau threefolds [14]. We focus on K-theoretic blowup equations which come from comparing the Nekrasov partition function of 5d gauge theory on a 5d Omega background and on its one-point blown-up. For a 5d gauge G theory with Coulomb parameter $\vec{a}$, mass parameter $\vec{m}_F$ and instanton counting q, the Nekrasov partition function $Z(\epsilon_1, \epsilon_2, \vec{a}, \vec{m}_F, q)$ with Omega deformation parameters ϵ_1, ϵ_2 satisfies the following form of K-theoretic blowup equations:

$$\sum_{\vec{k}\in Q^\vee_w(G)} Z(\epsilon_1, \epsilon_2 - \epsilon_1, \vec{a} + \epsilon_1\vec{k}, \vec{m}_F + \epsilon_1\vec{\lambda}_F, e^{\lambda_q\epsilon_1}q) Z(\epsilon_1 - \epsilon_2, \epsilon_2, \vec{a} + \epsilon_2\vec{k}, \vec{m}_F + \epsilon_2\vec{\lambda}_F, e^{\lambda_q\epsilon_2}q)$$
$$= \Lambda(\epsilon_1, \epsilon_2, \vec{m}_F, q)\, Z(\epsilon_1, \epsilon_2, \vec{a}, \vec{m}_F, q).$$

Here w is a vector in the coweight space $P^\vee_G$, and $Q^\vee_w(G)$ is the class in $(P^\vee/Q^\vee)_G$ containing w, where $Q^\vee$ denotes the coroot lattice. The gauge flux $\vec{k}$ takes over all coweights in $Q^\vee_w(G)$. Clearly, there are $r_c = |(P^\vee/Q^\vee)_G|$ non-equivalent choices of w. The $\Lambda(\epsilon_1, \epsilon_2, \vec{m}_F, q)$ is called the Λ factor of blowup equations. If $\Lambda(\epsilon_1, \epsilon_2, \vec{m}_F, q) \equiv 0$, blowup equations are called vanishing, otherwise unity. The nontrivial property of blowup equations lies in that the Λ factor is independent from the Coulomb parameter $\vec{a}$. This gives a strong constraint for the partition function and in many cases allows one to solve the partition function completely from blowup equations.

A K-theoretic blowup equation is specified by the r field–the triple

$$(\lambda_G, \lambda_F, \lambda_q) = (w, \vec{\lambda}_F, \lambda_q) \in \Big((P^\vee/Q^\vee)_G, P^\vee_F, \mathbb{Q}\Big).$$

We summarize all possible choices of w which are denoted by the associated representations for all simple Lie groups G in Table 1. For a 5d $\mathcal{N} = 1$ gauge theory, we aim to find all the r fields $(\lambda_G, \lambda_F, \lambda_q)$ such that the K-theoretic blowup equations are satisfied. We hope to find a general method to determine the full set of the r fields without even computing the partition function. The recent classification programs for 5d SCFTs [15, 16] and 5d gauge theories [17] from 6d inspire us to conjecture that all K-theoretic blowup equations can be reduced from elliptic blowup equations. More precisely, when compacitfying 6d SCFTs to 5d KK theories, the elliptic blowup equations are translated to K-theoretic blowup equations. Decoupling matter from 5d KK theories, one reaches various 5d gauge theories/5d SCFTs, in the meantime the K-theoretic blowup equations of 5d KK theories are reduced to the blowup equations of downstream theories. We find in many cases such as the rank-one case, this process can produce the full set of K-theoretic blowup equations for the downstream theories. We remark that a 5d gauge theory could receive r fields from more than one 5d KK theories.

Decoupling matter for 5d gauge theories means to make some of the hypermultiplet mass infinite. In the meantime, the instanton counting parameter needs to be redefined. In general, decoupling a hypermultiplet in representation $\mathbf{R}$ with mass m, the instanton counting parameter q gets multiplied by $m^{\pm I_2(\mathbf{R})/2}$, depending on whether sending $m \to -\infty$ or $+\infty$. With one mass parameter infinite, a unity blowup

Table 1 Some useful data for simple Lie algebra G including dual Coxeter number $h_G^\vee$, rank of center r_c and representations w associated to all classes in $(P^\vee/Q^\vee)_G$

G	$h_G^\vee$	r_c	w
A_{N-1}	N	N	$\mathbf{1}, \mathbf{N}, \Lambda^2, \Lambda^3, \ldots, \overline{\mathbf{N}}$
$B_{N\geq 3}$	$2N-1$	2	$\mathbf{1}, \mathbf{V}$
$C_{N\geq 2}$	$N+1$	2	$\mathbf{1}, \Lambda^{\mathbf{N}}$
$D_{N\geq 4}$	$2N-2$	4	$\mathbf{1}, \mathbf{V}, \mathbf{S}, \mathbf{C}$
G_2	4	1	$\mathbf{1}$
F_4	9	1	$\mathbf{1}$
E_6	12	3	$\mathbf{1}, \mathbf{27}, \overline{\mathbf{27}}$
E_7	18	2	$\mathbf{1}, \mathbf{56}$
E_8	30	1	$\mathbf{1}$

equation can become either a unity or a vanishing one, while a vanishing blowup equation can only become a vanishing one.

2 K-Theoretic Blowup Equations for 5d $Sp(1) + N_f\mathbf{F}$ Theories

E-string theory is a typical 6d $(1,0)$ SCFT which has E_8 flavor symmetry. In the Horava-Witten picture of $E_8 \times E_8$ heterotic string theory, E-strings are realized by M2-branes stretched between one M9-brane and one M5-brane. All elliptic blowup equations for E-string theory have been found in [10]. It is well-known E-string theory circle-compactified to 5d gives $Sp(1) + 8\mathbf{F}$ KK theory. We find all r fields for 5d $Sp(1) + 8\mathbf{F}$ KK theory which are summarized in Table 2.

Now we want to find the r fields of 5d $Sp(1) + N_f\mathbf{F}$ theories from the RG flow from 5d $Sp(1) + 8\mathbf{F}$ KK theory. Each time we decouple a fundamental, the instanton counting parameter is redefined by $q_{N_f-1} = q_{N_f}/\sqrt{m_{N_f}}$. Since there is only one vanishing r field, it will remain so during the flow. We find the RG flow gives the following unity r fields for 5d $Sp(1) + N_f\mathbf{F}$ theories in Table 3. The total numbers are the same with the analysis on local E_{N_f+1} del Pezzo surfaces in [10].

Table 2 All r fields for 5d $Sp(1) + 8\mathbf{F}$ KK theory. There exist two possible $Sp(1)$ gauge fluxes represented by trivial representation **1** and fundamental representation **2**

	$G = Sp(1)$	#	$F = SO(16)$
Vanish	**2**	1	**1**
Unity	**1**	128	**S**
Unity	**2**	112	Dominant orbit in **Adj**

Table 3 Number of unity r fields for 5d $Sp(1) + N_f\mathbf{F}$ theories

N_f	8	7	6	5	4	3	2	1	0	0_π
$\lambda_G = \mathbf{1}$	128	128	96	64	40	24	14	8	5	4
$\lambda_G = \mathbf{2}$	112	112	86	62	42	26	14	6	2	2
Total	240	240	182	126	82	50	28	14	7	6

3 Flavor Enhancement

When a 5d gauge theory flows to its SCFT point, perturbative flavor symmetry and instanton $U(1)$ symmetry often mix together to exhibit enhanced flavor symmetry, see some recent developments in [18–20]. To keep manifest the enhanced flavor symmetry, the Coulomb parameters $\vec{a}$ get shifted by mass parameters and the instanton counting parameter to become the invariant Coulomb parameter $\vec{\mathfrak{a}}$ [21]. For a 5d $\mathcal{N} = 1$ gauge theory flowing to a SCFT $\mathcal{T}$, we conjecture all its r fields form Weyl orbits of the (enhanced) global symmetry of $\mathcal{T}$ in the lattice of invariant Coulomb parameters. In fact, the r fields forming the Weyl orbits of the enhanced global symmetry is strong evidence that we find the full set of blowup equations.

Let us look at the RG flow and flavor enhancement for 5d $Sp(1) + N_f\mathbf{F}$ theories. For 5d $Sp(1) + N_f\mathbf{F}$ SCFTs, the invariant Coulomb parameter $\mathfrak{a}$ is related to the perturbative Coulomb parameter a by [21]

$$\mathfrak{a} = a + \frac{2q}{(8 - N_f)}, \quad \text{thus} \quad \lambda_{\mathfrak{a}} = \lambda_a + \frac{2\lambda_q}{(8 - N_f)}.$$

Let us choose the fundamental domain of $\lambda_{\mathfrak{a}}$ as $[-1/2, 1/2]$. Then we draw the RG flow of all r fields of 5d $Sp(1) + N_f\mathbf{F}$ SCFTs in Fig. 1.

4 5D Dualities

5d $\mathcal{N} = 1$ gauge theories exhibit many interesting dualities. From the viewpoint of 5d SCFT, the dual theories can be regarded as different mass deformations of the same SCFT. From the viewpoint of Calabi-Yau threefolds, the 5d dualities can be regarded as fibre-base duality inside the geometry [22]. Many 5d dualities can also be observed from the brane web diagrams, by rotating the brane webs, resolving $O7^-$ branes and Higgsing, see for example [23–25].

For two dual 5d $\mathcal{N} = 1$ gauge theories $\mathcal{T}_A$ and $\mathcal{T}_B$, all vanishing/unity r fields of $\mathcal{T}_A$ should one to one correspond to all vanishing/unity ones of $\mathcal{T}_B$. This fact is far from trivial. The reason is that the r fields take value in $(P^\vee/Q^\vee)_G \times P_F^\vee$, and two

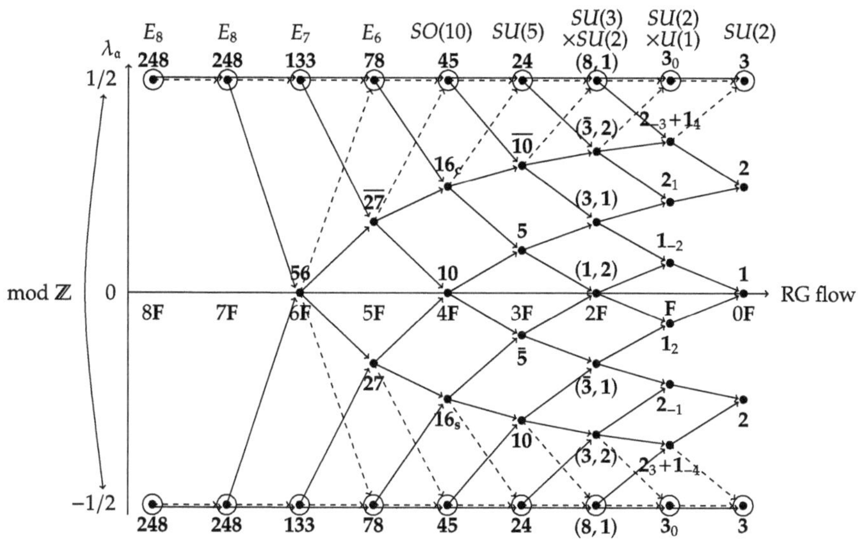

Fig. 1 RG flow of all r fields of 5d $Sp(1) + N_f\mathbf{F}$ SCFTs. The r fields are Weyl orbits of enhanced flavor group E_{N_f+1}. The circles represent vanishing r fields while the dots represent unity ones. The solid arrows represent the flowing to unity r fields, while the dashed ones represent the flowing to vanishing r fields

dual theories usually have completely different gauge and flavor groups. In many cases, the r fields can even help us to identify or rule out 5d dualities.

Acknowledgements The author would like to thank Kimyeong Lee for the collaboration on this project and Jie Gu, Hee-Cheol Kim, Sung-Soo Kim, Hiraku Nakajima and Xin Wang for helpful discussions. The author is supported by KIAS Grant QP081001.

References

1. K. Lee and K. Sun, "Blowup Equations with RG Flows, Dualities and Flavor Enhancements in 5d SCFTs," to appear.
2. H. Nakajima and K. Yoshioka, "Instanton counting on blowup. 1.," *Invent. Math.***162** (2005) 313–355.
3. H. Nakajima and K. Yoshioka, "Instanton counting on blowup. II. K-theoretic partition function," Transformation groups 10 (3), 489–519.
4. L. Göttsche, H. Nakajima and K. Yoshioka, "K-theoretic Donaldson invariants via instanton counting," Pure Appl. Math. Quart. **5**, 1029 (2009).
5. H. Nakajima and K. Yoshioka, "Perverse coherent sheaves on blowup, III: Blow-up formula from wall-crossing," Kyoto J. Math. **51**, no. 2, 263 (2011).
6. C. A. Keller and J. Song, "Counting Exceptional Instantons," JHEP **07** (2012), 085.
7. J. Kim, S. S. Kim, K. H. Lee, K. Lee and J. Song, "Instantons from Blow-up," JHEP **11**, 092 (2019) [erratum: JHEP **06**, 124 (2020)].

8. J. Gu, B. Haghighat, K. Sun and X. Wang, "Blowup Equations for 6d SCFTs. I," JHEP **1903**, 002 (2019).
9. J. Gu, A. Klemm, K. Sun and X. Wang, "Elliptic Blowup Equations for 6d SCFTs. II: Exceptional Cases," JHEP **12**, 039 (2019).
10. J. Gu, B. Haghighat, A. Klemm, K. Sun and X. Wang, "Elliptic Blowup Equations for 6d SCFTs. III: E-strings, M-strings and Chains," JHEP **07**, 135 (2020).
11. J. Gu, B. Haghighat, A. Klemm, K. Sun and X. Wang, "Elliptic blowup equations for 6d SCFTs. Part IV. Matters," JHEP **11** (2021), 090.
12. H. C. Kim, M. Kim, S. S. Kim and K. H. Lee, "Bootstrapping BPS spectra of 5d/6d field theories," JHEP **04**, 161 (2021).
13. Z. Duan, K. Lee, J. Nahmgoong and X. Wang, "Twisted 6d (2, 0) SCFTs on a circle," JHEP **07**, 179 (2021).
14. M. x. Huang, K. Sun and X. Wang, "Blowup Equations for Refined Topological Strings," JHEP **10**, 196 (2018).
15. P. Jefferson, S. Katz, H. C. Kim and C. Vafa, "On Geometric Classification of 5d SCFTs," JHEP **04** (2018), 103.
16. L. Bhardwaj, P. Jefferson, H. C. Kim, H. C. Tarazi and C. Vafa, "Twisted Circle Compactifications of 6d SCFTs," JHEP **12** (2020), 151.
17. L. Bhardwaj and G. Zafrir, "Classification of 5d $\mathcal{N} = 1$ gauge theories," JHEP **12** (2020), 099.
18. F. Apruzzi, C. Lawrie, L. Lin, S. Schäfer-Nameki and Y. N. Wang, "Fibers add Flavor, Part I: Classification of 5d SCFTs, Flavor Symmetries and BPS States," JHEP **11**, 068 (2019).
19. L. Bhardwaj, "Flavor Symmetry of 5d SCFTs, Part 1: General Setup," [arXiv:2010.13230 [hep-th]].
20. L. Bhardwaj, "Flavor Symmetry of 5d SCFTs, Part 2: Applications," [arXiv:2010.13235 [hep-th]].
21. V. Mitev, E. Pomoni, M. Taki and F. Yagi, "Fiber-Base Duality and Global Symmetry Enhancement," JHEP **04**, 052 (2015).
22. L. Bhardwaj, "Dualities of 5d gauge theories from S-duality," JHEP **07** (2020), 012.
23. H. Hayashi, S. S. Kim, K. Lee, M. Taki and F. Yagi, "A new 5d description of 6d D-type minimal conformal matter," JHEP **08** (2015), 097.
24. H. Hayashi, S. S. Kim, K. Lee and F. Yagi, "6d SCFTs, 5d Dualities and Tao Web Diagrams," JHEP **05** (2019), 203.
25. H. Hayashi, S. S. Kim, K. Lee, M. Taki and F. Yagi, "More on 5d descriptions of 6d SCFTs," JHEP **10**, 126 (2016).

Boundary $\mathcal{N} = 2$ Theory, Floer Homologies, Affine Algebras, and the Verlinde Formula

Meng-Chwan Tan

Abstract We explain how topologically-twisted $\mathcal{N} = 2$ gauge theory on a four-manifold with boundary, will allow us to furnish purely physical proofs of (i) the Atiyah-Floer conjecture, (ii) Muñoz's theorem relating quantum and instanton Floer cohomology, (iii) their monopole counterparts, and (iv) their higher rank generalizations. In the case where the boundary is a Seifert manifold, one can also relate its instanton Floer homology to modules of an affine algebra via a 2d A-model with target the based loop group. As an offshoot, we will demonstrate an action of the affine algebra on the quantum cohomology of the moduli space of flat connections on a Riemann surface, as well as derive the Verlinde formula. This presentation is based on the work [arXiv:1909.04058] published in Adv. Theor. Math. Phys. 25: 1–58, 2021.

1 Introduction

Four-dimensional (cohomological) topological quantum field theories (TQFTs), such as Donaldson-Witten (DW) theory and Seiberg-Witten (SW) theory, have been studied extensively over the years since they were first conceived in [1]. Such a TQFT possesses an underlying topological supersymmetry, which is generated by a scalar supercharge $\mathcal{Q}$. Observables of the theory are identified with the $\mathcal{Q}$-cohomology, and they are invariant under topological deformations. In the aforementioned theories, they generate the celebrated Donaldson and Seiberg-Witten invariants, respectively.

When DW theory is defined on a four-manifold with boundary Y_3, the corresponding Donaldson invariants are valued in the instanton Floer homology of Y_3 [1]. Likewise, for SW theory, the corresponding Seiberg-Witten invariants are valued in the monopole Floer homology of Y_3 [2].

One may also consider DW theory defined on a four-manifold of the form $M_4 = \Sigma \times C$, where Σ and C are compact Riemann surfaces. Upon shrinking C (or Σ), it is

Proceedings of the Nankai Symposium in Honour of S.S. Chern's 110th Birthday.

M.-C. Tan (✉)
Department of Physics, National University of Singapore, Singapore, Singapore
e-mail: mctan@nus.edu.sg

Y.-H. He et al. (eds.), *Nankai Symposium on Mathematical Dialogues*,
https://doi.org/10.1007/978-981-19-2328-9_40

known that the ensuing 2d theory is an A-twisted sigma model on Σ (or C) with target being the moduli space of flat connections on C (or Σ) [3]. The relevant observables of the A-model are identified with its $\mathcal{Q}$-cohomology, which in turn describe the quantum cohomology of the target space [3, 4]. Thus, the $\mathcal{Q}$-cohomology of DW theory must also be identified with quantum cohomology, i.e. Donaldson invariants on M_4 are related to the quantum cohomology of the space of flat connections on either Riemann surface [5].

In [5], it was suggested that upon shrinking C (or Σ), SW theory should also descend to an A-model on Σ (or C), albeit with target being the moduli space of vortices on C (or Σ). Likewise, the $\mathcal{Q}$-cohomology of SW theory must also be identified with quantum cohomology, i.e. Seiberg-Witten invariants on M_4 are related to the quantum cohomology of the space of vortices on either Riemann surface [5].

The main aim is to exploit properties of 4d $\mathcal{N} = 2$ TQFT in the manner described above, so that physical proofs of known mathematical conjectures and theorems as well as derivations of mathematically novel identities between 3d and 2d invariants, and more, can be obtained.

We shall now present a summary of the talk.

2 Summary of Presentation

First, we provide a relevant review of DW theory, and its relation to instanton Floer homology when the underlying four-manifold M_4 has boundary Y_3. In particular, if the moduli space of instantons is zero-dimensional, the partition function on M_4 is expressed as a sum of instanton Floer homology classes Ψ_{inst}

$$\boxed{Z_{M_4} = \langle 1 \rangle_{\Psi(\Phi_{Y_3})} = \sum_i \Psi_{\text{inst}}(\Phi^i_{Y_3})} \tag{1}$$

We will also take $M_4 = \Sigma \times C$, and shrink C, to obtain a sigma model on Σ with target $\mathcal{M}(C)$. This space is characterized by

$$\boxed{F_{ab} = 0} \tag{2}$$

which corresponds to the moduli space of flat connections on C, which we will denote by $\mathcal{M}_{\text{flat}}(C)$. Moreover, upon shrinking C, since the topological term of the form $F \wedge F$ leads to the pullback, i.e.

$$\boxed{\frac{1}{8\pi^2}\int_{M_4} \text{Tr}\,(F \wedge F) = \int_{\Sigma} X^*\omega_{\text{flat}}} \tag{3}$$

the ensuing 2d model on the remaining Riemann surface Σ must be an A-twisted sigma model with target $\mathcal{M}_{\text{flat}}(C)$, and action

$$\boxed{\begin{aligned} S'_{\text{DW}} = \frac{1}{e^2}\int_\Sigma d^2z\, \Bigg(G^{\text{flat}}_{I\overline{J}} \left(\frac{1}{2}\partial_z X^I \partial_{\overline{z}} X^{\overline{J}} + \frac{1}{2}\partial_{\overline{z}} X^I \partial_z X^{\overline{J}} + \rho_z{}^{\overline{J}} \nabla_{\overline{z}} \chi^I + \rho_{\overline{z}}{}^I \nabla_z \chi^{\overline{J}}\right) \\ - R_{I\overline{J}K\overline{L}} \rho_{\overline{z}}{}^I \rho_z{}^{\overline{J}} \chi^K \chi^{\overline{L}} \Bigg) + i\theta \int_\Sigma X^* \omega_{\text{flat}} \end{aligned}} \tag{4}$$

Second, we provide a review of SW theory, and its relation to monopole Floer homology when the underlying four-manifold M_4 has boundary Y_3. In particular, if the moduli space of monopoles is zero-dimensional, the partition function on M_4 is expressed as a sum of monopole Floer homology classes Ψ_{inst}

$$\boxed{Z_{M_4} = \langle 1 \rangle_{\Psi(\Phi_{Y_3})} = \sum_i \Psi_{\text{mono}}(\Phi^i_{Y_3})} \tag{5}$$

We will also take $M_4 = \Sigma \times C$, and shrink C, to obtain a sigma model on Σ with target $\mathcal{M}(C)$. This space is characterized by

$$\boxed{\begin{aligned} F_{w\overline{w}} &= \frac{i}{4}\left(1 - |\varphi|^2\right) \\ D_{\overline{w}}\varphi &= 0 \end{aligned}} \tag{6}$$

which corresponds to the moduli space of vortices on C, which we will denote by $\mathcal{M}^q_{\text{vort}}(C)$. Furthermore, upon shrinking C, a $\mathcal{Q}$-exact topological term S_{top}, which can be added inconsequentially, leads to the pullback, i.e.

$$\boxed{S_{\text{top}} = \frac{1}{8\pi^2}\int_{M_4} \mathcal{F} \wedge \mathcal{F} = \int_\Sigma X^* \omega_{\text{vort}}} \tag{7}$$

Hence, the ensuing 2d model on Σ must also be an A-model with target $\mathcal{M}^q_{\text{vort}}(C)$, and action

$$\boxed{\begin{aligned} S'_{\text{SW}} = \frac{1}{e^2}\int_\Sigma d^2z\, \Bigg(G^{\text{vort}}_{I\overline{J}} \left(\frac{1}{2}\partial_z X^I \partial_{\overline{z}} X^{\overline{J}} + \frac{1}{2}\partial_{\overline{z}} X^I \partial_z X^{\overline{J}} + \rho_z{}^{\overline{J}} \nabla_{\overline{z}} \chi^I + \rho_{\overline{z}}{}^I \nabla_z \chi^{\overline{J}}\right) \\ - R_{I\overline{J}K\overline{L}} \rho_{\overline{z}}{}^I \rho_z{}^{\overline{J}} \chi^K \chi^{\overline{L}} \Bigg) + i\theta \int_\Sigma X^* \omega_{\text{vort}} \end{aligned}} \tag{8}$$

Third, we provide physical proofs of various mathematical conjectures and theorems, by exploiting the fact that physical states of a TQFT are insensitive to topological deformations. We first prove the Atiyah-Floer conjecture, which relates instanton and Lagrangian intersection Floer homologies. This takes the form

$$\boxed{\text{HF}^{\text{inst}}_*(Y_3) \cong \text{HF}^{\text{Lagr}}_*(\mathcal{M}_{\text{flat}}(\Sigma), L_0, L_1)} \tag{9}$$

This is proved by considering DW theory on the four-manifold $M_4 = \mathbb{R}^+ \times Y_3 \cong \mathbb{R}^+ \times I \times_f \Sigma$ (LHS of (9)), and identifying its states with those of the A-model on $\mathbb{R}^+ \times I$ with target $\mathcal{M}_{\text{flat}}(\Sigma)$ (RHS of (9)). The second mathematical claim is Muñoz's theorem, which relates instanton and symplectic Floer cohomologies, whereby the result of [6] further relates symplectic Floer cohomology with quantum cohomology. It takes the form

$$\boxed{QH^*(\mathcal{M}_{\text{flat}}(\Sigma)) \cong \text{HF}^*_{\text{symp}}(\mathcal{M}_{\text{flat}}(\Sigma)) \cong \text{HF}^*_{\text{inst}}\left(\Sigma \times S^1\right)} \tag{10}$$

This is proved by considering DW theory on the four-manifold $M_4 = \Sigma \times S^1 \times \mathbb{R}^+$ (RHS of (10)), and identifying that with the A-model on $\mathbb{R}^+ \times S^1$ with target $\mathcal{M}_{\text{flat}}(\Sigma)$ (centre of (10)). The first equality follows from [6], as mentioned. The same analysis is carried out for SW theory, in which we consider monopole analogs of (9) and (10). In an analogous manner, we prove the monopole Atiyah-Floer conjecture, which takes the form

$$\boxed{\text{HF}^{\text{mono}}_*(q, Y_3) \cong \text{HF}^{\text{Heeg}}_*\left(\mathcal{M}^q_{\text{vort}}(\Sigma), L_0, L_1\right)} \tag{11}$$

Our analysis of SW theory also allows us to deduce the *mathematically novel* monopole analog of Muñoz's theorem, which takes the form

$$\boxed{QH^*(\mathcal{M}^q_{\text{vort}}(\Sigma)) \cong \text{HF}^*_{\text{symp}}\left(\mathcal{M}^q_{\text{vort}}(\Sigma)\right) \cong \text{HF}^*_{\text{mono}}\left(q, \Sigma \times S^1\right)} \tag{12}$$

These results can be generalized to higher rank gauge groups G, because the physical analysis is either independent of the choice of G, or simply involves a straightforward extension. In doing so, we can obtain higher rank generalizations of (9)–(12).

Fourth, we consider DW theory on $M_4 = \Sigma \times D \cong \Sigma \times S^1 \times \mathbb{R}^+$, with gauge group $G = SU(2)$. Instanton Floer homology can be defined on $Y_3 = \Sigma \times S^1$. Moreover, shrinking D allows us to identify the target of the A-model on Σ with the based loop group ΩG. Such an A-model is known to possess affine symmetry [7]. The corresponding A-model states form modules of an affine Lie algebra $\mathfrak{g}_{\text{aff}}$, which span the space of $\mathfrak{g}_{\text{aff}}$-modules on Σ that we denote by $\mathfrak{G}_{\text{mod}}(\Sigma)$. We can identify the corresponding partition functions, and hence establish the *mathematically novel* isomorphism

$$\boxed{\text{HF}^{\text{inst}}_*\left(\Sigma \times S^1\right) \cong \mathfrak{G}_{\text{mod}}(\Sigma)} \tag{13}$$

Next, we consider DW theory on $M_4 = \Sigma \times_f D \cong M_{g,p} \times \mathbb{R}^+$, where $M_{g,p}$ is a Seifert manifold, g is the genus of Σ, and p is the Chern number of the S^1 bundle which characterizes the nontriviality of the fibration. As such, we see that $\Sigma \times S^1$ is a trivially-fibered Seifert manifold–i.e. $\Sigma \times S^1 = M_{g,0}$. By inserting p copies of the fibering operator $\mathcal{P}$ [8] in the partition function over $M_{g,0}$, which has the effect of shifting the Chern number by p, the partition function over $M_{g,p}$ is obtained. We may then generalize (13) to the case where $\Sigma \times S^1$ is replaced by $M_{g,p}$. Denoting $\mathfrak{G}_{\text{mod},p}(\Sigma)$ as the space where each basis component is now acted upon by p copies of a suitable representation of $\mathcal{P}$, we similarly show that there is a *mathematically novel* isomorphism

$$\boxed{\text{HF}_*^{\text{inst}}\left(M_{g,p}\right) \cong \mathfrak{G}_{\text{mod},p}(\Sigma)} \tag{14}$$

It is straightforward to see, from (10) and (13), that there is yet another *mathematically novel* identity of the form

$$\boxed{QH^*(\mathcal{M}_{\text{flat}}(\Sigma)) \cong \mathfrak{G}_{\text{mod}}(\Sigma)} \tag{15}$$

Fifth, and last, as a preliminary step, we first consider an A-model on $D \cong \mathbb{R}^+ \times S^1$ with target $\mathcal{M}_{\text{flat}}(\Sigma)$. Since the theory is topological, we may further shrink S^1 so that we get a 1d sigma model which turns out to be a quantum mechanical model on $\mathcal{M}_{\text{flat}}(\Sigma)$ with action

$$\boxed{S_{\text{QM}} = \frac{1}{\hbar}\int d\tau\ \frac{1}{2}\dot{X}^I\dot{X}_I} \tag{16}$$

Thus, states of the A-model on D with target $\mathcal{M}_{\text{flat}}(\Sigma)$ are identified with quantum mechanical states on $\mathcal{M}_{\text{flat}}(\Sigma)$. Equation (16) also means we can write the commutator relations for the collective coordinates X as

$$\boxed{[\hat{X}^I, \hat{P}^J] = \hbar\delta^{IJ}} \tag{17}$$

which amounts to quantizing $\mathcal{M}_{\text{flat}}(\Sigma)$. Next, we make use of DW theory on $M_4 = \Sigma \times D$, to obtain, upon shrinking D, an A-model on Σ with target ΩG, and, upon shrinking Σ, an A-model on D with target $\mathcal{M}_{\text{flat}}(\Sigma)$. In doing so, we can derive Falting's definition of the Verlinde formula [9, 10]

$$\boxed{V_\ell(\Sigma) \cong H^0(\mathcal{M}_{\text{flat}}(\Sigma), \mathcal{L}^\ell)} \tag{18}$$

where $V_\ell(\Sigma)$ is the space of zero-point conformal blocks of $\mathfrak{g}_{\text{aff}}$ at level ℓ on Σ. The LHS of (18) is obtained from the A-model on Σ with target ΩG, and the RHS is obtained from the quantum mechanical model on $\mathcal{M}_{\text{flat}}(\Sigma)$, where H^0 is the space of holomorphic sections of the determinant line bundle $\mathcal{L}$. We also derive Pauly's definition of the Verlinde formula [11], which considers extra operator insertions in the theory. This derivation will proceed similar to the case with *no* operator insertions, because we can exploit the position-independence of operator insertions in an A-model. We are then able to derive the isomorphism

$$\boxed{V_\ell(\Sigma, \vec{p}) \cong H^0(\mathcal{M}_{\text{para}}(\Sigma, \vec{p}), \mathcal{L}^\ell)} \tag{19}$$

where $\vec{p} = (p_1, \ldots, p_n)$ are the operator insertion points on Σ, and $V_\ell(\Sigma, \vec{p})$ is the space of n-point conformal blocks of $\mathfrak{g}_{\text{aff}}$ at level ℓ on Σ.

References

1. E. Witten, "Topological quantum field theory," *Communications in Mathematical Physics*, vol. 117, no. 3, pp. 353–386, 1988.
2. M. Marcolli, *Seiberg Witten Gauge Theory*, vol. 17. Springer, 1999.
3. M. Bershadsky, A. Johansen, V. Sadov, and C. Vafa, "Topological reduction of 4D SYM to 2D σ-models," *Nuclear Physics B*, vol. 448, no. 1–2, pp. 166–186, 1995.
4. E. Witten, "Topological sigma models," *Communications in Mathematical Physics*, vol. 118, no. 3, pp. 411–449, 1988.
5. S. Gukov, "Surface operators and knot homologies," in *New Trends in Mathematical Physics*, pp. 313–343, Springer, 2009.
6. V. Sadov, "On equivalence of Floer's and quantum cohomology," *Communications in mathematical physics*, vol. 173, no. 1, pp. 77–99, 1995.
7. M. Ashwinkumar, J. Cao, Y. Luo, M.-C. Tan, and Q. Zhao, "Little strings, quasi-topological sigma model on loop group, and toroidal Lie algebras," *Nuclear Physics B*, vol. 928, pp. 469–498, 2018.
8. M. Blau and G. Thompson, "Chern-Simons theory on S^1-bundles: Abelianisation and q-deformed Yang-Mills theory," *Journal of High Energy Physics*, vol. 2006, no. 05, p. 003, 2006.
9. G. Faltings, "A proof for the Verlinde formula," *Journal of Algebraic Geometry*, vol. 3, no. 2, p. 347, 1994.
10. A. Beauville, "Conformal blocks, fusion rules and the Verlinde formula," *arXiv preprint alg-geom/9405001*, 1994.
11. C. Pauly, "Espaces de modules de fibrés paraboliques et blocs conformes," *Duke Mathematical Journal*, vol. 84, no. 1, pp. 217–235, 1996.
12. C. Lozano and M. Mariño, "Donaldson Invariants of Product Ruled Surfaces and Two-Dimensional Gauge Theories," *Communications in Mathematical Physics*, vol. 220, no. 2, pp. 231–261, 2001.
13. J. Labastida and M. Mariño, *Topological quantum field theory and four manifolds*, vol. 25. Springer, 2005.
14. G. Moore and E. Witten, "Integration over the u-plane in Donaldson theory," arXiv preprint hep-th/9709193, 1997.
15. M. Atiyah, N. Hitchin, and I. Singer, "Self-duality in four-dimensional Riemannian geometry," *Proceedings of the Royal Society of London. A. Mathematical and Physical Sciences*, vol. 362, no. 1711, pp. 425–461, 1978.
16. M. Atiyah, "New invariants of 3-and 4-dimensional manifolds," *The mathematical heritage of Hermann Weyl (Durham, NC, 1987)*, vol. 48, pp. 285–299, 1988.
17. S. Donaldson, *Floer homology groups in Yang-Mills theory*, vol. 147. Cambridge University Press, 2002.
18. A. Popov, "Loop groups in Yang–Mills theory," *Physics Letters B*, vol. 748, pp. 439–442, 2015.
19. K. Hori, S. Katz, C. Vafa, and R. Pandharipande, *Mirror symmetry*, vol. 1. American Mathematical Soc., 2003.
20. D. Salamon, "Notes on flat connections and the loop group," *Preprint, University of Warwick*, 1998.
21. M. Elbistan, "Weyl semimetal and topological numbers," *International Journal of Modern Physics B*, vol. 31, no. 29, p. 1750221, 2017.
22. Y. Shnir, *Magnetic monopoles*. Springer Science & Business Media, 2006.
23. E. Witten, "Monopoles and four-manifolds," arXiv preprint hep-th/9411102, 1994.

24. D. Tong, "TASI lectures on solitons," arXiv preprint hep-th/0509216, 2005.
25. R. Dey, "Geometric prequantization of the moduli space of the vortex equations on a Riemann surface," *Journal of mathematical physics*, vol. 47, no. 10, p. 103501, 2006.
26. A. Kapustin and E. Witten, "Electric-magnetic duality and the geometric Langlands program," arXiv preprint hep-th/0604151, 2006.
27. C. Kutluhan, Y.-J. Lee, and C. Taubes, "HF= HM I: Heegaard Floer homology and Seiberg–Witten Floer homology," arXiv preprint arXiv:1007.1979, 2010.
28. V. Muñoz, "Ring structure of the Floer cohomology of $\Sigma \times S^1$," *Topology*, vol. 38, no. 3, pp. 517–528, 1999.
29. V. Muñoz and B.-L. Wang, "Seiberg–Witten–Floer homology of a surface times a circle for non-torsion spinC structures," *Mathematische Nachrichten*, vol. 278, no. 7–8, pp. 844–863, 2005.
30. A. Floer, "An instanton-invariant for 3-manifolds," *Communications in mathematical physics*, vol. 118, no. 2, pp. 215–240, 1988.
31. S. Dostoglou and D. Salamon, "Self-dual instantons and holomorphic curves," *Annals of Mathematics*, vol. 139, no. 3, pp. 581–640, 1994.
32. Y. Luo, M.-C. Tan, P. Vasko, and Q. Zhao, "Four-dimensional N=2 supersymmetric theory with boundary as a two-dimensional complex Toda theory," *Journal of High Energy Physics*, vol. 2017, no. 5, p. 121, 2017.
33. E. Witten, "Quantum field theory and the Jones polynomial," *Communications in Mathematical Physics*, vol. 121, no. 3, pp. 351–399, 1989.
34. P. Francesco, P. Mathieu, and D. Sénéchal, *Conformal field theory*. Springer Science & Business Media, 2012.
35. G. Segal, "The definition of conformal field theory," in *Differential geometrical methods in theoretical physics*, pp. 165–171, Springer, 1988.

Inverse Algorithm and Triple Point Diagrams

Valdo Tatitscheff

Abstract Dimer models (also known as brane tilings) are special bipartite graphs on a torus $\mathbb{T}^2$. They encode the structure of the 4d $\mathcal{N} = 1$ worldvolume theories of D3 branes probing toric affine Calabi-Yau singularities. Constructing dimer models from a singularity can in principle be done via the so-called inverse algorithm, however it is hard to implement in practice. We discuss how combinatorial objects called triple point diagrams systematize the inverse algorithm, and show how they can be used to construct dimer models satisfying some symmetry or containing particular substructures. We present the construction of the Octagon dimer model which satisfies both types of constraints. Eventually we present a new criterion concerning possible implementations of symmetries in dimer models, in order to illustrate how the use of triple point diagrams could strengthen such statements.

1 Dimer Models

For our purposes a *dimer model* (or *brane tiling*) is a bipartite graph Γ embedded in a topological 2-torus $\mathbb{T}^2$ in such a way that each connected component of $\mathbb{T}^2 \setminus \Gamma$ is simply connected, and considered up to isotopy. An example of dimer model is given on the left of Fig. 1, where the square is a fundamental cell for $\mathbb{T}^2$. Fixing an orientation of the torus, it induces at each vertex a cyclic orientation of the edges incident to it, hence there is a globally well-defined notion of 'left' and 'right' at each vertex of Γ.

A *zig-zag path* (ZZP) on Γ is an oriented path of edges of Γ which turns maximally right at each black vertex, and maximally left at each white one. One can equivalently represent zig-zag paths as strands crossing edges at their middle. A zig-zag path and the corresponding strand are displayed respectively in the middle and on the right of Fig. 1. Note that this ZZP has a non-trivial winding around the torus: in the basis of homology induced by our choice of fundamental cell, its winding is $(1, 1)$.

V. Tatitscheff (✉)
Technical University of Munich, TUM School of Computation, Information and Technology, Boltzmannstraße 3, Garching 85748, Germany
e-mail: valdo.tatitscheff@normalesup.org

Y.-H. He et al. (eds.), *Nankai Symposium on Mathematical Dialogues*,
https://doi.org/10.1007/978-981-19-2328-9_41

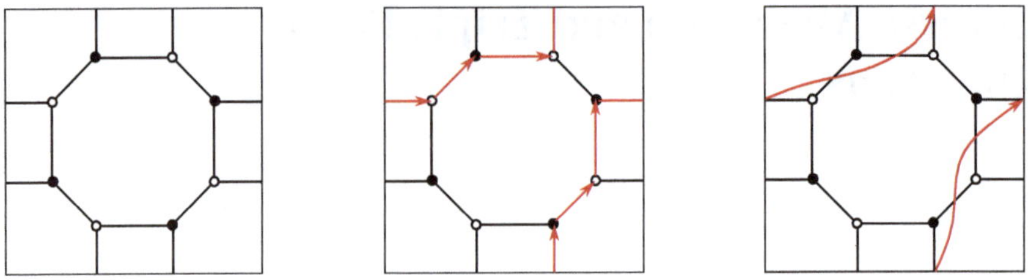

Fig. 1 A dimer model, a zig-zag path and a zig-zag strand

A dimer model is said to be *consistent* if (1) there is no topologically trivial ZZP (2) no lift of ZZP in the universal cover $\mathbb{R}^2$ of the torus self-intersects (3) any two distinct lifts of ZZPs in the universal cover of the torus never intersect more than once in the same direction. These technical conditions are equivalent to the existence of (possibly degenerate) isoradial embeddings of Γ in $\mathbb{T}^2$ [1–3]. In what follows we always assume dimer models to be consistent.

Dimer models have been introduced in string theory as a tool to study the world-volume quantum field theories (QFTs) of D3-branes probing affine toric Calabi-Yau threefolds (CY3) in type IIB string theory (see [4] and references therein). These setups have proved very fruitful in extending the gauge/gravity correspondence to $\mathcal{N} = 1$ conformal field theories and even non-conformal QFTs. A brane tiling encodes the structure of a quiver $\mathcal{N} = 1$ QFT as follows: each face is an $SU(N)$ gauge group, each edge is a matter field in the bifundamental representation of the faces of which it is the boundary, and each vertex encodes a monomial of the super-potential. Orientifolds can be implemented on dimer models as involutions of the bipartite map. They grant access to orthogonal and symplectic gauge groups, as well as matter in rank-2 tensor representations.

Each affine toric CY3 is in one-to-one correspondence with a class of lattice convex polygons (also called *toric diagrams*) up to $SL_2(\mathbb{Z})$, hence dimer models are also associated with lattice convex polygons up to $SL_2(\mathbb{Z})$. Since each edge of a dimer model belongs exactly to two ZZPs going in opposite directions, the sum of windings of all ZZPs in a brane tiling is always zero. Ordering the windings according to their slope, one builds the sides of a convex lattice polygon. The choice of a fundamental cell for $\mathbb{T}^2$ translates as the action of $SL_2(\mathbb{Z})$ on the corresponding polygon. In fact the lattice to which the vertices of the polygon belong can be naturally identified with the integral first homology group of the torus $H_1(\mathbb{T}^2, \mathbb{Z}) \simeq \mathbb{Z}^2$.

There are in general many dimer models corresponding to the same affine toric CY3. Going from a dimer model to the class of lattice polygons up to $SL_2(\mathbb{Z})$ as we explained before is the *forward algorithm* and it is in general not injective; going in the opposite direction and building consistent dimer models from a class of toric diagrams can be done via the *inverse algorithm* which is one-to-many in general [5]. Even if techniques have been developed to make the inverse algorithm more efficient [1, 5], it remains a set of instructions which has to be carried out case-by-case with a deft hand. Below we explain how Thurston's triple point diagrams [6] help

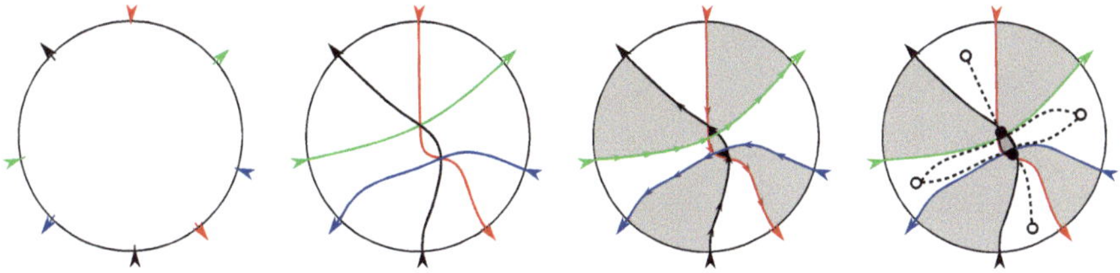

Fig. 2 From boundary data to bipartite graphs through triple crossing diagrams

systematizing the inverse algorithm and even allow the construction of dimer models satisfying combinatorial constraints.

There are interesting combinatorial transformations of consistent dimer models, as the so-called *spider moves* and the closely related *urban renewals*. From a physics perspective, they encode Seiberg dualities of the corresponding $\mathcal{N} = 1$ theories [7]. Mathematically, they are an instance of quiver mutations and link dimer models to cluster algebras and varieties. In particular, each class of consistent dimer models up to spider moves defines an algebraic integrable system whose phase space is a cluster $\mathcal{X}$-variety [8]. Not uncorrelated to this is the fact that dimer models encode double Bruhat cells in the affine Poisson-Lie groups $\widehat{\mathrm{PGL}(N)}$ [9].

Dimer models also appear in other fields of mathematics and physics–for instance crystal melting and topological strings [10].

2 Triple Points Diagrams and Fast-Inverse Algorithm

Triple point diagrams have been introduced by Thurston [6]. A triple point diagram is a collection of oriented dimension-one manifolds with boundary, mapped smoothly into a disk (the image of a connected component is called a *strand*) such that (1) exactly three strands cross at each intersection point (2) the endpoints of strands are distinct points on the boundary of the disk and no other point is mapped to the boundary (3) the orientations of the strands induce consistent orientations on the complementary regions. Triple point diagrams are considered up to isotopy. Theorem 3 of [6] states that given a disk with $2n$ points on the boundary labeled 'in' or 'out' and such that the labels alternate as one goes around the boundary, any matching (bijection) between in and out points can be realized by a triple point diagram.

On the left of Fig. 2 such a boundary data on a disk is shown–the pairing is displayed in colors. From it one easily constructs a triple point diagram. Deform the latter so that each crossing is replaced by a small triangle with counterclockwise oriented sides; one sees then that each connected component of the complement of the diagram in the disk has its boundary either oriented clockwise (white faces) or counterclockwise (black faces) or the orientation of edges along the boundary alternates (grey faces). Putting a black vertex in each black face, a white vertex in

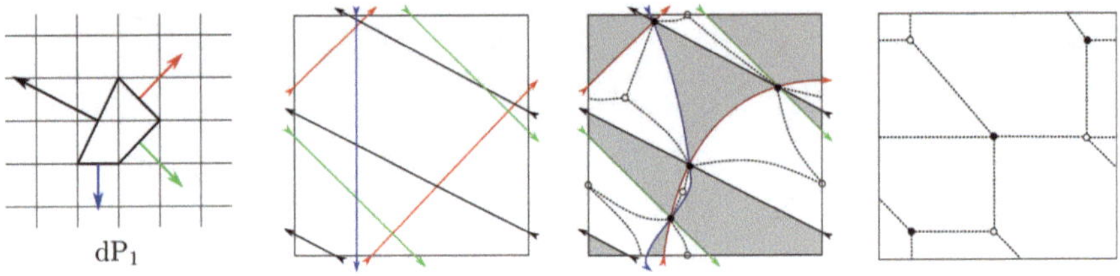

Fig. 3 From a toric diagram to a consistent dimer model through triple crossing diagrams

each white face and connecting vertices according to adjacency yields a bipartite graph on the disk. The so-called $2 \leftrightarrow 2$ moves of triple point diagrams correspond to spider moves of the graph.

Triple crossing diagrams can be used to systematically construct consistent dimer models from lattice polygons, as reviewed in [8]. One considers the simple outgoing normal vectors to the sides of a convex lattice polygon and for each of them one draws a straight line on a torus with winding the coordinates of the vector, such that the induced 'in' and 'out' points on the boundary alternate (it is always possible to do so [8]). Keeping only the boundary data together with the pairing between 'in' and 'out' points, forgetting momentarily the identifications between opposite sides of the square, one is left with a topological disk, boundary data and a pairing. Thus, one can construct a triple point diagram realizing the pairing and such that the number of triple crossings is minimal among such diagrams, deform it as before, and construct the corresponding bipartite graph. In the end we are left with a consistent dimer model on a torus that corresponds to the lattice polygon we started with.

An example for a polygon dubbed dP_1 is presented in Fig. 3. On the right of the same figure is the dimer model we obtained. It famously corresponds to dP_1 [7].

3 The Octagon Dimer Model

Supersymmetry (SUSY) has been introduced in the 80's in order to solve issues inherent to the Standard Model of particle physics, and grants a lot of control on the dynamics of the QFTs enjoying this property. However it is notoriously difficult to implement SUSY in a phenomenologically satisfactory way.[1] Gauge theories in which SUSY is broken by dynamical effects are essential in many of these putative efforts; they are called Dynamical Supersymmetry Breaking (DSB) models. Only a few classes of explicit DSB models are known explicitly.

The physical motivation for the Octagon dimer model is to embed DSB models in brane tilings, for if it is possible it provides a natural UV completion of these DSB field theories. For a QFT to be implementable in a dimer model it must contain matter fields in tensor representations of rank at most 2 of the gauge groups, which is the

[1] Many issues appear as one tries to explain how SUSY breaks spontaneously at low energies.

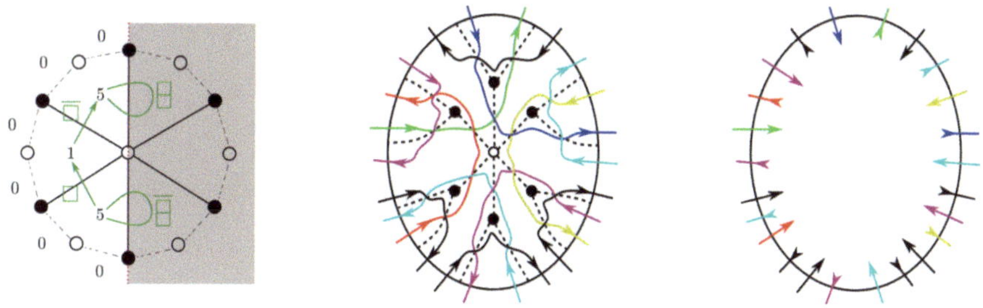

Fig. 4 The hexagonal cluster hosting the twin SU(5) model

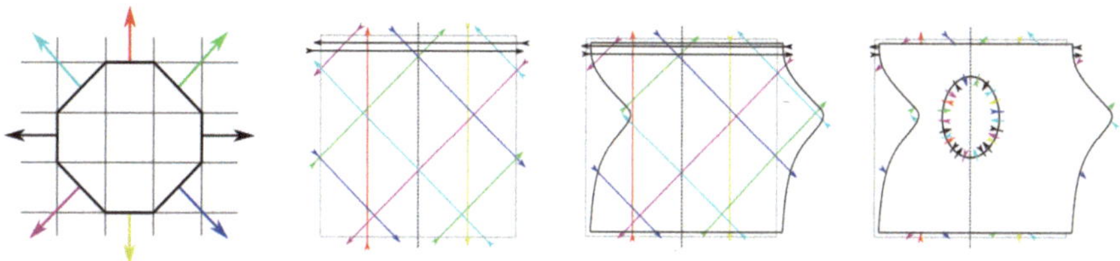

Fig. 5 From a polygon to a consistent dimer model through triple crossing diagrams

case for the so-called *uncalculable* SU(5) and the $3-2$ DSB models. Orientifolds are needed in order to construct dimer models with rank-2 tensor matter.

While instances of brane tilings hosting DSB models are known since long [11], it was shown more recently [12, 13] that the presence of $\mathcal{N}=2$ fractional branes at the toric affine CY3 actually spoils the presence of the DSB vacuum. There are $\mathcal{N}=2$ fractional branes whenever the CY3 singularity is not-isolated, or equivalently when the corresponding toric diagrams are convex, but not strictly convex.

It was shown in [14] that all possible substructures in dimer models leading to the SU(5) model imply the existence of $\mathcal{N}=2$ fractional branes, but the one shown on the left of Fig. 4. That led naturally to the question of whether it is possible at all to construct a dimer model corresponding to a strictly convex lattice polygon, symmetric with respect to two vertical axes in order to implement the orientifold one needs, and containing the hexagonal cluster of our interest, together with a few additional technical constraints carefully discussed in [14]. The hexagonal cluster can be inscribed in a disk; let us draw its ZZPs as in the middle of Fig. 4 and keep only the 'in' and 'out' insertions on the boundary together with the pairing, as displayed on the right of Fig. 4. By construction the hexagonal cluster of interest can be obtained from this data.

Our goal is to embed this circle (with in/out points and pairing) inside a consistent dimer model with the constraints listed above. One can show that a polygon that can possibly work must have at least 8 sides; the octagon shown on the left of Fig. 5 is a parsimonious choice. One places straight lines on a fundamental cell as before, in a symmetric way with respect to the dashed vertical axis. Deforming the boundary of the fundamental cell so that endpoints of strands are distinct, one only keeps the

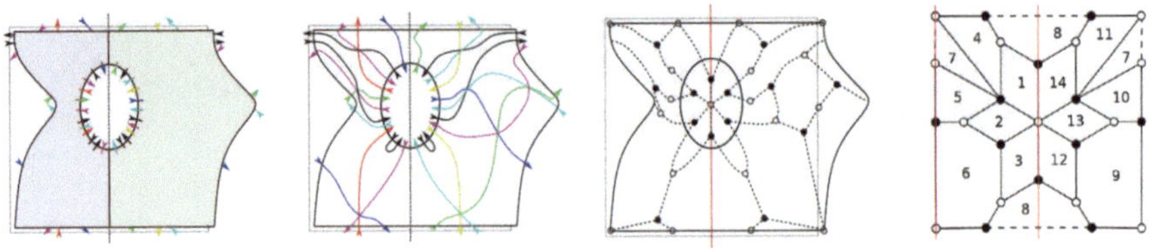

Fig. 6 The construction of the octagon dimer model

in/out points and inserts the circle corresponding to the hexagonal cluster on the symmetry axis in the middle of the cell. The different steps of this procedure are displayed in Fig. 5.

The bluish and greenish parts of the deformed fundamental cell on the left of Fig. 6 are topological disks with boundary data and pairing satisfying Thurston's conditions, hence one can construct a (symmetric) triple crossing diagram realizing the boundary data, and run the algorithm as before to obtain a dimer model satisfying all the constraints listed above. It is presented under a fancy guise on the right of Fig. 6. The physics of this Octagon dimer model is discussed in [15].

4 Symmetries in Dimer Models

Triple point diagrams are very likely to be useful in order to study general aspects of the inverse algorithm and its geometric counterparts. Let us consider a specific example for concreteness. In [16] it is shown that if a dimer model is symmetric with respect to a vertical axis mapping vertices to vertices of the same color, then the corresponding toric diagram must be symmetric with respect to a horizontal axis.[2] One can wonder about the reciprocal: given a toric diagram symmetric with respect to a horizontal axis, is there always a corresponding dimer model symmetric with respect to a vertical axis? One can argue that this is false by considering the toric diagram on the left of Fig. 7.

From the toric diagram it appears that no ZZP is mapped to itself under the reflection with respect to the dashed red line. Hence there must be edges on each symmetry axis in the dimer model, and subsequently it is always possible to choose a symmetric path as the black one displayed on the right of Fig. 7, such that the total number of intersections (counted with signs) between this path and ZZPs in the left half of the cell is zero. Let $n_i \in \mathbb{Z}$, $i = 1, \ldots, 6$ be the number of intersections (with signs) between the black path and the i-th ZZP in the left bluish half of the fundamental cell–and remember that we have chosen the path so that $\sum n_i = 0$.

If the dimer model under consideration is symmetric with respect to the dashed red lines, then the black path intersect (with signs) $-n_6$ times the ZZP (1) on the right half of the cell since the ZZPs (1) and (6) are exchanged under the symmetry.

[2] This follows by considering how the symmetry acts on ZZPs in the symmetric dimer model.

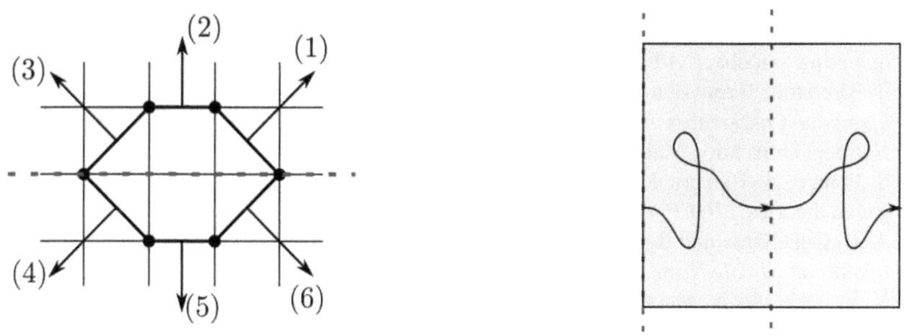

Fig. 7 From a polygon to a consistent dimer model through triple crossing diagrams

Then, from the toric diagram one infers that $n_1 - n_6 = 1$ since the winding of the black path is $(1, 0)$ and hence as it loops around the torus it crosses once (with sign) the ZZPs (1) whose winding is $(1, 1)$. Similarly, $n_2 - n_5 = 1$ and $n_3 - n_4 = 1$. Hence $\sum n_i = 2(n_4 + n_5 + n_6) + 3$, which can not be zero since the n_i's are integers, and we are led to the conclusion that there can not be any symmetric dimer model corresponding to this toric diagram and this symmetry. The example of above can be generalized straightforwardly to many other toric diagrams, and answers negatively the question of whether there always exist symmetric dimer models when the toric diagram is compatible with this symmetry. Thus one presumably needs intrinsically non-perturbative sectors [17] to describe the corresponding orientifolds, if any.

As one wonders about the existence of constrained dimer models, ZZPs-based arguments provide in general merely *necessary* criteria–such as the one we just derived, whereas the existence theorems for triple point diagrams let one hope for *sufficient* conditions. The knowledge of these would certainly deepen our understanding of orientifolds of affine toric CY3 singularities.

⋆ ⋆ ⋆

It is a pleasure to thank Riccardo Argurio, Matteo Bertolini, Sebastián Franco, Eduardo García-Valdecasas, Shani Meynet and Antoine Pasternak for countless inputs and illuminating discussions, as well as comments on the manuscript. I would also like to thank Mo-Lin Ge, Cheng-Ming Bai, Yang-Hui He, Hong-Qin Li, Jiakang Bao, Edward Hirst and Suvajit Majumder for organizing this very nice symposium.

References

1. A. Hanany and D. Vegh, "Quivers, tilings, branes and rhombi," *JHEP*, vol. 10, p. 029, 2007.
2. N. Broomhead, *Dimer models and Calabi-Yau algebras*, vol. 1011. 2012.
3. A. Ishii and K. Ueda, "A note on consistency conditions on dimer models," arXiv preprint arXiv:1012.5449, 2010.
4. M. Yamazaki, "Brane Tilings and Their Applications," *Fortsch. Phys.*, vol. 56, pp. 555–686, 2008.

5. B. Feng, Y.-H. He, K. D. Kennaway, and C. Vafa, "Dimer models from mirror symmetry and quivering amoebae," *Adv. Theor. Math. Phys.*, vol. 12, no. 3, pp. 489–545, 2008.
6. D. Thurston, "From dominoes to hexagons," in *Proceedings of the 2014 Maui and 2015 Qinhuangdao Conferences in Honour of Vaughan FR Jones' 60th Birthday*, pp. 399–414, Australian National University, Mathematical Sciences Institute, 2017.
7. S. Franco, A. Hanany, K. D. Kennaway, D. Vegh, and B. Wecht, "Brane dimers and quiver gauge theories," *JHEP*, vol. 01, p. 096, 2006.
8. A. B. Goncharov and R. Kenyon, "Dimers and cluster integrable systems," in *Annales scientifiques de l'École Normale Supérieure*, vol. 46, pp. 747–813, 2013.
9. V. V. Fock and A. Marshakov, "Loop groups, Clusters, Dimers and Integrable systems," in *Geometry and quantization of moduli spaces*, pp. 1–65, Springer, 2016.
10. A. Iqbal, N. Nekrasov, A. Okounkov, and C. Vafa, "Quantum foam and topological strings," *JHEP*, vol. 04, p. 011, 2008.
11. S. Franco, A. Hanany, D. Krefl, J. Park, A. M. Uranga, and D. Vegh, "Dimers and orientifolds," *JHEP*, vol. 09, p. 075, 2007.
12. G. Buratti, E. García-Valdecasas, and A. M. Uranga, "Supersymmetry Breaking Warped Throats and the Weak Gravity Conjecture," *JHEP*, vol. 04, p. 111, 2019.
13. R. Argurio, M. Bertolini, S. Meynet, and A. Pasternak, "On supersymmetry breaking vacua from D-branes at orientifold singularities," *JHEP*, vol. 12, p. 145, 2019.
14. R. Argurio, M. Bertolini, S. Franco, E. García-Valdecasas, S. Meynet, A. Pasternak, and V. Tatitscheff, "Dimers, Orientifolds and Stability of Supersymmetry Breaking Vacua," *JHEP*, vol. 01, p. 061, 2021.
15. R. Argurio, M. Bertolini, S. Franco, E. García-Valdecasas, S. Meynet, A. Pasternak, and V. Tatitscheff, "The Octagon and the Non-Supersymmetric String Landscape," *Phys. Lett. B*, vol. 815, p. 136153, 2021.
16. A. Retolaza and A. Uranga, "Orientifolds of Warped Throats from Toric Calabi-Yau Singularities," *JHEP*, vol. 07, p. 135, 2016.
17. I. García-Etxebarria and B. Heidenreich, "Strongly coupled phases of $\mathcal{N} = 1$ S-duality," *JHEP*, vol. 09, p. 032, 2015.

Connections Between Reflected Entropies and Hyperbolic String Vertices

Haitang Yang

We establish connections between the reflected entropies of multipartite mixed states in CFT_2 and hyperbolic string vertices of closed string field theory (CSFT). We show that the reflected surfaces, which are bulk duals of the reflected entropies, share the same Riemann surfaces with the hyperbolic string vertices. This observation enables us to build quantitative relations between the reflected entropies and hyperbolic string vertices.

Both the boundaries of string vertices and reflected surfaces are simple closed geodesics in hyperbolic surfaces. According to theorems in hyperbolic geometry, there is a unique geodesic perpendicular to two ultra-parallel geodesics; and given any three positive numbers l_a, l_b, l_c, there exists a unique Y-piece whose boundary geodesics have lengths l_a, l_b, l_c. Since the Y-pieces are the fundamental building blocks for higher order string vertices and reflected surfaces, it is reasonable to expect there are some close connections between them. However, the connections are not straightforward. We summarize the differences in Fig. 1 and the table.

One might be puzzled why we introduce the string vertices to complicate the story. It looks that the Y-pieces and surfaces built on them are sufficient to construct the reflected surfaces, without the two strict constraints imposed by the string vertices. The reasons are three folds:

1. The string vertices are physical and respect the dynamical equation, namely the BV equation. So, the string vertices provide us a practical approach to figure out dynamical properties of the reflected surfaces and entropies.
2. Though the classical string vertices allow any boundary geodesic length, they are only applicable to tree diagrams. To see this, let us recall the BV equation

$$\partial \mathcal{V} + \hbar \Delta \mathcal{V} + \frac{1}{2}\{\mathcal{V}, \mathcal{V}\} = 0.$$

H. Yang (✉)
College of Physics, Sichuan University, Chengdu, China
e-mail: hyanga@scu.edu.cn

Y.-H. He et al. (eds.), *Nankai Symposium on Mathematical Dialogues*,
https://doi.org/10.1007/978-981-19-2328-9_42

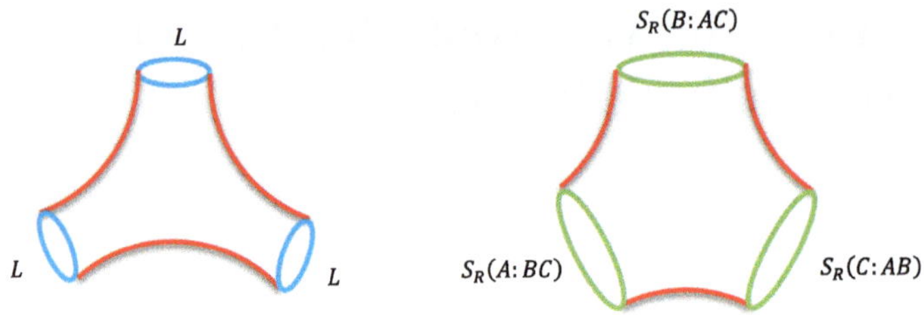

	Closed string vertices $\tilde{\mathcal{V}}_{0,3}(L)$	Reflected entropies $S_R(\bullet:\bullet\bullet)$
Constraint 1	Equal lengths of boundaries	No requirement
Constraint 2	$L \le L_*$ for quantum vertices $L > 0$ for classical vertices	$S_R(\bullet:\bullet\bullet) > 2L_*$

Fig. 1 Differences of Y-pieces in two theories

Fig. 2 Gluing two string Y-pants to get X-piece

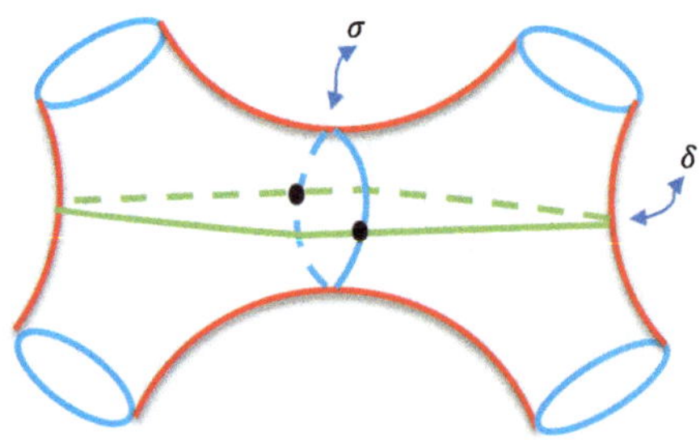

In the classical limit $\hbar \to 0$, the second term $\hbar\Delta\mathcal{V}$ disappears, and $\mathcal{V} = \sum_{g,n} \hbar^g \mathcal{V}_{g,n} \to \mathcal{V} = \sum_{n\ge 3} \mathcal{V}_{0,n}$. So, we will mainly discuss the quantum vertices and mention the classical limits when needed.

3. Through the connection, it is possible to build in CFT the holographic duals of the operations Δ and $\{,\}$.

We illustrate the connections with several examples. We then address the equal length restriction of the string vertices.

Example 1

Let us start with Fig. 2, the X-piece $\tilde{\mathcal{V}}_{0,4}(L)$, constructed by gluing two Y-pieces along the blue geodesic boundary σ.

The X-piece is uniquely fixed by the blue geodesics, and we have:

Corollary 1 Considering two closed geodesics with lengths σ and δ on a X-piece, if σ and δ intersect twice, we have $max\,\{\sigma, \delta\} > 2L_*$, with quantitative relation:

$$\cosh\frac{\sigma}{2} = \sinh\frac{\delta}{4}\sinh\frac{\sigma}{4}. \tag{1}$$

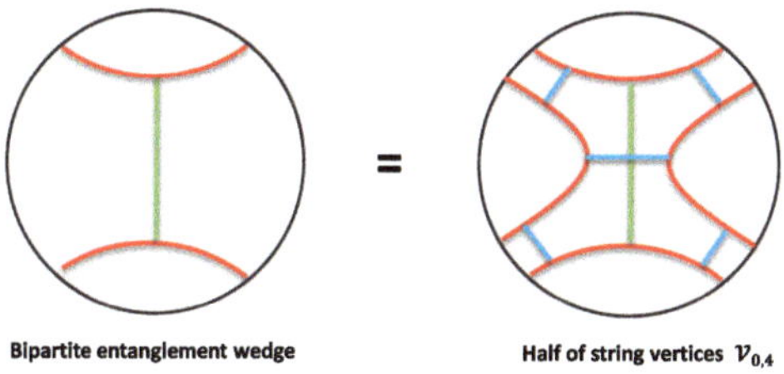

Fig. 3 A quantum X-piece represents a bipartite reflected surface. Since the length of the blue geodesics is smaller than $L_*/2$, they can not be reflected surfaces and no entanglement wedge exists. The only legitimate reflected surface is the green geodesic which indicates an up-down entangling

This result naturally combines the bounds of two theories together: $L \leq L_*$ in string vertices and $S_R > 2L_*$ in entanglement, and it makes it possible to establish connections between two theories. It is easy to derive that δ has a minimum,

$$\delta = 4\text{arc}\sinh\left(\frac{\cosh\sigma/2}{\sinh\sigma/4}\right) \geq 4\text{arc}\sinh(2\sqrt{2}) \equiv \delta_{min} > S_{Rmin} \equiv 4\text{arc}\sinh(1) = 2L_*. \quad (2)$$

In the classical theory, if the length of σ is also larger than $2L_*$, Fig. 2 basically represents a quadripartite reflected surface with four $S_R\,(\bullet : \bullet\bullet\bullet) = \sigma$, one $S_R\,(\bullet\bullet : \bullet\bullet) = \sigma$ and one $S_R\,(\bullet\bullet : \bullet\bullet) = \delta$.

Note that in quantum theory, blue geodesics $\sigma < L_*$, so they do not contribute to the reflected entropy and the corresponding entanglement wedges disappear. We are therefore only left with the up-down bipartite entanglement $S_R\,(\bullet : \bullet) = \delta$, namely the green geodesic, as illustrated in Fig. 3, where the right panel collapses to the left panel. From Eq. (2), a quantitative relation between the boundary length of the string vertices $L = \sigma$ and the reflected entropy $S_R(A : B) = \delta$ is

$$S_R(A : B) = 4\text{arc}\sinh\frac{\cosh\frac{L}{2}}{\sinh\frac{L}{4}}, \quad (3)$$

$$L = 4\text{arc}\sinh\left[\frac{1}{4}\left(\sinh\frac{S_R}{4} \pm \sqrt{\sinh^2\frac{S_R}{4} - 8}\right)\right]. \quad (4)$$

<u>The gap between δ_{min} and S_{Rmin}, *generalized* reflected entropy</u>

In Eq. (2), there exists a nontrivial gap between the minimal values of $\delta_{min} = 4\text{arc}\sinh(2\sqrt{2})$ and $S_{Rmin} = 2L_* = 4\text{arc}\sinh(1)$. It is very instructive to study the origin of this gap.

The reason is that the hyperbolic string vertices with explicit local coordinates provide non-perturbative corrections to the reflected entropies. The hyperbolic three-string vertex reduces to the naive three-string vertex as the geodesic boundaries L's approach zero. It is not difficult to generalize this result to the four-string vertex. Under this limit, the outer four geodesic boundaries L's of $\mathcal{V}_{0,4}$ go to zero, one then

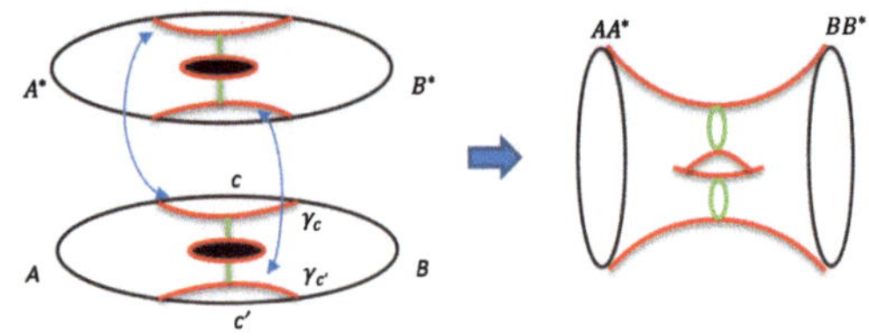

Fig. 4 Canonical purification for a thermal state

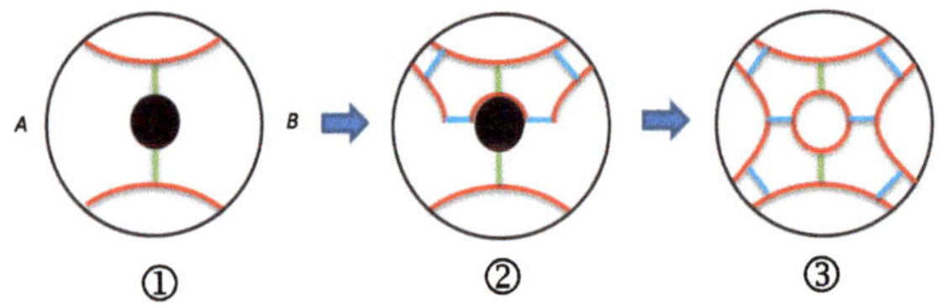

Fig. 5 Bipartite canonical purification with a black hole can be built by $\mathcal{V}_{1,4}$

gets $\delta_{min} \to S_{Rmin} = 2L_*$. It therefore indicates that the real bulk dual of the reflected entropy is not the reflected surface (hyperbolic string vertex), but the Moosavian-Pius surface! Since the Moosavian-Pius surface is a limit of the hyperbolic string vertex, the hyperbolic string vertex should be dual to a *generalized* reflected entropy in the CFT, which has the reflected entropy as a limit.

Example 2
This example is a bipartite system for a thermal state of CFT_2. The bulk dual of this state intrinsically has a black hole located at the center of the Poincare disk. The black hole introduces a genus after purification, see Fig. 4, which invalidates classical constructions. So, this reflected surface can only be constructed quantum mechanically.

To construct the bulk geometry with string vertices, we have the following steps as illustrated in Fig. 5:

1. Two disconnected green geodesics in the left panel has the total length $S_R(A:B) = 2s$.
2. From example 1, we can replace each green geodesic by a string vertex $\tilde{\mathcal{V}}_{0,4}(L)$.
3. Gluing two $\tilde{\mathcal{V}}_{0,4}(L)$ along blue boundaries, we see that the bulk geometry can be built by a string vertex $\mathcal{V}_{1,4}(L)$.

From Eq. (1), we have

$$S_R(A:B) = 2s = 8\text{arc}\sinh\frac{\cosh\frac{L}{2}}{\sinh\frac{L}{4}},$$
$$\mathcal{V}_{1,4}(L) = \mathcal{V}_{1,4}\left(4\text{arc}\sinh\left[\frac{1}{4}\left(\sinh\frac{s}{4} - \sqrt{\sinh^2\frac{s}{4} - 8}\right)\right]\right). \tag{5}$$

Moreover, the area A of the black hole's horizon also can be fixed by the string vertices:

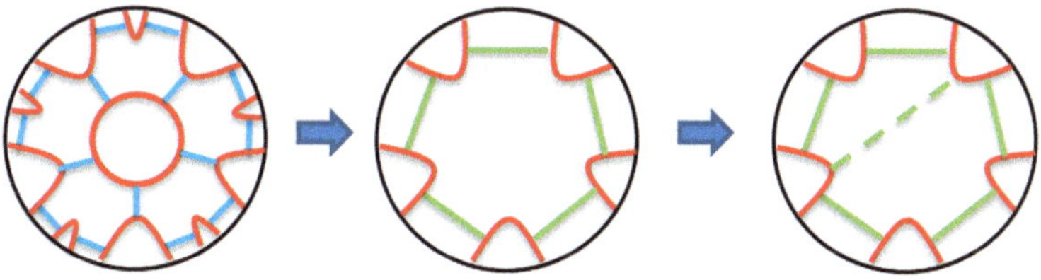

Fig. 6 The blue geodesics disappear when we calculate the entanglement wedge cross sections since $L < L_*$. In addition to the near boundary bipartite reflected surfaces $S_R\,(\bullet : \bullet \bullet \bullet\bullet)$ (solid green lines), a new class of bipartite reflected entropies (green dashed line) emerges $S_R\,(\bullet\bullet : \bullet \bullet \bullet) \neq S_R\,(\bullet : \bullet \bullet \bullet\bullet)$

Fig. 7 The sides of all hexagons are equal. By prolonging the sides to the boundary, each hexagon corresponds to a different region on the boundary. One thus can get different reflected entropies by selecting different hexagons

$$A = 2\text{arc}\cosh \frac{\cosh \frac{L}{2}}{\cosh \frac{L}{2} - 1}. \tag{6}$$

Remove the equal length restriction

Since the boundary geodesics of string vertex $\widetilde{\mathcal{V}}_{0,3}(L)$ has equal length $L < L_*$, one may worry the reflected surfaces built on it are too restricted. To clarify this problem, we consider the canonical purification of a pentapartite mixed state. The dual bulk geometry can be rebuilt by five X-pieces, as sketched in Fig. 6. Since the geodesic lengths of string vertices have the same value L, all bipartite reflected entropies $S_R\,(\bullet : \bullet \bullet \bullet\bullet)$ (solid green lines) near the boundary are identical. Remarkably, there emerges a new class of bipartite reflected entropies (green dashed line) which have different values $S_R\,(\bullet\bullet : \bullet \bullet \bullet) \neq S_R\,(\bullet : \bullet \bullet \bullet\bullet)$.

It is easy to understand that by adding more X-pieces, various reflected surfaces can be constructed. Particularly, one general situation is exhibited by Fig. 7, where all pieces are right-angled equilateral hexagons. The sides of these hexagons are all equal. Obviously, each hexagon relates different disk boundary regions when we prolong the sides to the boundary of the Poincare disk. By selecting appropriate hexagons, one can get, say, arbitrary bipartite surfaces.